Device Circuit Co-Design Issues in FETs

This book provides an overview of emerging semiconductor devices and their applications in electronic circuits, which form the foundation of electronic devices. *Device Circuit Co-Design Issues in FETs* provides readers with a better understanding of the ever-growing field of low-power electronic devices and their applications in the wireless, biosensing, and circuit domains. The book brings researchers and engineers from various disciplines of the VLSI domain together to tackle the emerging challenges in the field of engineering and applications of advanced low-power devices in an effort to improve the performance of these technologies. The chapters examine the challenges and scope of FinFET device circuits, 3D FETs, and advanced FET for circuit applications. The book also discusses low-power memory design, neuromorphic computing, and issues related to thermal reliability. The authors provide a good understanding of device physics and circuits, and discuss transistors based on the new channel/dielectric materials and device architectures to achieve low-power dissipation and ultra-high switching speeds to fulfill the requirements of the semiconductor industry.

This book is intended for students, researchers, and professionals in the field of semiconductor devices and nanodevices, as well as those working on device-circuit co-design issues.

Materials, Devices, and Circuits: Design and Reliability

Series Editor: Shubham Tayal, K. K. Paliwal, Amit Kumar Jainy

Tunneling Field Effect Transistors
Design, Modeling and Applications
Edited by T.S. Arun Samuel, Young Suh Song, Shubham Tayal, P. Vimala and Shiromani Balmukund Rahi

Quantum-Dot Cellular Automata Circuits for Nanocomputing Applications
Edited by Trailokya Sasamal, Hari Mohan Gaur, Ashutosh Kumar Singh, Xiaoqing Wen

Device Circuit Co-Design Issues in FETs
Edited by Shubham Tayal, Billel Smaani, Shiromani Balmukund Rahi, Samir Labiod and Zeinab Ramezani

For more information about this series, please visit: https://www.routledge.com/Materials-Devices-and-Circuits/book-series/MDCDR#

Device Circuit Co-Design Issues in FETs

Edited by
Shubham Tayal, Billel Smaani,
Shiromani Balmukund Rahi, Samir Labiod
and Zeinab Ramezani

CRC Press
Taylor & Francis Group
Boca Raton London New York

CRC Press is an imprint of the
Taylor & Francis Group, an **informa** business

Designed cover image: © Shutterstock

First edition published 2024
by CRC Press
6000 Broken Sound Parkway NW, Suite 300, Boca Raton, FL 33487-2742

and by CRC Press
4 Park Square, Milton Park, Abingdon, Oxon, OX14 4RN

CRC Press is an imprint of Taylor & Francis Group, LLC

Library of Congress Cataloging-in-Publication Data
Names: Tayal, Shubham, editor. | Smaani, Billel, editor. | Rahi, Shiromani
Balmukund, editor. | Labiod, Samir, 1981- editor. | Ramezani, Zeinab, editor.
Title: Device circuit co-design issues in FETs / edited by Shubham Tayal,
Billel Smaani, Shiromani Balmukund Rahi, Samir Labiod and Zeinab Ramezani.
Other titles: Device circuit co-design issues in field-effect transistors
Description: Boca Raton : CRC Press, 2024. | Series: Materials, devices,
and circuits | Includes bibliographical references and index.
Identifiers: LCCN 2023007962 (print) | LCCN 2023007963 (ebook) |
ISBN 9781032414256 (hardback) | ISBN 9781032416823 (paperback) |
ISBN 9781003359234 (ebook)
Subjects: LCSH: Field-effect transistors. | Electronic circuit design. | Semiconductors.
Classification: LCC TK7871.95 .D48 2024 (print) | LCC TK7871.95 (ebook) |
DDC 621.3815/284--dc23/eng/20230503
LC record available at https://lccn.loc.gov/2023007962
LC ebook record available at https://lccn.loc.gov/2023007963

ISBN: 9781032414256 (hbk)
ISBN: 9781032416823 (pbk)
ISBN: 9781003359234 (ebk)

DOI: 10.1201/9781003359234

Typeset in Times
by Deanta Global Publishing Services, Chennai, India

Contents

Preface

Our daily lifestyle is a witness to the importance of CMOS science and technology. CMOS changes our daily life. Nearly all modern industries and other cutting-edge developments including artificial intelligence (AI), autonomous systems, 5G communications, and quantum computing rely on it. The world population's use of electronics, communications, computers, and information technology (IT) applications has increased dramatically in recent years. Users can easily be identified in most places in our daily activities. Mobile technology is the best example of its applications. Scaling is the main and most important feature of CMOS devices. This feature of CMOS devices has continuously helped to develop various types of circuits and systems for our daily life uses, medical sciences, aerospace, and military-based development over the past four decades. Only due to the scaling of conventional MOSFET, various types of circuits and systems have developed for mankind's better life. The journey of scaling MOSFET technology is continuous and has reached the 5 nm range.

Scaling of the CMOS channel length below 0.5 u/m and increasing chip density to the VLSI range has placed power dissipation on an equal footing with performance as a figure of merit in digital circuit design. Portability and reliability have also played a major role in the emergence of low-power, low-voltage, digital circuits, and system designs. In this regard, the need to extend the battery life, have inexpensive packaging and cooling systems, and reduce the weight and size of the equipment have been the driving forces. The demand for low-power and high-speed FET devices has grown exponentially to meet the requirements of these applications. FETs are the leading electronics technology and will continue to advance in future CMOS. MOSFET has played a leading role in the development of VLSI circuits and systems. This device was a basic building block of CMOS technology and, as a consequence, the predominant device of integrated circuits and system development. The demand for small, portable, and affordable electronic equipment is growing by the day. In order to satisfy consumer demand, researchers are constantly searching for new semiconductor devices. This book, *Device Circuit Co-Design Issues in FETs*, provides industry professionals and beginners with an overview of emerging semiconductor devices and their applications in VLSI circuits and systems. The goal of this book is to provide readers with a better understanding of the ever-growing field of low-power electronic devices and their applications.

Through detailed derivations, discussions, layout, and simulation examples, Chapter 1 provides a concise summary of the thought process and practical implementation of CMOS IC design for the reader. The foundation of all digital designs is the inverter. Designing increasingly complex structures like NAND gates, adders, multipliers, and microprocessors becomes significantly easier once their operations and attributes are thoroughly understood. By extrapolating the conclusions found for inverters, it ispossibletoextracttheelectricalbehaviorsoftheseintricatecircuitsalmos tentirely.Inverteranalysisis frequently expanded to explain the operations of more complicated gates, such as NAND, NOR, or XOR, which serve as the basis for components like multipliers and processors. The static CMOS inverter, also known as the CMOS inverter, is the only iteration of the inverter gate on which this chapter focuses. The inverter, NAND, and NOR gates are implemented and simulated using the LT-spice computer-aided design (CAD) tool. Similar to other CAD tools, this method can be used to simulate more complex circuits by combining or altering these fundamental elements.

The design and simulation of CMOS integrated circuits are covered in Chapter 2. Beginning with a brief overview of CMOS process integration, this section focuses on interconnects, providing information on propagation delay and simulating certain parasitic effects. In the chapter, some fundamental understanding concepts for the analog design, layout techniques, and simulation of current mirror are presented. Moreover, DC characteristics and dynamic behavior have been analyzed. In addition to this, the layout of the basic CMOS static logic gates (inverter, NAND, and NOR) and arithmetic functions such as the full adder are presented. SPICE simulations have been performed on both 50 nm (short-channel) and 1 μm (long-channel) technologies.

The existing limitations of conventional CMOS technology are presented in Chapter 3. Conventional CMOS technology has reached its physical and technological limits, according to semiconductor experts, and as a result, numerous field-effect transistor (FET) architectures have been developed. The junctionless (JL) gate-all-around (GAA) MOSFET has attracted a great deal of research interest. In addition, compact models of FETs need to be incorporated into circuit simulators using a hardware description language (HDL), such as VHDL and Verilog-A. This is for the potential use of emerging transistors in various integrated circuits. The compact modeling of JL GAA MOSFET as an important issue is addressed here. The chapter begins with a discussion of the characteristics of compact models for the development of new electronic systems and applications. The value of hardware description language for the design and simulation of circuits is then demonstrated. In addition to the theoretical basis and main approach for developing compact models of JL GAA MOSFET, the charge-based, surface-potential-based, and threshold-voltage-based models are also presented. The most significant compact models are surface-potential-based and charge-based, specially dedicated to circuit simulation and design. Furthermore, the challenges of compact modeling of JL GAA MOSFET are also discussed.

Chapter 4 presents different variations of the novel gate-overlap tunnel field-effect transistors (GOTFETs) and their applications in analog, digital, and ternary logic circuits. For benchmarking their device and circuit performance with the industry-standard 45 nm CMOS technology, the presented GOTFETs have an effective channel length of 45 nm, commensurate with the technology node. These devices have a higher band-to-band generation rate than the conventional TFET devices, due to the gate fully overlapping on the source side, resulting in excellent improvement in the I_{on} levels while maintaining very low I_{off}. Introduction of an epi-layer between the source and oxide layers, the proposed variant of GOTFET, the line TFET (LTFET), exhibits almost flat drain current saturation characteristics, leading to very high R_{out}

for superior analog circuit applications. Optimization of the LTFET device has been done by changing critical parameters such as epi-thickness, gate-to-source overlap, and doping concentration, and has shown its influence on analog performance. Therefore, the proposed LTFET has a two-order improvement in r_o leading to a two-order improvement in the intrinsic gain A_{vo} over the MOSFET. Due to lower connection, smaller chip footprint, and faster-operating speeds, the GOTFET structure has been further modified for ternary logic circuit applications. The intended LVT and HVT GOTFET shave been designed such that low threshold voltage $V_{TL} \approx V_{DD}/3$ and high threshold voltage $V_{TH} \approx 2V_{DD}/3$ for the unique voltage levels $\{0-V_{DD}/3\}$, $\{V_{DD}/3-2V_{DD}/3\}$, $\{2V_{DD}/3-V_{DD}\}$ correspond to ternary logic states 0, 1, and 2. The proposed LVT and HVT TFET devices will be the starting point for all applications involving ternary logic. This chapter is, in essence, a comprehensive review of the GOTFET devices and their circuits performance, such that the readers of this book chapter will learn about specialized TFETs (GOTFETs), which perform much better than conventional CMOS when switched on while consuming less power than conventional TFETs when switched off. Consequently, complementary GOTFET (or CGOT) technology combines the robustness and high performance of CMOS with the low-power benefits of TFET in a single-device technology.

Chapter 5 is devoted to the few years when the development of ultra-low-power oxide electronics devices has been facilitated by abrupt, ultrafast, nanoscale switching caused by an insulator-to-metal transition in phase transition materials. These transitions, particularly those caused by electrical triggering, aid in the achievement of dimensional scaling at the lower technology node. The unique electrical properties of these materials can be used to create innovative devices and circuits for next-generation electronics. This chapter examines the history of the phase transition materials family, including its origin, history, modeling, and application in cutting-edge devices. There is a focus on various applications of phase transition materials in low-power electronics, such as steep switching devices, digital circuits, memory, and non-Boolean computing.

Chapter 6 describes the extensive use of semiconductor devices in the electronic systems of satellites. In the outer atmosphere, natural radiation is the major threat to semiconductor devices. The radiation raises there liability issues of these types of devices when the irradiation accumulation of trap charges is found in the oxides and semiconductor/insulator interfaces. These trap charges are well capable of shifting the threshold voltage towards negative and increasing the leakage current. The radiation effects are classified into two categories: total ionizing dose (TID) effects and single event effects (SEEs). The impact of TID on SOI-FinFET with the spacer technique is investigated. At a higher radiation dose of 2000 krad, the high-k dielectric (HfO_2) spacer maintains lower leakage current and positive threshold voltage. The proposed engineering technique enhances the OFF-state device performance after and before the irradiation of the device. For the pre-radiation condition, a 48% improvement in OFF-state current (I_{OFF}) is observed for the SiO_2 spacer-based device and an 83% improvement is obtained for the HfO_2 spacer. The HfO_2 spacer-based device shows 4.2%, 2.6%, 2.5%, and 2.4% lower subthreshold swing (SS) after the 2000 krad dose as compared to SiO_2, Si_3N_4, Al_2O_3, and AlN, respectively. An improvement of 23% in I_{OFF} and a 42% lower shift in threshold voltage is observed for the HfO_2 spacer SOI-FinFET as compared to SiO_2 spacer-based SOI-FinFET. This investigation shows that SOI-FinFET with HfO_2 spacer-based device is best suited for electronic systems in space applications.

Chapter 7 is dedicated to the rapid development of technology that has increased the density, speed, and performance of transistors embedded in modern chips. According to the presented literature, FinFET technology down to 7 nm has shown more acceptable performance than others. However, further scaling down to 5 and 3 nm scales imposes undeniable challenges to this technology. Thus, the community of semiconductor designers, in order to introduce a suitable alternative to FinFET, proposed the technology of nanosheet FET (NSFET) to overcome these challenges. Fundamentally NSFET is an advanced version of FinFET. The prominent feature of the NSFET is having a horizontal gate stacked around the channel in all directions. This feature gives the gate more control over the channel. Therefore, it significantly improves the performance and ON current of the NSFET compared to other FETs. This has made the NSFET more popular than other devices, especially for scaling down to 3 nm. On the other hand, the successful fabrication of NSFET by Samsung/ IBM for sub-7-nm technology has pushed the semiconductor industry towards these devices. In this regard, to deal with the performance of NSFET in integrated circuits, it is of particular importance to investigate the electrical characteristics of these devices from the perspective of the circuit. In this chapter of the book, NSFET is introduced, and two key challenges of nanodevices, the self-heating and short channel effects, are investigated. Subsequently, in more detail, the behavior and challenges of this device have been analyzed from the circuit point of view.

In Chapter 8, the authors provided a brief introduction to tunneling FET. Tunneling FET surpasses the subthreshold swing limitation and off-state current issues of conventional CMOS devices. The structure of TFET, its characteristics, and its scope with specific applications are discussed in this chapter. This will be useful for researchers who have just started their research on TFET. Only certain applications are explored in this chapter. Still, there are many more applications in the research to explore.

In Chapter 9, the demand for memory is increasing day by day, and the downscaling of conventional 1T-1C DRAM in sub-10 nm technology is becoming a topic of concern. The fabrication and scalability of 3-D cell storage capacitors are extremely difficult. To overcome this issue, the concept of capacitorless 1T-DRAM is introduced. The silicon-on-insulator (SOI) transistor will store the charges in the floating body of the metal oxide semiconductor field-effect transistor (MOSFET) by impact ionization. Further scaling of MOSFET devices is approaching its boundary, and it is giving rise to short-channel effects. Hence multi-gate transistors (such as FinFET, GAA FET, and RFET), in which more than one gate surrounds the channel are introduced. In this chapter, FinFET-based capacitorless 1T-DRAM is introduced, due to its simplicity in fabrication. In FinFET, the gate controls the channel from three sides; hence the electrostatic control over the channel increases and the leakage current also reduces.

In Chapter 10, the majority of current embedded systems use microprocessors equipped with volatile cache memory based on static random access memory (SRAM) technology. As part of a core computing component, its performance is critical and needs to have more attention. Actual systems-on-chips (SoCs) need to be more performant because less than 20% of the globally integrated transistors are used for arithmetic and logic operations, and the rest of the transistors, about 80%, are mostly used for the cache memory. Additionally, modern implantable electronic components and devices, for specific and general use, are based on artificial intelligence (AI) and require efficient and reliable SRAM circuits designed for having enhanced and fast responses to compute-in-memory (CIM). In order to reach desired performances, reliability should be maintained, especially with regard to the most recent technological areas. In this chapter, the authors have cited, for example, embedded systems using low power supplies, which may pose a risk to the stability of the SRAM circuits and also their unavailability. In sophisticated devices, the process variations change the transistor design parameters and consequently the design integrity. Additionally, sensitive information treatment, environmental conditions (such as temperature variation, shocks, and vibration), and static charge emission from adjacent integrated circuits can affect SRAM reliability. Fin field-effect transistor (FinFET) technology has been used to design SRAMs to enhance the overall performance, which takes into account efficiency, power, and area. In this work, we have reviewed various colossal challenges to SRAM design after classifying them into five distinct categories and each one will be presented with viable solutions.

In Chapter 11, FinFET technology is discussed, which is the slogger of today's semiconductor world. However, the demand for further scaling with a desire for ultra-low power and high-speed applications leads to undesired short-channel effects, where new transistors are required for the next generation. Thanks to science and technological innovation, different transistors from the GAA (gate-all-around) FET family and their competitive benefits have been brought together. This chapter tries to answer why and how 3D devices emerge for future computing paradigms. In addition to the limitation of FinFET, it further discusses the scope and challenges of different members of the GAAFET family, such as nanowire FET, nanosheet FET, junctionless nanosheet FET, complementary FET, and forksheet FET.

Editor biographies

Shubham Tayal is an assistant professor in the Department of Electronics and Communication Engineering at SR University, Warangal, India. He has more than six years of academic/research experience teaching at undergraduate and postgraduate levels. He received his Ph.D. in microelectronics and VLSI design from the National Institute of Technology, Kurukshetra, M.Tech (VLSI Design) from YMCA University of Science and Technology, Faridabad, and B.Tech (Electronics and Communication Engineering) from MDU, Rohtak.

He has published more than 40 research papers in various international journals and conferences of repute, and many papers are under review. He is on the editorial and reviewer panel of many SCI/SCOPUS-indexed international journals and conferences. He is editor/co-editor of eight books published by CRC Press (Taylor & Francis Group, USA) and Springer Nature. He acted as a keynote speaker and delivered professional talks on various forums. He is a member of various professional bodies including IEEE and IRED. He is on the advisory panel of many international conferences. He is a recipient of the Green ThinkerZ International Distinguished Young Researcher Award 2020. His research interests include the simulation and modeling of multi-gate semiconductor devices, device-circuit co-design in the digital/analog domain, machine learning, and IoT.

Billel Smaani received his Ph.D. degree from the University of Frère Mentouri, Constantine, Algeria, in 2015. He joined the Centre Universitaire Abdelhafid Boussouf, Mila, Algeria, in 2021, where he has been an associate professor since June 2022. From 2015 to 2021, he was with the University of M'hamed Bougara Boumerdes, Algeria. His current research interests include the study, analysis, and compact modeling of advanced nanoscale field-effect transistors for analog and digital circuit co-design.

Shiromani Balmukund Rahi received a B.Sc. (Physics, Chemistry Mathematics) in 2002, an M.Sc. (Electronics) from Deen Dyal Upadhyaya Gorakhpur University, Gorakhpur, in 2005, M. Tech. (Microelectronics) from Panjab University Chandigarh in 2011, and a Doctorate of Philosophy in 2018 from the Indian Institute of Technology, Kanpur, India. He completed his Master project (M.Sc.) at the Central Electronics Engineering Research Institute (CEERI, 2005), Pilani Rajasthan, under the supervision of Dr. P C

Panchariya (Director and Chief Scientist, CEERI, Pilani) and thesis (M. Tech.) under Prof. RenuVig (director and Professor, UIET Panjab University Chandigarh), post-doctoral research (Department of Computer Science, Korea Military Academy, Seoul, Republic of Korea). He has 25 international publications and ten book chapters. He has edited two books for CRC publication. He is associated with research with the Indian Institute of Technology Kanpur, India, and in the electronics department at the University Mostefa Benboulaid of Algeria, developing ultra-low power devices such as tunnel FETs, NC TFET, negative capacitance, and nanosheet FETS.

Samir Labiod was born in Constantine, Algeria, on January 5, 1981. He received electrical engineering and magister degrees in electronics from Constantine University, Algeria, in 2005 and 2008, respectively. He also received his Ph.D. from Constantine University Institute of Sciences and Technology, Constantine, Algeria, in 2013. His current research interests include the numerical modeling of electromagnetic compatibility of semiconductor devices.

Zeinab Ramezani received her Ph.D. in Electrical and Computer Engineering in 2017. She worked as an assistant professor at IAU University from 2017 to 2019 and as a research scientist at Northeastern University in Boston, MA, USA, from 2019 to 2021.She is a scientist with over ten years of experience in modeling, simulation, and characterization of novel structures; micro-and nanoelectronics; nanotechnology; nanophotonic and nanomagnetic power semiconductor devices; wide bandgap semiconductors; optoelectronic devices; plasmonic devices; and bioelectronics and biosensors. Her current research at the University of Miami, FL, USA, is focused on developing and modeling new materials and electronic tools to enable leapfrog advancements in health applications.

Contributors

Atefeh Rahimifar Department of Electrical Engineering, Karoon Institute of Higher Education, Ahvaz, Iran

Abdelmalek Mouatsi Boumerdes University, Faculté de Technologie, Laboratoire d'Ingénierie des Sytémes et des Telecommunication, Boumerdes, Algeria

Abhishek Acharya S.V. National Institute of Technology, Surat, India

Abhishek Kumar Upadhyay 3X-FAB Semiconductor Foundries, Erfurt, Germany

Abhishek Ray Department of Electronics and Communication Engineering, National Institute of Technology Raipur, Chhattisgarh, India

Alain Tshipamba University of Connecticut, Storrs, CT, USA

Alok Naugarhiya Department of Electronics and Communication Engineering, National Institute of Technology Raipur, Chhattisgarh, India

B. Karthikeyan Velammal College of Engineering and Technology, Madurai, Tamilnadu, India

Bhaskar Awadhiya Department of Electronics and Communication, Manipal Institute of Technology, Manipal Academy of Higher Education, Manipal, Udupi, Karnataka, India

Billel Smaani Centre Universitaire Abdel Hafid Boussouf, Mila, Algeria

Bouchra Nadji Department of Automation University Mhamed Bougara Boumerdes Algeria

Fares Nafa Boumerdes University, Faculté de Technologie, Laboratoire d'Ingénierie des Sytémes et des Telecommunication, Algeria

Guru Prasad Mishra Department of Electronics and Communication Engineering, National Institute of Technology Raipur, Chhattisgarh, India

Hamza Akroum Laboratoire d'Automatique Appliquée, Université M'Hamed Bougara de Boumerdes, Algeria

Husien Salama University of Connecticut, Storrs, CT, USA

Ismahan Mahdi Department of Electrical Systems Engineering, University Mhamed Bougara Boumerdes, Algeria

Khalifa Ahmed University of Turkish Aeronautical Association, Ankara, Turkey

Maya Lakhdara Département d'Electronique, Faculté des Sciences et de la Technologie, Laboratoire Hyperfréquences et Semi-conducteurs, Université des frères Mentouri Constantine 1, Constantine, Algeria

Mitali Rathi Department of Electronics and Communication Engineering, National Institute of Technology Raipur, Chhattisgarh, India

Mohamed Salah Benlatreche Centre Universitaire Abdelhafid Boussouf, Mila, Algeria

P. Anand Velammal College of Engineering and Technology, Madurai, Tamilnadu, India

P. Suveetha Dhanaselvam Velammal College of Engineering and Technology, Madurai, Tamilnadu, India

P.N Kondekar Department of Electronics and Communication Engineering, PDPM-Indian Institute of Information Technology, Design & Manufacturing, Jabalpur, India

P. Vanitha SRMIST Ramapuram Campus, Chennai, India

Ramakant Yadav Electrical & Electronics Engineering Department, Mahindra University, Hyderabad, India

Sameer Yadav Department of Electronics and Communication Engineering, PDPM-Indian Institute of Information Technology, Design & Manufacturing, Jabalpur, India

Samir Labiod Department of Physics, Faculty of Sciences, Universite 20 Aout 1955, Skikda, Algeria

Shiromani Balmukund Rahi Department of Electrical Engineering, Indian Institute of Technology Kanpur, Kanpur 208016, India

Shobhit Srivastava S.V. National Institute of Technology, Surat, India

Simhadri Hariprasad BITS Pilani, Hyderabad Campus, India

Surya Shankar Dan BITS Pilani, Hyderabad Campus, India

Yasmine Guerbai Department of Electrical Systems Engineering University Mhamed Bougara Boumerdes, Algeria

Yassine Meraihi Department of Electrical Systems Engineering University Mhamed Bougara Boumerdes, Algeria

Zakaria Hadef Department of Physics, Faculty of Sciences, Universite 20 Aout 1955, Skikda, Algeria

Zeinab Ramezani Electrical and Computer Engineering Department, University of Miami, Miami, FL, USA

1 Modeling for CMOS circuit design

Husien Salama, Alain Tshipamba,
and Khalifa Ahmed

CONTENTS

1.1 CMOS DEVICES

1.1.1 INTRODUCTION

A CMOS transistor consists of a P-channel MOS (PMOS) and an N-channel MOS (NMOS) [1]. The operation of a CMOS device is like other types of field effect transistors (FET) except it depends on an added oxide layer between the gate and the substrate. CMOS are active devices, meaning they require external power sources

DOI: 10.1201/9781003359234-1

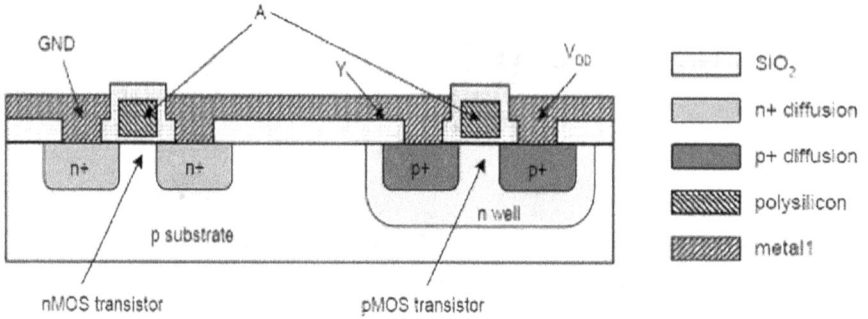

FIGURE 1.1 CMOS (complementary metal oxide semiconductor). [1]

to operate. For this reason, shown in Figure 1.1, CMOS devices are designed with a power supply, input voltage terminal (V_{IN}), output voltage (V_{OUT}), gate, drain, and PMOS and NMOS transistors which are connected to the gate and the drain terminals [2].

The main advantage of CMOS over NMOS and PMOS technology is a much smaller power dissipation, which has become a crucial element for scalability in IC design. Unlike NMOS, PMOS, or bipolar circuits, a CMOS circuit has almost no static power dissipation. Power is only barely dissipated if the circuit switches between high and low power levels or states. This superior performance of CMOS technology enables the integration of more CMOS gates on an IC than with NMOS or bipolar technology.

1.1.2 SWITCH OF CMOS

In basic CMOS concepts, we see the use of transistors for designing logic gates. The same approach can be used to design other blocks (such as flip-flops or memories). Ideally, a transistor behaves like a switch [3] for logic implementation. For NMOS transistors, if the input is a logic high, the switch is ON; otherwise, it is OFF. On the other hand, for the PMOS, if the input is a logic low the transistor is ON; otherwise, the transistor is OFF [4].

A graphic representation is shown in Figure 1.2.

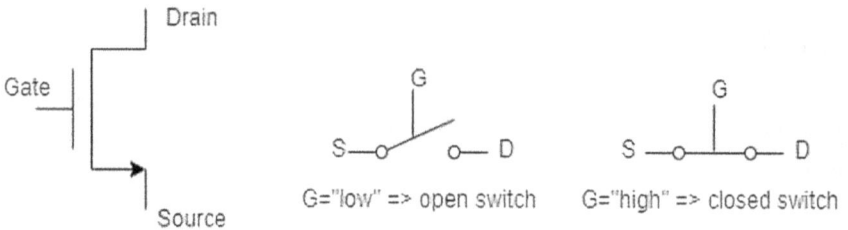

FIGURE 1.2 MOS as a switch

For the NMOS in Figure 1.1, the gate (G) can be thought of as the switch's handle and the signal flowing through the gate as the force acting on it. Considering an initially inactivated switch, if one does not apply enough force on the switch (i.e., a logic low is applied), then the switch remains open (Figure 1.2 on G = low). However, if sufficient force is applied (i.e., a logic high is acting on the switch), the switch closes (Figure 1.2 on G = "high"), and electric contact is established between the drain (D) and the source (S). The PMOS operates similarly with the main difference being that its activation is a logic low.

1.1.3 THE IMPLEMENTATION AND OPERATION OF THE CMOS INVERTER

We have seen that a CMOS device is a combination of NMOS and PMOS technology. To understand the basics of operation, we will design a simple inverter gate. Figure 1.3 shows the CMOS implementation of the inverter and how it works for different inputs (1 and 0) [5]. The symbol VDD is the source voltage (or logic 1), and GND is the ground (or logical 0).

The CMOS inverter operation is simple and straightforward. Referring to Figure 1.3, when the low input voltage (0) is given to the CMOS inverter's gate, the PMOS transistor is switched ON, whereas the NMOS transistor is switched OFF. Since the PMOS is connected to VDD, this facilitates the provision of a low resistance path for electrons from VDD to the output through the PMOS, hence the generation of a logic high output for a low input [6].

Similarly, when the high input voltage (1) is given to the CMOS inverter's gate, the PMOS transistor is OFF, whereas the NMOS transistor is now switched ON. Since the NMOS is connected to the ground, this renders the ground a low resistance path for electrons and, consequently, from the output to the ground, hence the generation of a logic low output for a high input voltage [7]. As a result, Figure 1.3 works like an inverter.

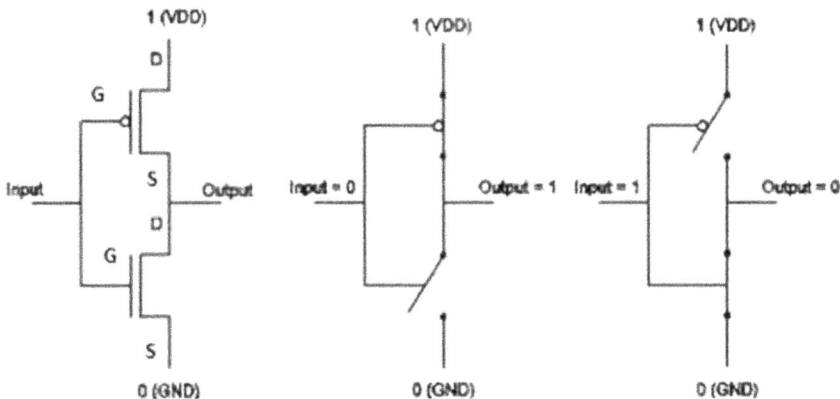

FIGURE 1.3   CMOS inverter and switch equivalent [2]

1.2   THE CMOS IC DESIGN PROCESS

The CMOS circuit design process consists of defining circuit inputs and outputs, hand calculations, circuit layout, simulations including parasitic revaluation of circuit inputs and outputs, fabrication, and testing [8]. The circuit specifications are usually set at the beginning of the design and adjusted as the process evolves and matures. This will be the result of trade-offs made between cost and performance, changes within the marketability of the chip, or just changes in the customer's needs. However, in most cases, major changes after the chips have gone into production are impossible [9].

1.2.1   BACKGROUND

The CMOS concept is relatively newer to other semiconductor technologies and was first introduced to the semiconductor community around 1963. The thought that a circuit could be made with discrete complementary MOS devices– an NMOS (N-channel MOSFET) transistor and a PMOS (P-channel) transistor – was quite novel at the time given the immaturity of MOS technology and the rising popularity of the bipolar junction transistor (BJT) as a replacement for the vacuum tube [10].

The inverter circuit shown in Figure 1.4 consists of PMOS and NMOS FET. The input is the gate voltage for both transistors.

1.2.2   CMOS INVERTER CHARACTERISTICS

CMOS inverters are the most widely used and adaptable MOSFET inverters used in chip design. They operate with little to no power loss and at relatively high speeds. Furthermore, the CMOS inverter has good logic buffer characteristics: its capacitance in both low and high states is large. A CMOS inverter consists of a set of PMOS and an NMOS transistor connection [11]. In this setting, the supply voltage VDD is placed at the PMOS drain terminal, and the NMOS source terminal is connected to the ground. V_{IN} is connected to the gate terminals of both transistors, and V_{OUT} is set between the drain of the NMOS and the source of the PMOS (Figure 1.4). It is important to note that the CMOS inverter does not contain any resistors, which makes it more power efficient than a regular resistor of a MOSFET inverter. Because the voltage at the input of the CMOS device varies between 0 and VDD, the state of the NMOS and PMOS varies accordingly. Figure 1.5 and Figure 1.6 show the characteristics and modes of CMOS [12].

To constitute an operating point, the currents via the NMOS and PMOS bias must be equal. This indicates graphically that the DC points must be situated at the intersection of the relevant load-lines. A few of those points (for VIN = 0, 0.5, 1, 1.5, 2, and 2.5 V) are indicated on the graph. It is evident that every operating point lies either at the upper or lower end of the line.

The voltage transfer characteristic (VTC) of the inverter hence exhibits a narrow transition zone. This zone results from the high gain during the switching flash when

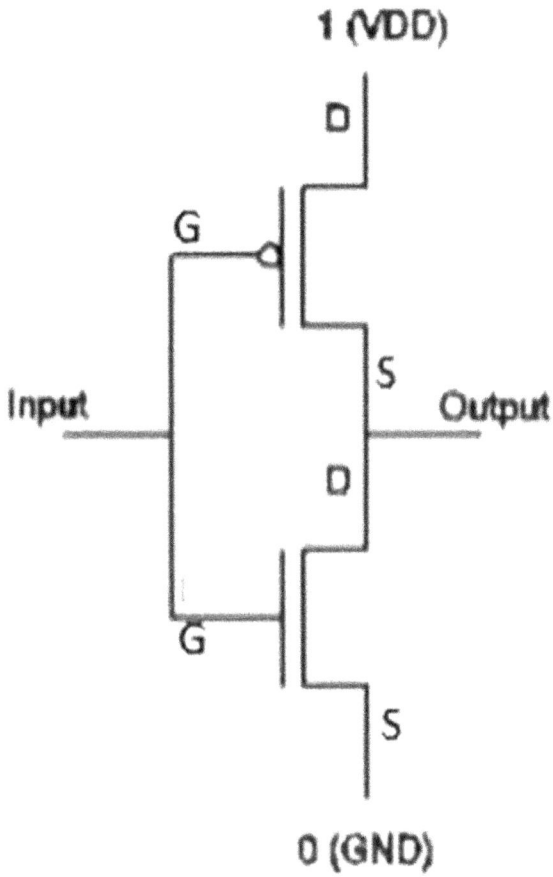

FIGURE 1.4   CMOS inverter [2]

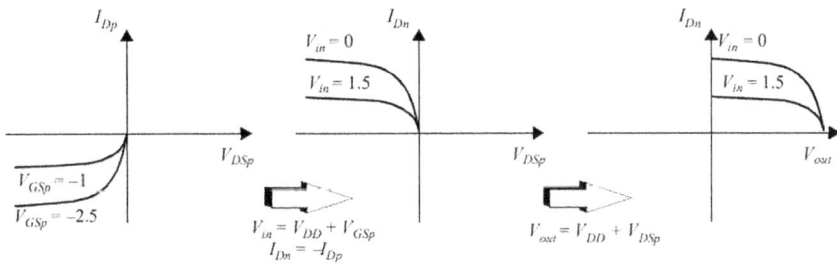

FIGURE 1.5   Transforming PMOS I-V characteristic to a common coordinate set (assuming VDD = 2.5 V)

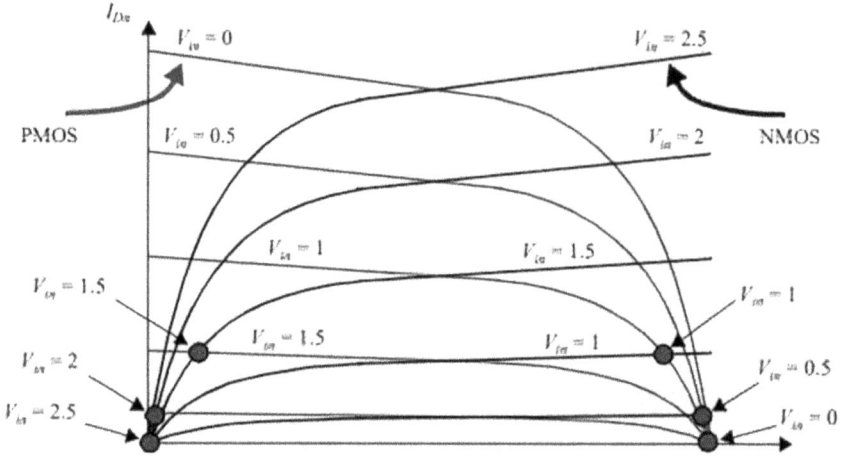

FIGURE 1.6 Load curves for the static CMOS inverter's NMOS and PMOS transistors (VDD = 2.5 V). The dots represent various input voltages' dc operation points [12].

both NMOS and PMOS are temporarily ON. In that operation region, a small change in the input voltage results in a large variation [13].

The VTC shown in Figure 1.7 looks like an inverted step function that specifies accurate switching between ON and OFF. However, in real bias, a gradual transition region exists. The voltage transfer characteristic specifies that for lower input voltage V_{IN}, the circuit generates high voltage V_{OUT}, whereas, for high input, it generates 0 volts.

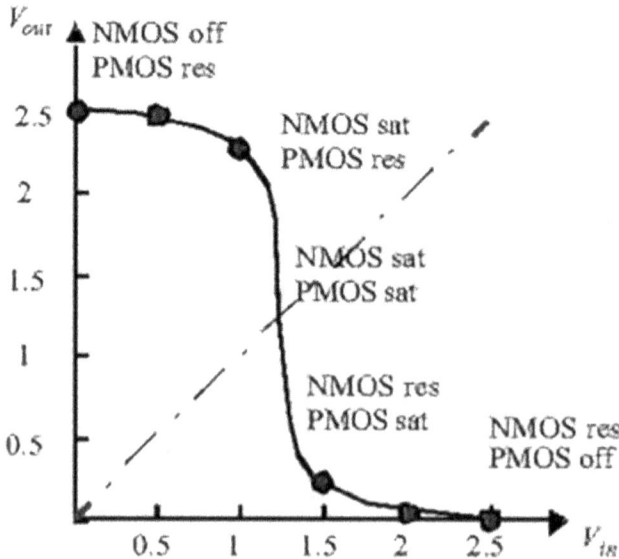

FIGURE 1.7 Characteristics of an inverter [12]

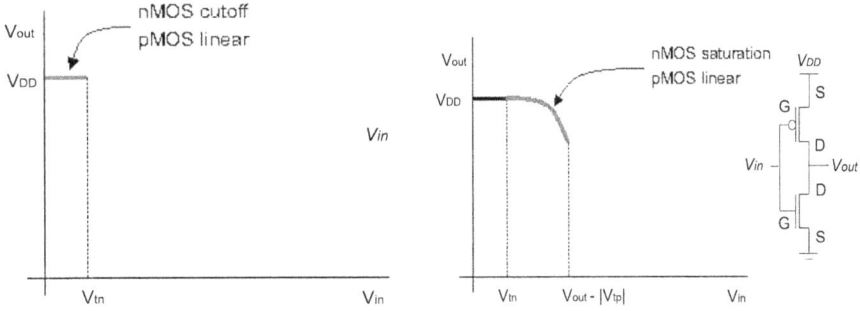

FIGURE 1.8 CMOS inverter VTC

The transition region pitch is a measure of quality steep pitches that exact switching. The noise can be calculated by assessing the difference between the input to the output for every region of the ON-OFF mode of operation [12].

Figure 1.8 shows the mode equations for PMOS and NMOS.

Below are the mode equations for PMOS and NMOS. Figures 1.9a and 1.9b show the operation modes of CMOS.

- Setting PMOS linear IDS equal to NMOS saturation IDS

$$k_n\left(\frac{\left(V_{in}-V_{tn}\right)^2}{2}\right)=k_p\left(\left(V_{in}-V_{DD}-V_{tp}\right)\left(V_{out}-V_{DD}\right)-\frac{\left(V_{out}-V_{DD}\right)^2}{2}\right) \qquad (1)$$

$$\frac{\left(V_{out}-V_{DD}\right)^2}{2}-\left(V_{in}-V_{DD}-V_{tp}\right)\left(V_{out}-V_{DD}\right)+\frac{k_n\left(V_{in}-V_{tn}\right)^2}{2}=0 \qquad (2)$$

$$\left(V_{out}-V_{DD}\right)=\left(V_{in}-V_{DD}-V_{tp}\right)+\sqrt{\left(V_{in}-V_{DD}-V_{tp}\right)^2-\frac{k_n}{k_p}\left(V_{in}-V_{tn}\right)^2} \qquad (3)$$

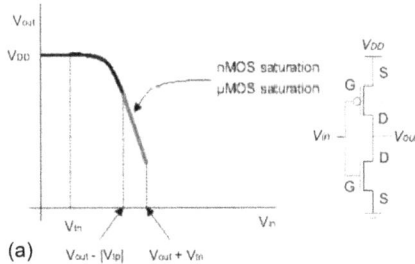

FIGURE 1.9A CMOS saturation mode

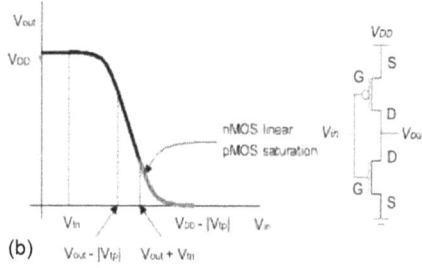

(b)

FIGURE 1.9B   NMOS linear and PMOS saturation modes

$$V_{out} = \left(V_{in} - V_{tp}\right) + \sqrt{\left(V_{in} - V_{DD} - V_{tp}\right)^2 - \frac{k_n}{k_p}\left(V_{in} - V_{tn}\right)^2} \qquad (4)$$

- Setting NMOS linear I_{DS} equal to PMOS saturation I_{DS}

$$k_p\left(\frac{\left(V_{in} - V_{DD} - V_{tp}\right)^2}{2}\right) = k_n\left(\left(V_{in} - V_{tn}\right)V_{out} - \frac{V_{out}^2}{2}\right) \qquad (5)$$

$$\frac{V_{out}^2}{2} - \left(V_{in} - V_{tn}\right)V_{out} + \frac{k_p\left(V_{in} - V_{DD} - V_{tp}\right)^2}{2} = 0 \qquad (6)$$

$$V_{out} = \left(V_{in} - V_{tn}\right) - \sqrt{\left(V_{in} - V_{tn}\right)^2 - \frac{k_p}{k_n}\left(V_{in} - V_{DD} - V_{tp}\right)^2} \qquad (7)$$

Figure 1.10 shows the linear and cutoff modes of CMOS.
   All modes are summarized in Table 1.1.

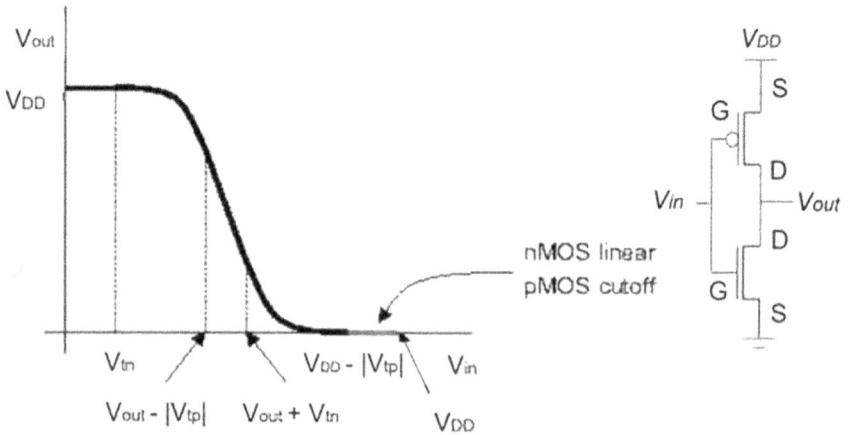

FIGURE 1.10   NMOS linear and PMOS cutoff modes

TABLE 1.1
CMOS modes

V_{IN}	PMOS mode	NMOS mode	V_{OUT}
$V_{in} < V_{tn}$	Linear	Cutoff	V_{DD}
$V_t < V_{in} < V_{out} - V_{tp}$	Linear	Saturation	$V_{out} = \left(V_{in} - V_{tp}\right) + \sqrt{\left(V_{in} - V_{DD} - V_{tp}\right)^2 - \dfrac{k_n}{k_p}\left(V_{in} - V_{tn}\right)^2}$
			Transition, V_{OUT} drops
$V_{out} - \left|V_{tp}\right| < V_{in} < V_{out} + V_{tn}$	Saturation	Saturation	
$V_{out} + V_{tn} < V_{in} < V_{DD} - V_{tp}$	Saturation	Linear	$V_{out} = \left(V_{in} - V_{tn}\right) - \sqrt{\left(V_{in} - V_{tn}\right)^2 - \dfrac{k_p}{k_n}\left(V_{in} - V_{DD} - V_{tp}\right)^2}$
$V_{in} > V_{DD} - \left|V_{tp}\right|$	Cutoff	Linear	0

1.3   THE LOGIC CIRCUIT OF CMOS

Combinational logic gates with one or more inputs and one output make up static CMOS circuits. Here are some significant CMOS logic gates [14].

1.3.1   THE INVERTER

1.3.1.1   Overview

The inverter is the most basic logic gate. Understanding how an inverter works and its characteristics will make it much simpler to examine other logic gates, adders, and other components of digital design and memory devices [15]. Figure 1.11 shows the operation of a CMOS inverter in low (0) and high (1) inputs.

The NOT gate is another name for the CMOS inverter. The circuit shown above demonstrates that an N-channel MOSFET (NMOS) and a P-channel MOSFET make up a CMOS inverter (PMOS). The NMOS transistor is OFF and the PMOS transistor is ON when the input A is LOW, or logic 0. The VDD has a path to the output thanks to the P-channel MOSFET. The output is HIGH as a result.

Logic 1 as is NMOS is ON and PMOS is OFF when the input is HIGH. The output signal is LOW and connected to GND.

The operation of the inverter is summarized in Table 1.2.

1.3.1.2   Simulation

We can use computer-aided design (CAD) software to simulate the operation of the CMOS inverter described above. Figure 1.12 shows the circuit implementation of a CMOS inverter in LTspice for simulation.

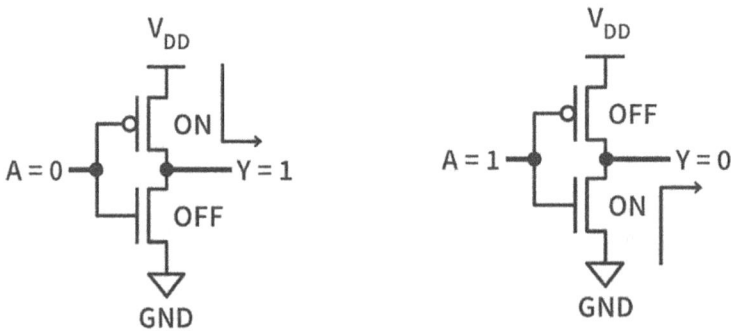

FIGURE 1.11   CMOS inverter

TABLE 1.2
CMOS inverter truth table

Input	Output
0	1
1	0

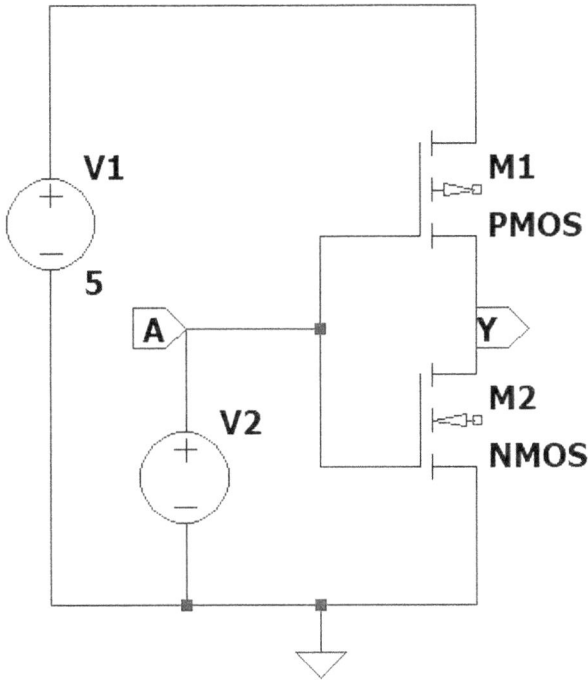

FIGURE 1.12 CMOS inverter schematic

Now that we have demonstrated how a CMOS device operates and how to design basic logic CMOS gates, we will see how one can use CAD tools to simulate our design and confirm its feasibility. There are many tools that designers use, from freeware like LTspice to licensed tools such as Cadence [16].

In this chapter, we use LTspice for both its user-friendly interface and availability (free of charge).

To design a device, there is a set of parameters that should be provided in a ".txt" file to define the device. Without a template, designing can be challenging for both experienced and junior users. Alternatively, one can use predefined NMOS and PMOS devices provided in LTspice and modify parameters to meet the needs, requirements, and specifications of one's device.

The inverter in Figure 1.12 uses two voltage sources: the biasing power (V1) and the signal to be inverted (V2). This signal is presented to the input as a pulsed voltage source with high and low levels representing 1s and 0s, respectively.

Figure 1.13 shows the input voltage V(a) and its corresponding inverted output V(y). This confirms the proper operation of the circuit as the inverter of CMOS with results matching the truth table (see Table 1.2 for context).

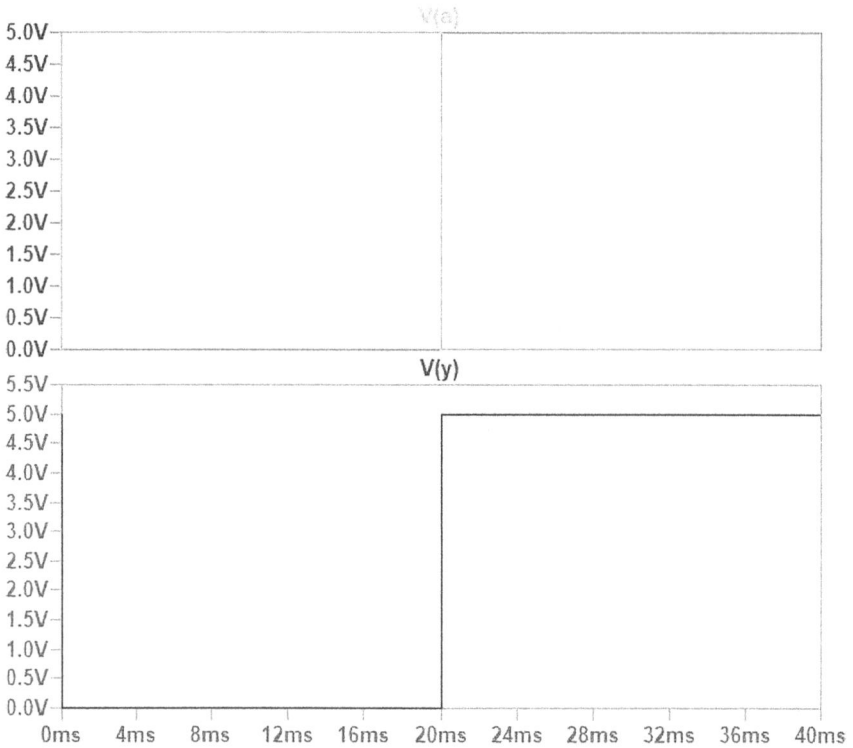

FIGURE 1.13   Simulation results of CMOS inverter

1.3.2   CMOS NAND GATE

1.3.2.1   Overview

Two N-channel MOSFETs are connected in series between Y (the output) and GND, and two P-channel MOSFETs are connected in parallel between VDD and Y in a two-input NAND gate. A NAND gate's schematic is shown in Figure 1.14 in LTspice for simulation.

At least one of the NMOS transistors will be OFF if either input A or B is low (logic 0). Since the NMOS transistors are wired in series and lead to the GND, this disrupts the path from Y to GND. To complete a path from Y to VDD in this instance, however, at least one of the PMOS transistors will be VDD. This makes the output Y high (logic 1) [17].

If A and B are high (logic 1), both NMOS transistors are ON. This completes the path from Y to GND. This makes Y low (logic 0). The output Y will be high for all other combinations of inputs A and B. The truth table of the NAND logic gate is given below [18]. The truth table below shows the inputs/output of the NAND gate.

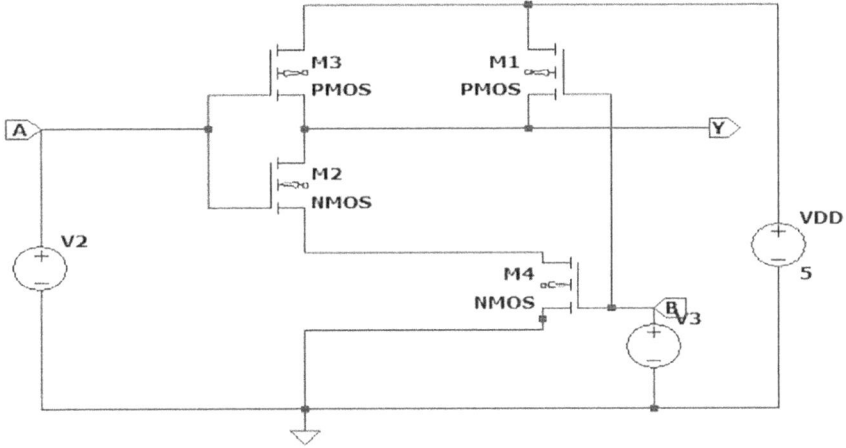

FIGURE 1.14 CMOS NAND circuit

1.3.2.2 Simulation

The circuit of Figure 1.14 is an LTspice implementation of the NAND logic gate. Upon providing pulsed inputs V(a) and V(b), as may be seen, the result of V(y) matches the NAND function behavior demonstrated in Table 1.3 and depicted in the simulation results of Figure 1.15.

1.3.3 CMOS NOR GATE

1.3.3.1 Overview

The NMOS transistors and PMOS transistors are coupled in series and parallel, respectively, in a two-input NOR gate. At least one NMOS transistor pulls the output low when at least one of the inputs is high. Only when both inputs are low does the output become high. Figure 1.16 represents the schematic of a NOR gate in LTspice for simulation. The truth table of the NOR logic gate is presented in Table 1.4 (Figure 1.17).

TABLE 1.3
NAND truth table

Input A	Input B	Output
0	0	1
0	1	1
1	0	1
1	1	0

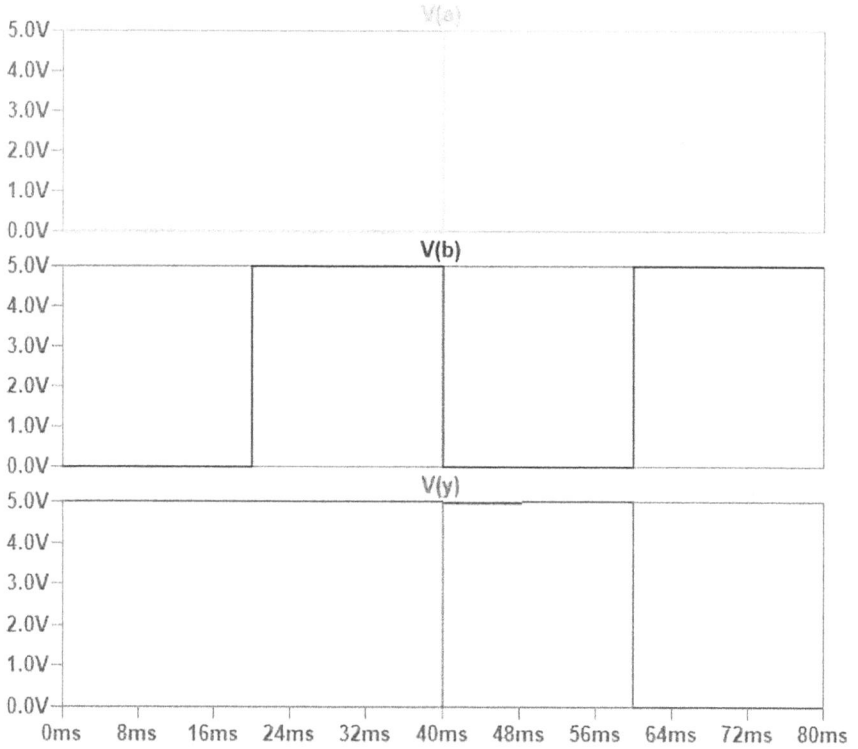

FIGURE 1.15   CMOS NAND simulation

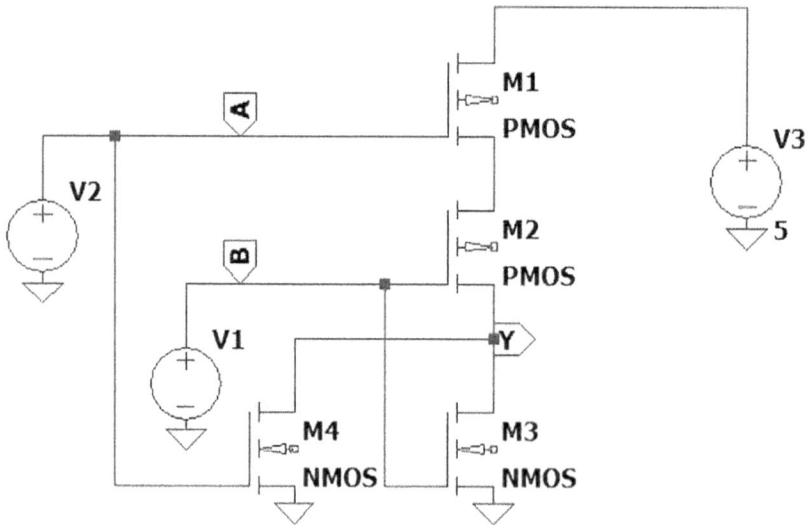

FIGURE 1.16   CMOS NOR circuit

TABLE 1.4
NOR truth table

Input A	Input B	Output
0	0	1
0	1	0
1	0	0
1	1	0

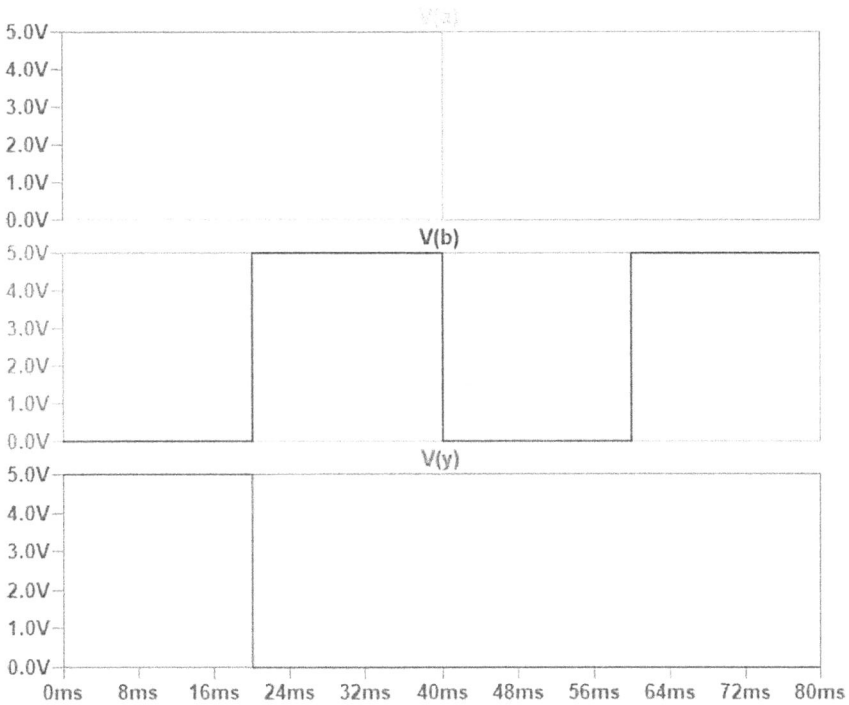

FIGURE 1.17 CMOS NOR simulation

1.3.3.2 Simulation

The simulated results for the NOR circuit design are shown in Figure 1.16 with their corresponding response to given inputs matching the NOR truth table of Table 1.4.

1.4 CMOS TECHNOLOGY AND APPLICATIONS

Due to its adaptability and efficiency in the use of electricity, CMOS is the technology of choice for the manufacture of integrated circuits (ICs). The low power design is the most dependable of the current technologies and offers the benefit of little heat dissipation [19]. Depending on the circuit layout, P- and N-type transistors can be set up to create logic gates.

CMOS technology is one of the most popular technologies in the computer chip design industry. This technology makes use of both the P-channel and N-channel semiconductor biases. CMOS is one of the most popular MOSFET technologies available [20]. This is the dominant semiconductor technology for all semiconductor devices including memory devices, volatile and non-volatile, and logic gate circuits.

The N-channel MOSFET and the P-channel MOSFET are both made with matching properties thanks to their design (ON and OFF). The primary benefit of CMOS technology over bipolar or the formerly common NMOS technologies is its exceptionally low power consumption in static settings because it only consumes power during switching operations [21].

When compared to bipolar or NMOS technology, this enables the integration of a significantly greater number of sensing gates on the VLSI IC. It is simpler to create various logic functions when NMOS and PMOS bias are combined in CMOS logic gates. The size of the transistor can be varied and further shrunk thanks to developments in CMOS IC production methods [22].

By scaling down the transistor, it is possible to incorporate more logical operations into the same IC without sacrificing performance. CMOS IC technology was initially employed to create digital logic ICs. CMOS technology is now used in analog ICs and mixed-signal designs because of its low cost and greater functionality [23]. CMOS logic has two different modes: low power dispersion and high noise perimeters. In both modes, it operates over a wide range of source and input voltages.

1.5   LAYOUT OF CMOS

The CMOS design layouts are based on the following components:

(a) Substrates.
(b) Wells: for NMOS and PMOS devices, respectively, wells are P-type and N-type.
(c) Diffusion areas: in these regions, the transistors are produced and are referred to as an active layer. For NMOS and PMOS transistors, respectively, these are denoted by n+ and p+.

Figure 1.18 shows a layout for a CMOS with P-substrate and N-well.

The circuit architecture (mask layout) and initial transistor scaling are the first steps in the iterative process of designing the physical (mask layout) of CMOS logic gates (to realize the desired performance specifications). Based on the fan-out, the number of devices, and the anticipated length of the interconnection lines, the designer can only make an estimate of the overall parasitic load at the output node at this time [24].

If the logic gate contains more than four transistors, the ideal ordering of the transistors in logic gates with more than four transistors. Now, it is possible to design

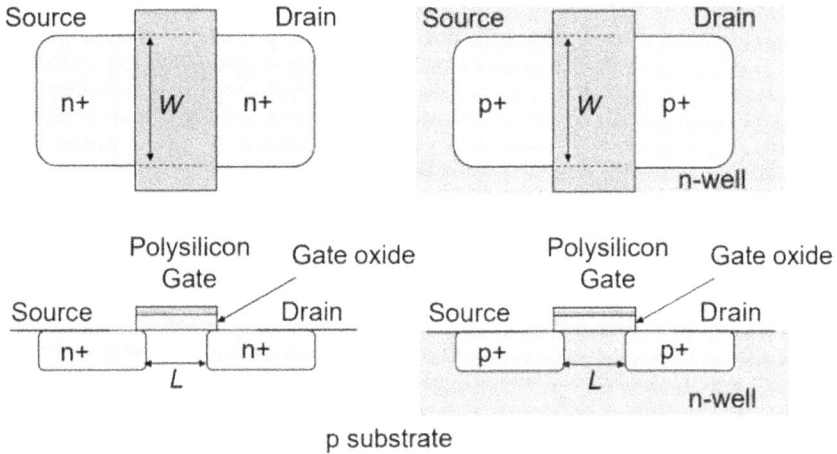

FIGURE 1.18   CMOS Inverter layout

a straightforward stick diagram layout that displays the contacts' locations, as well as the transistors' local interconnections [25]. Figure 1.19 depicts a mask layout for a CMOS inverter architecture.

The layout's topology is created by drawing the mask layers in accordance with the layout design guidelines (using a layout editor tool). To account for all design principles, this technique may need to go through multiple tiny iterations, but the fundamental topology shouldn't be significantly altered. The finished layout is subjected to a circuit extraction technique after the final design rule check (DRC) to

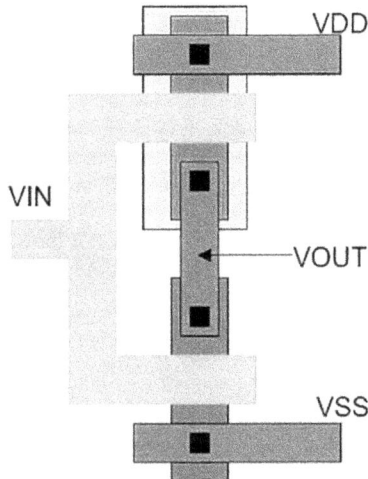

FIGURE 1.19   Mask layout of CMOS inverter

ascertain the true transistor sizes and, more crucially, the parasitic capacitances at each node. The extraction step's outcome is often in-depth [26].

1.6 CONCLUSION

The CMOS circuit configuration process comprises characterizing circuit information sources and results, hand computations, circuit recreations, circuit format, reconsideration of circuit sources of info and results, creation, and testing. The analysis of inverters is often extended to elucidate the behaviors of more complex gates such as NAND, NOR, or XOR, which successively form the building blocks for modules like multipliers and processors. In this chapter, we specialize in one single incarnation of the inverter gate, the static CMOS inverter, or the CMOS inverter short. We implement and simulate the inverter, NAND, and NOR gates using the LTspice CAD tool. The process is like other CAD tools and can be used to simulate more complex circuits by combining or modifying these basic elements. A green revolution in the CMOS domain using low-cost MOSFET is not far from being realized [27, 28].

REFERENCES

1. B. Hoeneisen and C. A. Mead, "Fundamental limitations in microelectronics—I. MOS technology." *Solid-State Electronics*, 15(7), 819–829 (1972).
2. T. Sakurai and A. R. Newton, "Alpha-power law MOSFET model and its applications to CMOS inverter delay and other formulas." *IEEE Journal of Solid-State Circuits*, 25(2), 584–594 (1990). doi: 10.1109/4.52187.
3. J.-H. Wang, H.-H. Hsieh, and L.-H. Lu, "A 5.2-GHz CMOS T/R switch for ultra-low-voltage operations." *IEEE Transactions on Microwave Theory and Techniques*, 56(8), 1774–1782 (2008).
4. R. H. Caverly, "Linear and nonlinear characteristics of the silicon CMOS monolithic 50-ohm microwave and RF control element." *IEEE Journal of Solid-State Circuits*, 34(1), 124–126 (1999).
5. I. Baturone, S. Sanchez-Solano, A. Barriga, and J. L. Huertas, "Implementation of CMOS fuzzy controllers as mixed-signal integrated circuits." *IEEE Transactions on Fuzzy Systems*, 5(1), 1–19 (1997). doi: 10.1109/91.554443.
6. R. A. Blauschild, P. A. Tucci, R. S. Muller, and R. G. Meyer, "A new NMOS temperature-stable voltage reference." *IEEE Journal of Solid-State Circuits*, SC-13, 767–774 (1978).
7. X. Zhang and E. I. EI-Masry, "A novel CMOS OTA based on body-driven MOSFETs and its applications in OTA-C filters." *IEEE Transaction on Circuits and Systems Part I: Fundamental Theory and Applications*, 54(6), 1204–1212 (2007).
8. A. A. Cherepanov, I. L. Novikov, and V. Y. Vasilyev, "An evaluation of CMOS inverter operation under cryogenic conditions." In *2018 19th International Conference of Young Specialists on Micro/Nanotechnologies and Electron. Devices (EDM)*, 2018, pp. 40–43. doi: 10.1109/EDM.2018.8435038.
9. H. Salama, B. Saman, E. Heller, R. H. Gudlavalleti, R. Mays, and F. Jain, "Twin drain quantum well/quantum dot channel spatial wave-function switched (SWS) FETs for multi-valued logic and compact DRAMs." *International Journal of High Speed Electronics and Systems*, 27(03n04), 1840024 (2018).

10. S. Mahapatra, V. Vaish, C. Wasshuber, K. Banerjee, and A. M. Ionescu, "Analytical modeling of single electron transistor for hybrid CMOS-SET analog IC design." *IEEE Transactions on Electron Devices*, 51(11), 1772–1782 (2004). doi: 10.1109/TED.2004.837369.

11. L. Bisdounis, S. Nikolaidis, and O. Koufopavlou, "Analytical transient response and propagation delay evaluation of the CMOS inverter for short-channel devices." *IEEE Journal of Solid-State Circuits*, 33(2), 302–306 (1998).

12. M. R. Bruce and V. J. Bruce, "ABCs of photon emission microscopy." *Electronic Device Failure Analysis*, 5(3), 13–22 (2003).

13. C.-C. Huang, C.-Y. Wang, and J.-T. Wu, "A CMOS 6-bit 16-GS/s time-interleaved ADC using digital background calibration techniques." *IEEE Journal of Solid-State Circuits*, 46(4), 848–858 (2011). doi: 10.1109/JSSC.2011.2109511.

14. C. Chen et al., "Integrating poly-silicon and InGaZnO thin-film transistors for CMOS inverters." *IEEE Transactions on Electron Devices*, 64(9), 3668–3671 (2017). doi: 10.1109/TED.2017.2731205.

15. C. Schlünder, F. Proebster, J. Berthold, K. Puschkarsky, G. Georgakos, W. Gustin, and H. Reisinger, "Circuit relevant HCS lifetime assessments at single transistors with emulated variable loads." In *2017 IEEE International Reliability Physics Symposium (IRPS)*, 2017, pp. 2D-2. IEEE.

16. H. Salama, B. Saman, R. H. Gudlavalleti, P. Y. Chan, R. Mays, B. Khan, E. Heller, J. Chandy, and F. Jain, "Simulation of stacked quantum dot channels SWS-FET using multi-FET ABM modeling." *International Journal of High Speed Electronics and Systems* 28(03n04), 1940025 (2019).

17. R. H. Gudlavalleti, B. Saman, R. Mays, H. Salama, E. Heller, J. Chandy, and F. Jain, "A noveladdressing circuit for SWS-FET based multi-valued dynamic random access memory array." *International Journal of High Speed Electronics and Systems*, 29, 2040009 (2020).

18. A. Boglietti, P. Ferraris, M. Lazzari, and M. Pastorelli, "Influence of the inverter characteristics on the iron losses in PWM inverter-fed induction motors." *IEEE Transactions on Industry Applications*, 32(5), 1190–1194 (1996).

19. B. G. Lee, A. V. Rylyakov, W. M. J. Green, S. Assefa, C. W. Baks, R. Rimolo-Donadio, D. M. Kuchta et al., "Monolithic silicon integration of scaled photonic switch fabrics, CMOS logic, and device driver circuits." *Journal of Lightwave Technology*, 32(4), 743–751 (2013).

20. B. Lu and S. K. Sharma, "A literature review of IGBT fault diagnostic and protection methods for power inverters." *IEEE Transactions on Industry Applications*, 45(5), 1770–1777 (2009). doi: 10.1109/TIA.2009.2027535.

21. H. T. Bui, Y. Wang, and Y. Jiang, "Design and analysis of low-power 10-transistor full adders using novel XOR-XNOR gates." *IEEE Transactions on Circuits and Systems. Part II: Analog and Digital Signal Processing*, 49(1), 25–30 (2002). doi: 10.1109/82.996055.

22. M. G. Johnson, "A symmetric CMOS NOR gate for high-speed applications." *IEEE Journal of Solid-State Circuits*, 23(5), 1233–1236 (1988).

23. S. Kang, B. Choi, and B. Kim, "Linearity analysis of CMOS for RF application." *IEEE Transactions on Microwave Theory and Techniques*, 51(3), 972–977 (2003). doi: 10.1109/TMTT.2003.808709.

24. H. Salama, B. Saman, R. Gudlavalleti, R. Mays, E. Heller, J. Chandy, and F. Jain, "Compact 1-bit full adder and 2-bit SRAMs using n-SWS-FETs." *International Journal of High Speed Electronics and Systems*, 29(01n04), 2040013 (2020).

25. R. J. Baker, et al. *CMOS: Circuit Design, Layout, and Simulation*. Rev. 4th ed., 2019. IEEE Press.

26. K. H. Cirne and S. P. Gimenez, "Layout design of CMOS inverters with circular and conventional gate MOSFETs by using IC station mentor." In: *Proc. 9th Microeletronics Students Forum*, SBmicro 2009 (2009).

27. L. Zhang, R. Fakhrabadi, M. Khoshkhoo, and S. Husien, "A robust second-order conic programming model with effective budget of uncertainty in optimal power flow problem.", *Journal of Energy and Power Technology*, (4) 1–15, (2022).

28. H. Salama, "Quantum dot solar cells." arXiv Preprint ArXiv:2211.06898 (2022).

2 Conventional CMOS circuit design

Samir Labiod, Abdelmalek Mouatsi,
Zakaria Hadef, and Billel Smaani

CONTENTS

2.1 INTRODUCTION

Digital CMOS (complementary metal-oxide semiconductor) integrated circuits (ICs) have been the driving force behind very large-scale integration (VLSI) for high-performance computing and engineering applications [1–3]. Low power, reliable performance, and circuit techniques for high speed, such as using dynamic circuits and ongoing improvements in processing technology, are prominent features and the reason behind the constant demand for digital CMOS ICs. With this kind of technology, it can be seen that the level of integration which once existed in several millions of transistors for logic chips and reached an even higher level in the case of memory

DOI: 10.1201/9781003359234-2

21

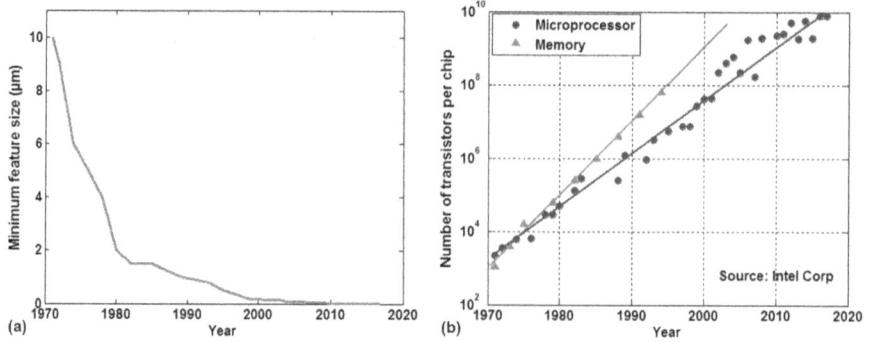

FIGURE 2.1 Evolution of technologic in integrated circuits versus time: (a) minimum feature size and (b) level of memory and logic chips integrations

ones, is now available in only one chip. It presents an immense challenge for chip developers in processing, methodology, design, testing, and project management.

Advances in device manufacturing technology allow the steady reduction of minimum feature size. Figure 2.1a shows the progress of the minimum feature size of transistors in integrated circuits since the late 1970s. In 1980, at the beginning of the VLSI era, the typical minimum feature size was 2 μm, and a feature size of 14 nm was expected around the year 2017. The actual development of the technology, however, has far exceeded these expectations. A minimum feature size of 0.25 μmn had been reached by 1995.

When the integration density of circuits is inspected, there is clearly a distinction between the memory and the logic chips. Figure 2.1b shows the level of integration through time for memory and logic circuits, beginning in 1970. The increase of transistors number has continued at an exponential rate over the last three decades, effectively confirming Gordon Moore's prediction on the growth rate of chip complexity, which was made in the early 1960s (Moore's Law) [4]. In this chapter, we present a brief overview of CMOS process integration. Process integration refers to the well-defined collection of semiconductor processes required to fabricate CMOS integrated circuits [5, 6]. We provide more information and examples related to the layout of the different interconnections, and MOSFETs. Design rules and the main fundamental layout techniques have been covered [7].

To illustrate the effect of miniaturization, different simulations were performed for long and short-channel MOSFETs models. The parasitic effect plays a very important role in the field of CMOS circuit design [8], and for this, it is essential to do some simulations to illustrate these effects.

A current mirror is a basic building block for analog circuit design [9]. Fundamental understanding and layout techniques for analog circuit design have been presented and simulated.

Finally, the main concepts for digital circuit design have been presented, such as the basic CMOS static logic gates inverter, NAND, and NOR) and arithmetic function (full adder). SPICE simulations have been performed for DC and dynamic characteristics.

2.2 CMOS FABRICATION TECHNOLOGY

The CMOS (complementary metal-oxide silicon) fabrication technology is recognized as the leader of VLSI systems technology.

2.2.1 WELL FORMATION

Figure 2.2 shows cross-sections of the wafer after each processing step involved in forming the n-well [10, 11]. Figure 2.3a illustrates the bare substrate before processing. Changing the substrate from p-type to n-type in the region of the well by adding enough dopants into it is what is required to form the n-well. The growth of a protective layer of oxide over the entire wafer, and relocating it to where we want the wells, is necessary for us to define what regions receive n-wells.

In a high-temperature (typically 900–1200 °C) furnace, the wafer is first oxidized, due to Si and O_2 reacting and becoming SiO_2 on the wafer surface (Figure 2.2b). The pattern of oxide is obligated to define the n-well. An organic photoresist that softens when exposed to light is spun onto the wafer (Figure 2.2c). The photoresist is exposed through the n-well mask (Figure 2.3b), which allows light to pass through only where the well should be. The softened photoresist is removed to expose the oxide (Figure 2.2d).

The oxide part that is not protected by the photoresist is etched with hydrofluoric acid (HF) (Figure 2.2e), then the acids mixture called piranha etch is for cleaning out the remaining photoresist (Figure 2.2f). The well is formed where the substrate is not covered with oxide. Two ways to add dopants are diffusion and ion implantation. In the diffusion process, the wafer is placed in a furnace with a gas containing the dopants. When heated, dopant atoms diffuse into the substrate. Notice how the well is wider than the hole in the oxide on account of lateral diffusion (Figure 2.2g). With ion implantation, dopant ions are accelerated through an electric field and blasted into the substrate. In either method, the atoms are prevented from entering the substrate where no well is intended. Finally, stripping the remaining oxide with HF leaves the bare wafer with wells in the appropriate places.

2.2.2 MOSFET FABRICATION PROCESS

A general CMOS process flow is demonstrated in Figure 2.3. Fabrication of NMOS and PMOS devices is detailed in [13, 14]. The first step, Figure 2.3a, is to grow a thin pad oxide on the surface of the entire wafer. Depositing nitride and photoresist layers follow this. The photoresist is then patterned using the active mask. The remaining photoresist, seen in Figure 2.3a, ultimately defines the openings in the field oxide (FOX) [15].

In Figure 2.3b, the areas not covered by the photoresist are etched. The etching extends down into the wafer so that shallow trenches are formed. In Figure 2.3c, the shallow trenches are filled with SiO_2. These trenches isolate the active areas and form the field regions (FOX). This type of device isolation is called shallow trench isolation (STI).

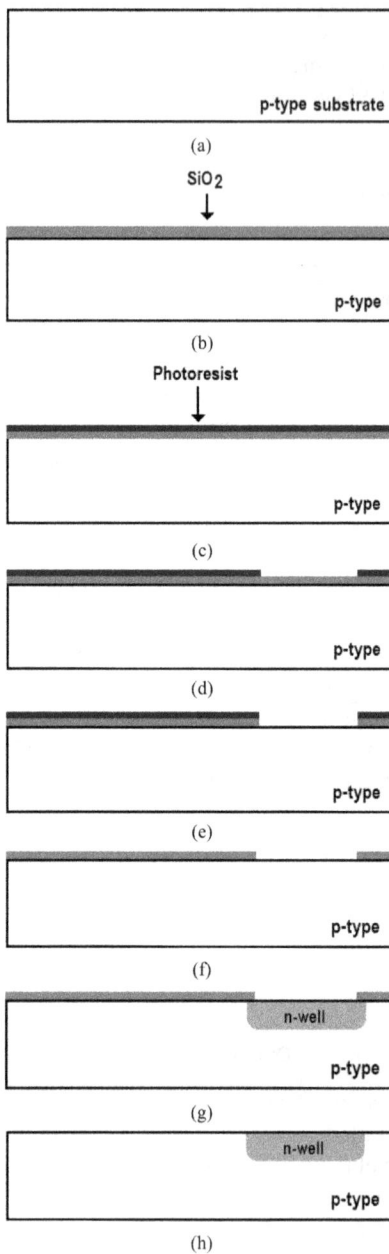

FIGURE 2.2   Cross-section while manufacturing the n-well [12].

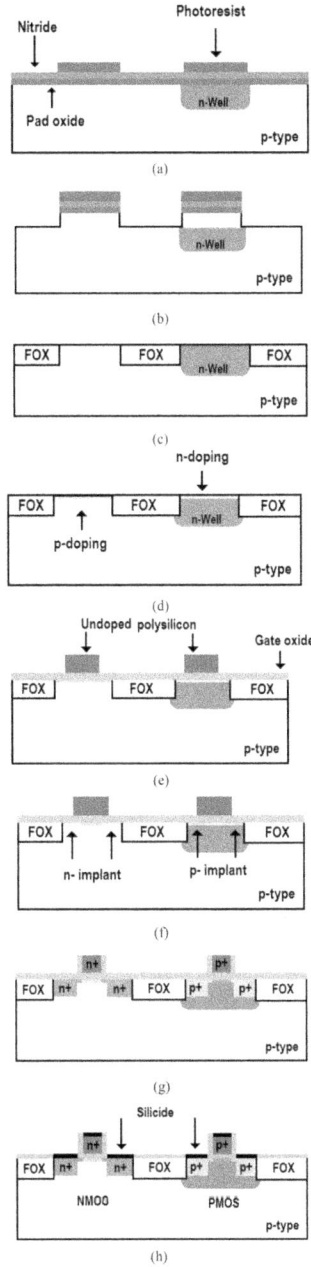

FIGURE 2.3 General CMOS process flow [16].

In Figure 2.3d, adjusting the threshold voltages of the devices is performed with two separate implants. A photoresist is patterned (twice) to select the areas for threshold voltage adjustment.

Figure 2.3e shows the results after the deposition and patterning of polysilicon. This is followed by various implants. In Figure 2.3f, we see a shallow implant forming the MOSFET's lightly doped drains (LDD). The LDD implants prevent the electric field directly next to the source/drain regions from becoming too high. Note that the poly gate is used as a mask during this step.

The next step is to grow a spacer oxide on the sides of the gate poly (Figure 2.3g). After the spacer is grown, the n+/p+ implants are performed. This implant dopes the areas used for the source and drain of the MOSFETs as well as the gate poly. The last step is to silicide the source and drain regions of the MOSFET. Finally, note that the process sequence seen in Figure 2.3 is often referred to, in the manufacturing process, as the front-end-of-line (FEOL). The fabrication of the metal layers and associated contacts/vias is referred to as the back-end-of-line (BEOL).

2.2.3   INTERCONNECTIONS

Typical metalization is used with aluminum, while contact holes are filled by a plug of tungsten. Low resistivity and low dielectric isolation layers have been used to minimize the RC time constants of the interconnection lines. Metalization level and a via are established at the same time in the dual damascene technique. Figure 2.4 shows an example layout and cross-section view. The vial layer connects metal and metal2. In the location indicated, the via layer specifies that the insulator will be removed. Once metal2 is set, the two metals are connected by the plug [17]. Notice

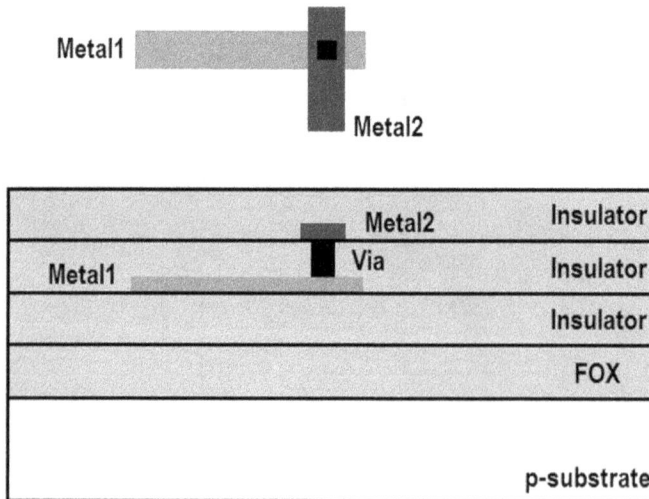

FIGURE 2.4   Cross-section and layout views of metal1/metal2 interconnection [14]

that in the case of using more than two layers of metal, via2 would connect metal2 to meta3, and via3 would connect metal3 to metal4.

The contact layer connects metal1 to either active (n+/p+) or poly. Unless we want to form a rectifying contact (a Schottky diode), we never connect metal directly to the substrate or well.

Further, we won't connect metal to poly without having the silicide in place. Never put a suicide block around a contact to poly [18].

Figure 2.5a shows a layout and corresponding cross-sectional view of the layers metal1, contact, and poly (a contact to poly). Figure 2.5b shows a connection to n+ and p+.

2.2.4 LAYOUT OF MOS TRANSISTOR

Layout and cross-section views of the NMOS device are shown in Figure 2.6. We recall that the MOSFET is a four-terminal device drain, source, gate, and substrate. Figure 2.6 shows the bulk connection in the layout and in the schematic. In an n-well process, the bulk is tied to ground, so the bulk connection is normally not shown in the schematic symbol. Source and drain are interchangeable.

Cross-section and layout views for the NMOS device are shown in Figure 2.7. Note how we lay the device out in an n-well. Also seen in the figure is the schematic symbol for the PMOS device. Again, the source and drain of the MOSFET are interchangeable. The n-well is normally tied to the highest potential, V_{DD}, in the circuit to keep the parasitic n-well/p-substrate diode from forward biasing.

2.2.5 LONG AND SHORT-CHANNEL MOSFETS

In this section, we will perform electrical simulations to present the main results for long and short-channel MOSFET. The typical parameters for the sizes and electrical parameters used for long and short-channel CMOS are shown in Table 2.1.

Figure 2.8a shows the I-V curves for a 50/2 NMOS device (actual size of 2.5 μm/100 nm device) with different values of V_{GS}=300 mV, 350 mV, and 400 mV. The current shows significant variation as V_{DS} changes. When I_D change with V_{DS} and V_{GS}=350 mV, the drain current is 10 μA as an approximation. Figure 2.8b presents the output resistance for a 50/2 NMOS device. We get the output resistance by taking the reciprocal of the drain current's derivative in Figure 2.8a. To calculate the V_{DSsat}, we can look at the point where the output resistance starts to increase. In the case where V_{GS}=350 mV, we have V_{DSsat}=50 mV. However, notice that if we use larger V_{DS}, we get considerably higher output resistances.

Figure 2.9a shows a plot of drain current versus V_{GS} for a MOSFET in the short-channel process. The threshold voltage is calculated by the linear extrapolation back to the axis of the gate voltage. In Figure 2.9b, we take the derivative of (a) to get the g_m of the device. We can linearly extrapolate the threshold voltage back to the gate voltage axis. The two methods give different results; from (a) the threshold voltage is V_{THN}=358.3 mV, while from (b) V_{THN}=281.4 mV.

Figure 2.10a depicts the I-V curves for a 10/2 NMOS device (actual size of 10 μm/2 μm device) with different values of V_{GS}=1V, 1.05V, and 1.1V. The current shows

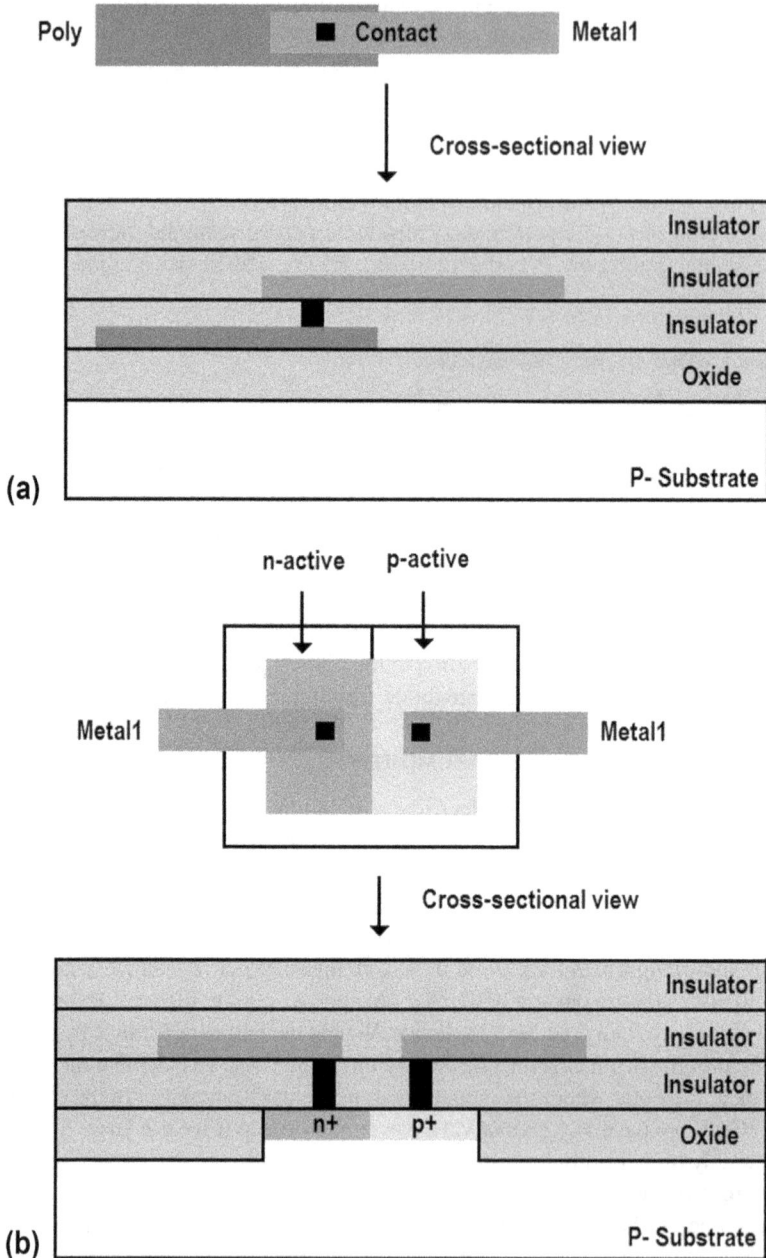

FIGURE 2.5 Metal connections: (a) metal connecting to poly and (b) metal connecting to n and p active [14]

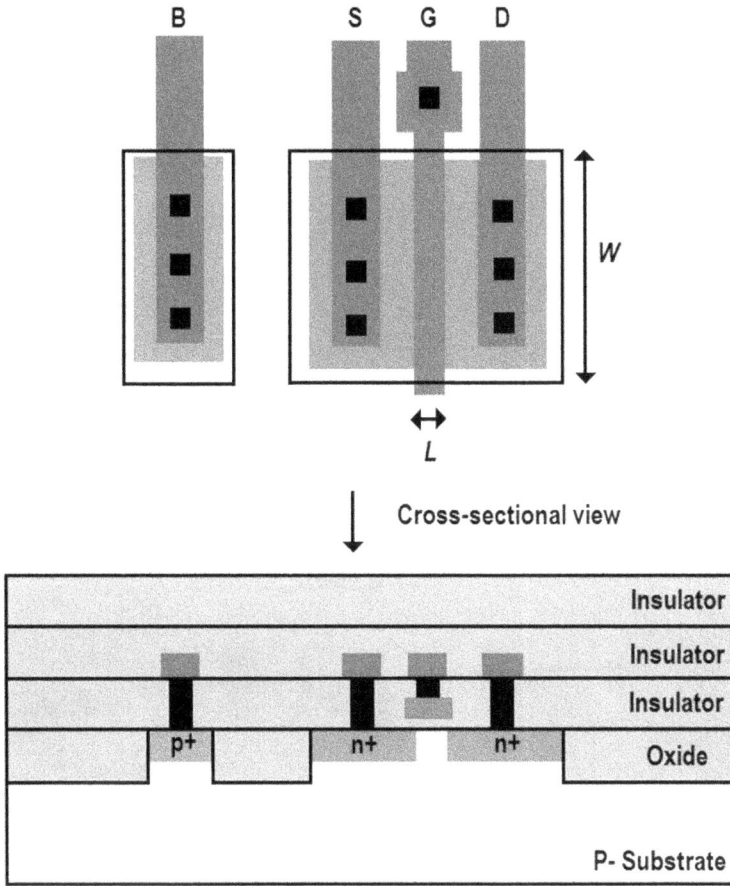

FIGURE 2.6 Cross-section and layout views for the NMOS device [19]

significant variation as V_{DS} changes. From the I-V characteristics, *the* drain current is approximately 20 μA for V_{GS}=1.05V. The output resistance for a 10/2 NMOS device is presented in Figure 2.10b for different values V_{GS}=1V, 1.05V, and 1.1V.

Figure 2.11a and Figure 2.11b depict the variation of drain current and transconductance against V_{GS} respectively for a MOSFET in the long-channel process. From (a), the threshold voltage is V_{THN}=1.12V, while from (b) V_{THN}=800mV.

2.3 PARASITICS ASSOCIATED WITH CMOS TECHNOLOGY

2.3.1 Rc delay through the n-well

In this part, we notice that the n-well can be used as a diode and as a resistor. Parasitic resistance and capacitance related to the n-well are shown in Figure 2.12.

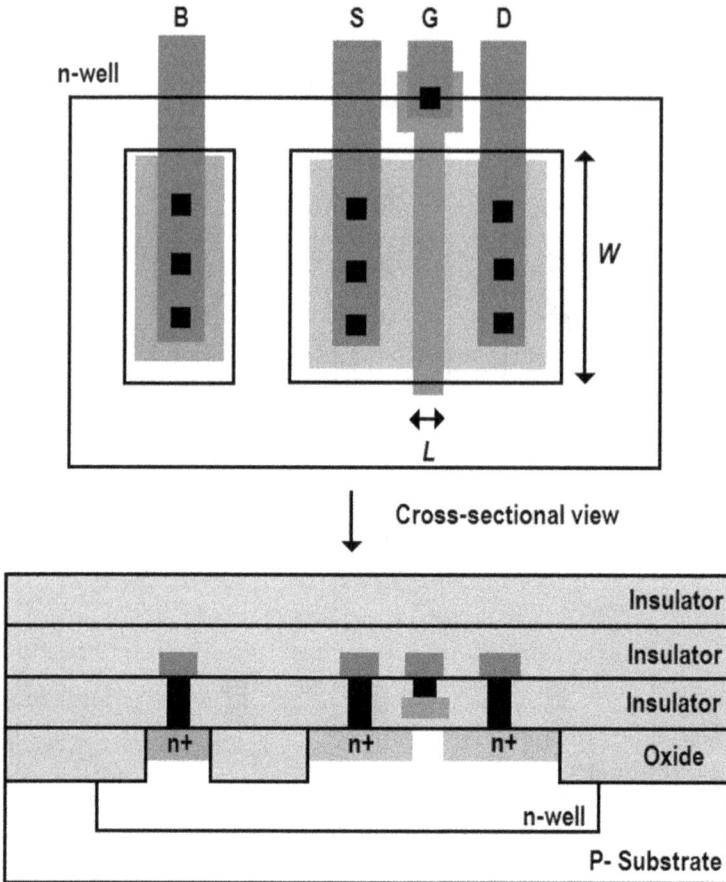

FIGURE 2.7 Cross-section and layout views for the PMOS device [19]

If we apply a voltage pulse to one side of the n-well resistor and measure the delay time at the 50% points of the pulses, the pulse will occur after the delay time [20–23].

Calculation of the delay through a distributed RC of the circuit is shown in Figure 2.13. The delay to node A is estimated using

$$t_{dA} = 0.7R_{sheet}C_{sheet} \tag{1}$$

The delay to the second node is the sum of the delay to the first node plus the delay associated with charging the capacitance at the second node through $2R_{square}$ or

$$t_{dB} = 0.7\left(R_{sheet}C_{sheet} + 2R_{sheet}C_{sheet}\right) \tag{2}$$

TABLE 2.1

The main parameters for long and short-channel MOSFET [19]

	NMOS		PMOS	
Parameters	**Short channel**	**Long channel**	**Short channel**	**Long channel**
Scale	50 *nm*	1 *μm*	50 *nm*	1 *μm*
W/L	50/2	10/2	100/2	30/2
V_{DSsat}	50 *mV*	250 *mV*	50 *mV*	250 *mV*
V_{GS}	350 *mV*	1.05 *V*	350 *mV*	1.15 *V*
V_{TH}	280 *mV*	800 *mV*	280 *mV*	900 *mV*
I_D	10 *μA*	20 *μA*	10 *μA*	20 *μA*
$\partial V_{TH} / \partial T$	−0.6 *mV/C°*	−1 *mV/C°*	−0.6 *mV/C°*	−1.4 *mV/C°*
C'_{ox}	25 *fF/μm²*	1.75 *fF/μm²*	25 *fF/μm²*	1.75 *fF/μm²*
C_{ox}	6.25 *fF*	35 *fF*	12.5 *fF*	105 *fF*
C_{gs}	4.17 *fF*	23.3 *fF*	8.34 *fF*	6 *fF*
C_{gd}	1.56 *fF*	2 *fF*	3.7 *fF*	6 *fF*
Λ	0.6 *V⁻¹*	0.01 *V⁻¹*	0.6 *V⁻¹*	0.0125 *V⁻¹*
g_m at I_D=10 μA	150 *μA/V*	150 *μA/V*	150 *μA/V*	150 *μA/V*

We use the distributed *RC* delay to write the general delay for a great number of sections, *l*, as

$$t_d = 0.7 R_{sheet} C_{sheet} \left(1 + 2 + 3 + \ldots + l\right) \tag{3}$$

then

$$t_d \approx 0.35 R_{sheet} C_{sheet} l^2 \tag{4}$$

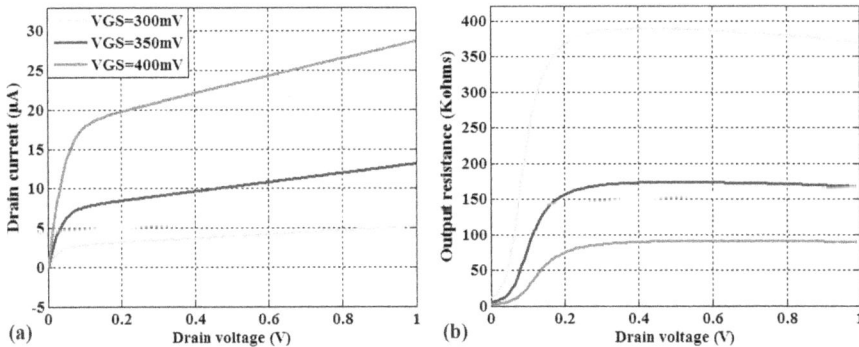

FIGURE 2.8 Output properties for a *50/2* NMOS in *50nm* technology process with V_{GS}=*300mV*, V_{GS}=*350mV*, and V_{GS}=*400mV*: (a) rain current and (b) output resistance

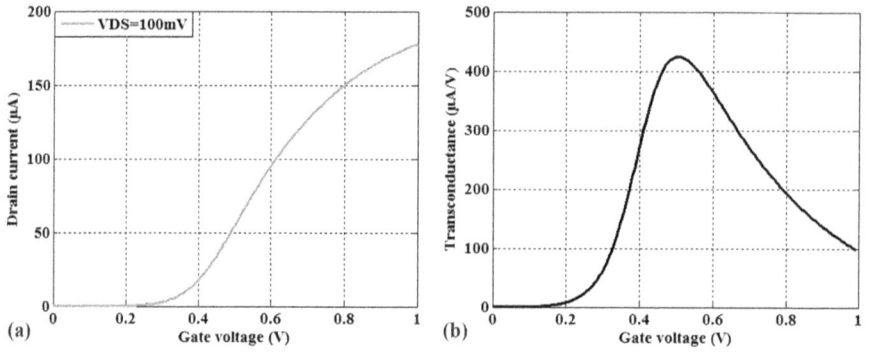

FIGURE 2.9 Input electric characteristics for a *50/2* NMOSin *50nm* technology process: (a) drain current plotted against gate-source and (b) transconductance plotted against gate-source voltage

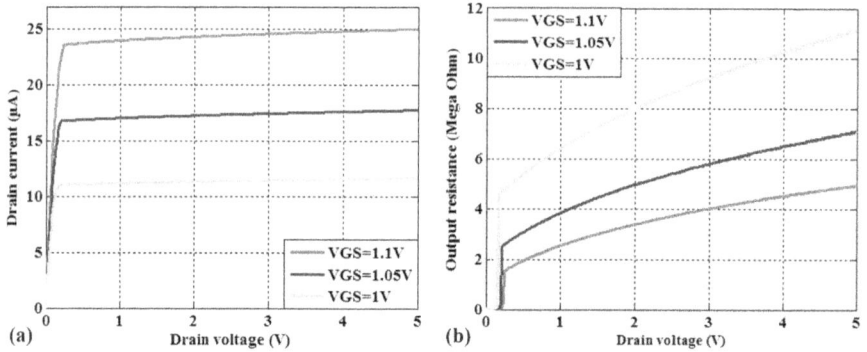

FIGURE 2.10 Output electric characteristics for a *10/2* NMOS in *1μm* technology process with V_{GS}=1V, V_{GS}=1.05V, and V_{GS}=1.1V: (a) drain current and (b) output resistance

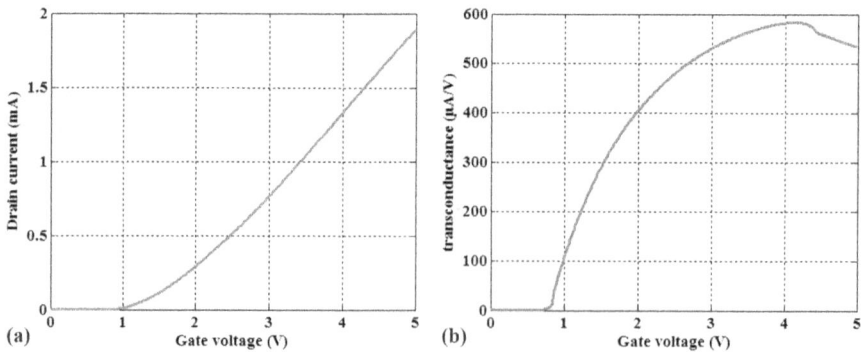

FIGURE 2.11 Input electric characteristics for a 10/2 NMOSin *1μm* technology process: (a) drain current plotted against gate-source and (b) transconductance plotted against gate-source voltage

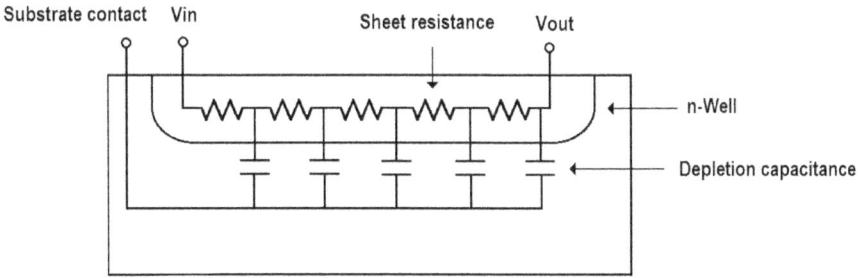

FIGURE 2.12   Parasitic resistance and capacitance of the n-well [23]

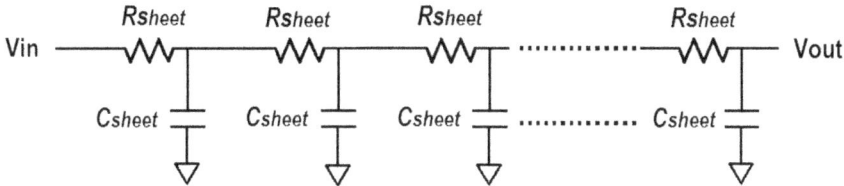

FIGURE 2.13   The use of a distributed RC delay to determine the delay [23]

Where the C_{sheet} and R_{sheet} are the sheet capacitance resistance of each distributed RC line.

To estimate the RC delay through the n-well resistor, a SPICE loss transmission line is used to model the distributed effects of the different n-well resistors, R_1=100 $k\Omega$, R_2=200 $k\Omega$, and R$_3$=300 $k\Omega$ with a width of 10 and a length of 200. The n-well to substrate capacitance of a 10 10×10 square is $5fF$.

We split the n-well into 20 squares where the size of each square is10×10 and resistances of 5 $k\Omega$, 10 $k\Omega$, and 15 $k\Omega$. The delay through different resistors is then$t_{d1} \approx 0.35(5k\Omega)(5fF)(20)^2 = 3.5ns$, $t_{d2} \approx 7ns$ and $t_{d3} \approx 10.5ns$

Figure 2.14 presents the delay through different resistors using SPICE simulation.

2.3.2   DEPLETION CAPACITANCE

We can form a p-n junction when we set an n-well in the p-substrate. It is important to know how to model a p-n junction for analytical calculations and space simulations [24].

The expression of the diode current I_D is given by

$$I_D = I_S \left(e^{\frac{V_D}{nV_T}} - 1 \right)$$
(5)

Note that $V_T = \dfrac{kT}{q}$, I_s is the scale (saturation) current, V_D is the voltage across the n-well diode, V_T is the thermal voltage, n is the emission coefficient, k is Boltzmann's constant, T is the temperature, and q is the elementary charge.

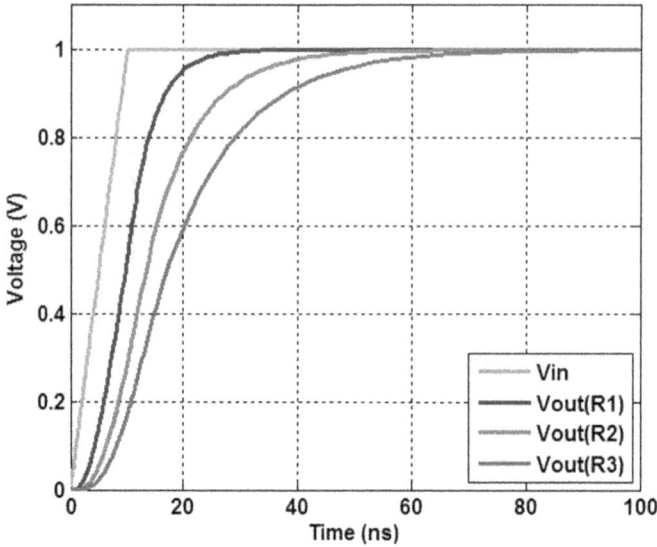

FIGURE 2.14   Delay through the n-well resistors using SPICE simulation

As shown in Figure 2.15, depletion capacitance is composed of bottom capacitance and sidewall capacitance.

The zero bias of the bottom capacitance, C_{B0}, is be calculated using

$$C_{B0} = C_A B_{area} (Sc)^2 \tag{6}$$

where Sc presents the scale, C_A is the capacitance area, and B_{area} is the bottom area.

The zero bias of the sidewall capacitance, C_{S0}, is calculated using

$$C_{S0} = C_A.(Depth\,of\,the\,well).(Perimeter\,of\,the\,well)(Sc)^2 \tag{7}$$

The overall depletion capacitance of the n-n junction is the parallel mixture between the bottom and sidewall capacitances, or

$$C_j = \frac{C_{B0} + C_{S0}}{\left[1 - \dfrac{V_D}{V_{bi}}\right]^m} \tag{8}$$

V_D is the voltage across the diode, m is the grading coefficient (showing how the silicon changes from n- to p-type), and V_{hl} is the built-in potential given by

$$V_{bi} = \frac{kT}{q} \ln\left(\frac{N_D N_A}{n_i^2}\right) \tag{9}$$

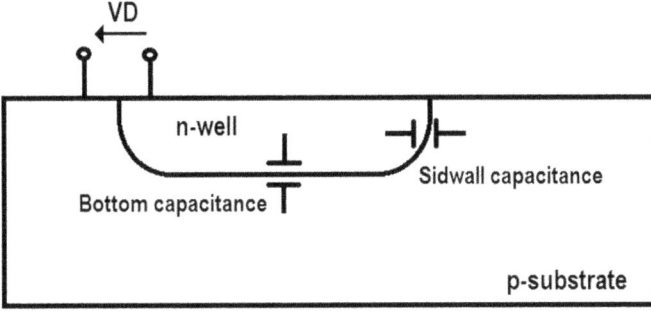

FIGURE 2.15 Bottom capacitance and sidewall capacitance of the p-n junction

where N_A is the substrate doping, and N_D is the n-well doping.

We consider an n-well/p-substrate diode with a 100 x 100 square at a scale factor of 1 μm, as the doping of the substrate is 10^{16} atoms/cm³ and the doping of the well is 10^{19} atoms/cm³. The measured zero-bias depletion capacitance of the junction is 80 aF/μm^2 (=100 x 10^{-18}F/μm^2), and the grading coefficient is 0.333. Assume the depth of the n-well is 4 μm.

The built-in potential is calculated using

$$V_{bi} = 26mV \times \ln\left(\frac{10^{16}10^{19}}{\left(14.5\ 10^9\right)^2}\right) = 303.39mV$$

From Equation (6), we get

$$C_{B0} = \left(80aF\ /\ \mu m^2\right) \times \left(100\right)^2 \times \left(1\mu m\right)^2 = 0.8pF$$

From Equation (7), we get

$$C_{S0} = \left(80aF\ /\ \mu m^2\right) \times \left(4\right) \times \left(400\right) \times \left(1\mu m\right)^2 = 0.128pF$$

Substituting numbers in Equation (8), we get

$$C_j = \frac{1.120pF}{\left[1 - \dfrac{V_D}{0.759}\right]^{0.33}}$$

Figure 2.16 depicts the capacitance changes of the n-well with reverse potential.

Generally, we consider the depletion capacitance of the p-n junction in the case of the reverse bias. Injecting electrons from the p junction and holes from the n-well

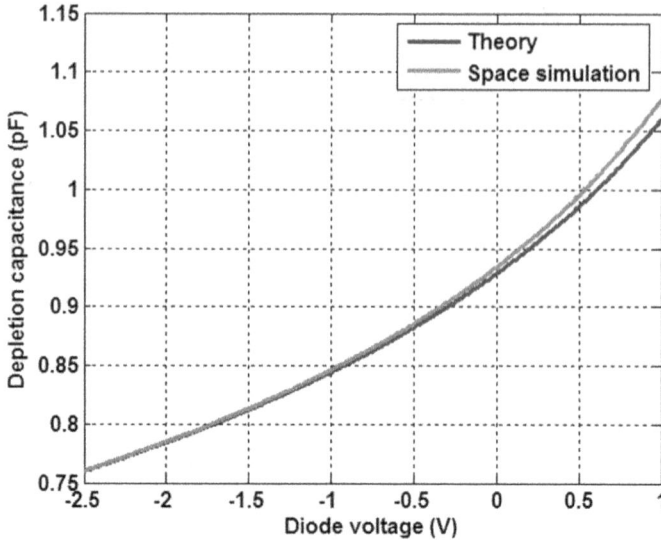

FIGURE 2.16   Handy and space simulation of depletion capacitance

across causes the diode to become a forward-biased minority carrier, which results in a storage capacitance. This capacity is typically much greater than the exhaustion capacity.

2.3.3   STORAGE CAPACITANCE

We can characterize the storage capacitance, C_s, in terms of the minority carrier lifetime. Within DC operating conditions, the expression of the storage capacitance is stated as

$$C_s = \frac{I_D}{nV_T}\tau_t \qquad (10)$$

where τ_t is carrier transit time. Notice that the diode capacitance is very functional for analog AC small signals. However, for digital applications, we have focused on the large signal-switching performance of the diode. In general, it can be said that it is undesirable to forward bias the p-n junction.

Consider the diode circuit in Figure 2.17. At the time t_1, the input voltage source makes an abrupt transition from a forward voltage of V_F to a reverse voltage of V_R, causing the current to change from $\dfrac{V_F - 0.7}{R}$ to $\dfrac{V_R - 0.7}{R}$. The diode voltage remains at 0.7 V because the diode contains a stored charge that must be removed. At time t_2, the stored charge is eliminated. At this point, the diode is similar to the voltage-dependent capacitor.

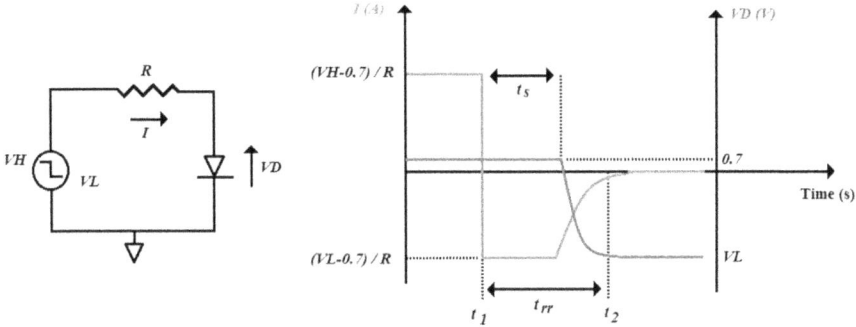

FIGURE 2.17   Forward-reverse diode circuit

The storage time t_s is clearly the difference between t_2 and t_1, then

$$t_s = t_2 - t_1 \tag{11}$$

We can also write

$$t_s = \tau_t \ln\left(\frac{i_F - i_R}{-i_R}\right) \tag{12}$$

where $i_F = \dfrac{V_F - 0.7}{R}$ and $i_R = \dfrac{V_R - 0.7}{R}$

To illustrate our understanding, SPICE simulation is used to model the circuit appearing in Figure 2.18, knowing that the carrier lifetime of the diode is 20 ns.

From Equation 12, the storage time is calculated as

$$t_s = 20\ln\left(\frac{4.35 + 5.65}{5.65}\right) = 11.4186\text{ns}$$

Figure 2.19 shows the diode current versus time, the input voltage step V_{IN}, and the voltage across the diode V_D. From the obtained results, storage time is t_s=11.214 ns which is close to the handy calculation.

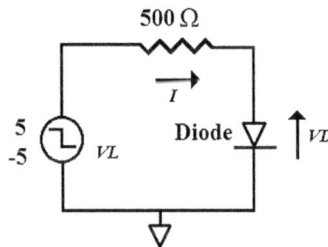

FIGURE 2.18   Diode circuit used in space simulation

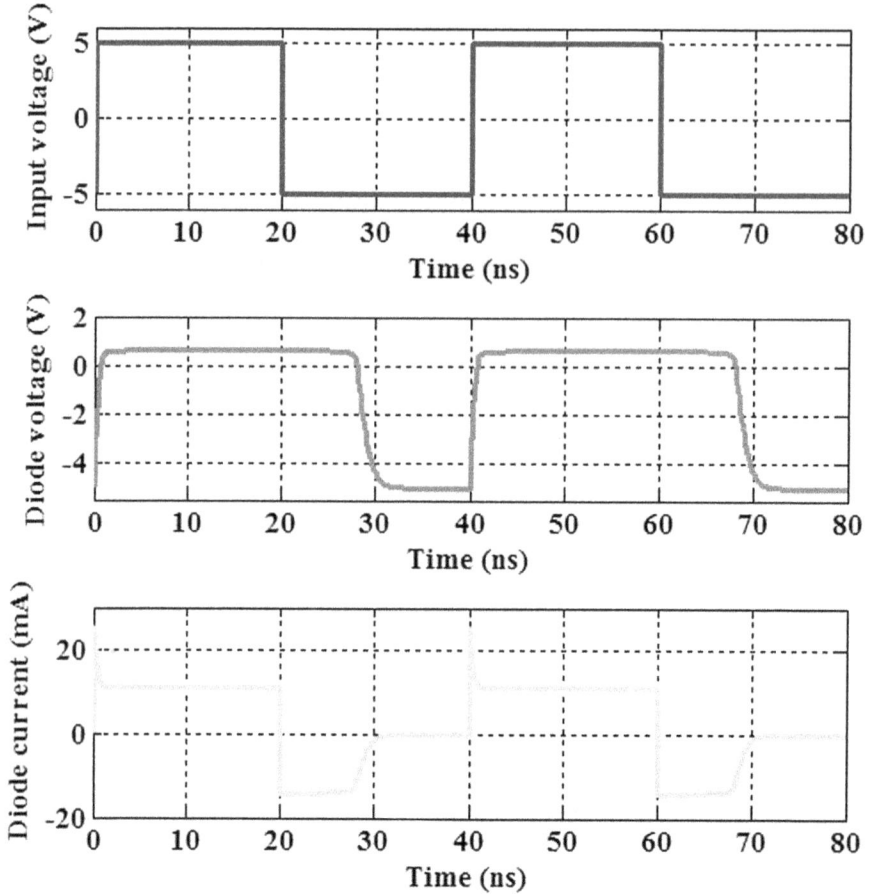

FIGURE 2.19   Voltage and diode current versus time

2.3.4   METAL-SUBSTRATE CAPACITANCE

The fundamental size of the bonding pad defined by MOSIS (MOS implementation system) is 100 μm by 100 μm. For a probe buffer, the size must be greater than 6 μm by 6 μm. The substrate is connected at ground, and then it can be supposed as an equipotential plane. This is important because we have to drive this capacitance to get a signal off the chip. Parasitic capacitance values are presented in Table 2.2 for CMOS process technology.

For example, the capacitance associated with the 100 μm by 100 μm square pad is the sum of the plate (or bottom) capacitance and the edge capacitance. We can write

$$C_{pad,m2-sub} = area\, C_{plate} + perimeter\, C_{fringe} \qquad (13)$$

TABLE 2.2
Capacitance parasitic values in CMOS technology process [14]

	Plate cap (aF/μm^2)			Fring cap (aF/μm)		
Poly1 to subs	53	58	63	85	88	92
Metal1 → poly1	35	38	43	84	88	93
Metal1 → subs	21	23	26	75	79	82
Metal1 → diffusion	35	38	43	84	88	93
Metal2 → poly1	16	18	20	83	87	91
Metal2 → subs	13	14	15	78	81	85
Metal2 → diffusion	16	18	20	83	87	91
Metal2 → metal1	31	35	38	95	100	104

The pad surface is 100 μm^2 square, whereas its perimeter is 400 μm. The use of typical capacity values for metal2 substrates in Table 2.2 gives

$$C_{pad,m2-sub} = (10000) \times 14aF + (400) \times 81aF = 0.172pF$$

2.4   LAYOUT DESIGN RULES

Design rules of layout are defined according to the size of the characteristics, separations, and overlaps. Feature size defines the dimensions of constructs, such as the channel length and the width of wires. Separation defines the distance between two constructs on the same layer. Overlap defines the necessary overlap of two constructs on adjacent layers in a physical construction, such as a contact connecting a poly wire with a metal1 wire, in which the metal1 wire must overlap with the poly wire below.

Mead and Conway [25, 26] popularized scalable design rules based on a single parameter, λ, that characterizes the resolution of the process. Λ is mostly half of the minimum channel length MOS transistor. The channel length is set by the minimum width of a polysilicon wire.

For example, a 50 nm process has a minimum polysilicon width of 0.05 μm and uses design rules with $\lambda = 0.025$ μm. Lambda-based rules are necessarily stable because they round up dimensions to an integer multiple of λ.

It is important to exercise caution when using lambda-based design guidelines in submicron geometries. We provide a sample set of the lambda-based layout design rules developed for the MOSIS [27] CMOS process in the following and demonstrate the effects of these rules on a piece of a basic layout that has two transistors (see Table 2.3 and Figure 2.20).

TABLE 2.3

MOSIS layout λ design rules [28]

Rule code	Description	λ-Rule
Active area		
R1	Minimum active area width	3λ
R2	Minimum active area spacing	3λ
Polysilicon		
R3	Minimum polysilicon width	2λ
R4	Minimum polysilicon spacing	2λ
R5	Minimum gate addition of polysilicon over active area	2λ
R6	Minimum polysilicon – active area spacing	1λ
R7	Minimum polysilicon – active area spacing	3λ
Metal		
R8	Minimum metal width	3λ
R9	Minimum metal spacing	3λ
Contact		
R10	Polysilicon contact size	2λ
R11	Minimum polysilicon contact spacing	2λ
R12	Minimum polysilicon contact to poly edge spacing	1λ
R13	Minimum polysilicon contact to metal edge spacing	1λ
R14	Minimum poly contact to active edge spacing	3λ
R15	Active contact size	2λ
R16	Minimum active contact spacing	2λ
R17	Minimum active contact to active edge spacing	1λ
R18	Minimum active contact to metal edge spacing	1λ
R19	Minimum active contact to polysilicon edge spacing	3λ
R20	Minimum active contact spacing	6λ

2.5 ANALOG AND DIGITAL CMOS CIRCUIT DESIGN

2.5.1 CURRENT MIRRORS

The current mirror is a very important unit for analog integrated circuit design [29]. It is mainly used to copy a current circulating from one active device to another active device. Figure 2.21 presents the fundamental NMOS current mirror, fabricated using two identical MOSFET, M_1 and M_2. If the two drain resistors are equal, then $V_{DS1} = V_{DS2}$.

From M_1, the expression of the reference current is

$$I_{REF} = I_{D1} = \frac{KP_n W_1}{2L_1}\left(V_{GS1} - V_{THN}\right)^2 \left(1 + \lambda\left(V_{DS1} - V_{DS1sat}\right)\right) \quad (14)$$

We have $V_{DS1} = V_{GS1}$ and $V_{DS1} = V_{GS1} - V_{THN}$.

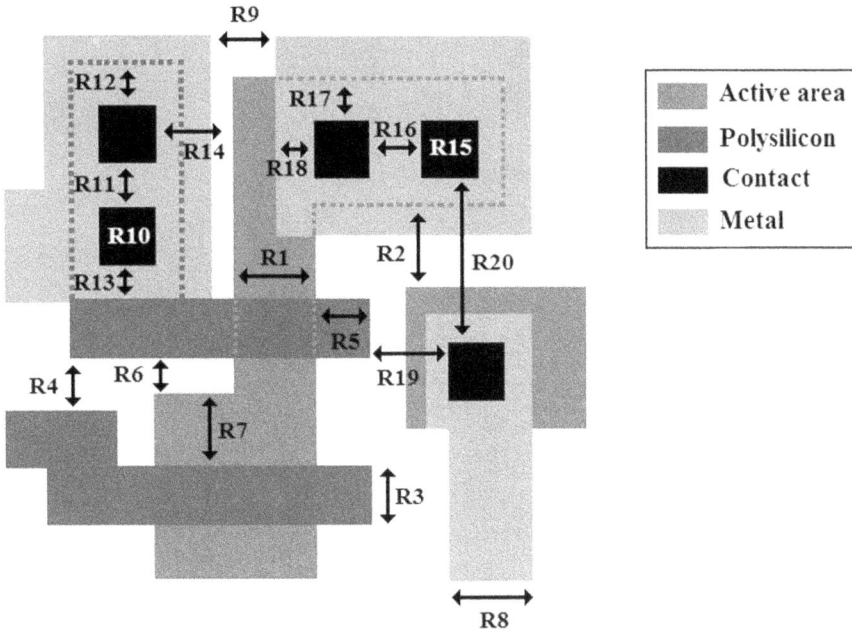

FIGURE 2.20 Clarification of some of the typical MOSIS layout design rules [28]

Then the drain current flowing through M_2 is

$$I_0 = I_{D2} = \frac{KP_nW_2}{2L_2}\left(V_{GS1} - V_{THN}\right)^2\left(1 + \lambda\left(V_0 - V_{DS1sat}\right)\right) \qquad (15)$$

Note that $V_{GS2} = V_{GS2}$ and $V_{DS1sat} = V_{DS2sat}$, and V_0 is the voltage across the current source. Looking at the ratio of the drain currents, we get

$$\frac{I_0}{I_{REF}} = \frac{W_2L_1}{W_1L_2}\frac{1 + \lambda\left(V_0 - V_{DS1sat}\right)}{1 + \lambda\left(V_{DS1} - V_{DS1sat}\right)} \qquad (16)$$

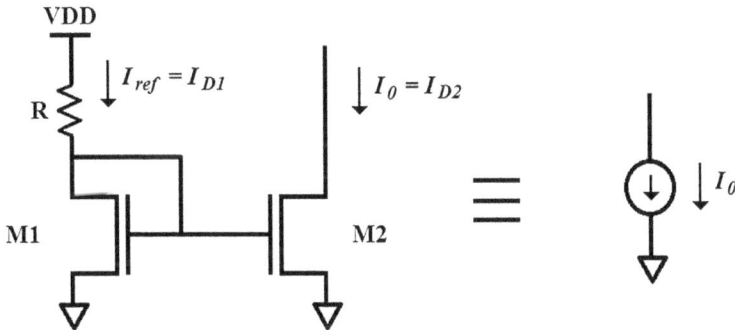

FIGURE 2.21 A basic current mirror

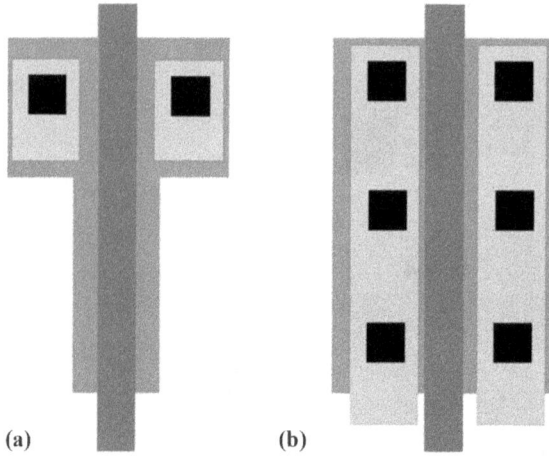

FIGURE 2.22 MOSFET contact in analog design: (a) with a single contact and (b) with more contacts [23]

Figure 2.22a depicts a basic MOSFET device with a high ratio of W/L.

Adding many contacts along the width of the drain and source contacts (as seen in Figure 2.22b) results in reduced resistance and high drain current.

Splitting the devices into parallel devices and interdigitating them can distribute process gradients across both devices and thus improve matching [30]. As seen in Figure 2.23a, each MOSFET in Figure 2.24b is divided into four MOSFETs. If the W/L of each MOSFET in Figure 2.23b is 100/2, then the size of each MOSFET (finger) in Figure 2.23a is 25/2.

We consider a circuit mirror presented in Figure 2.24. M1 and M2 are 10/2 NMOS with a scale of 1 μm.

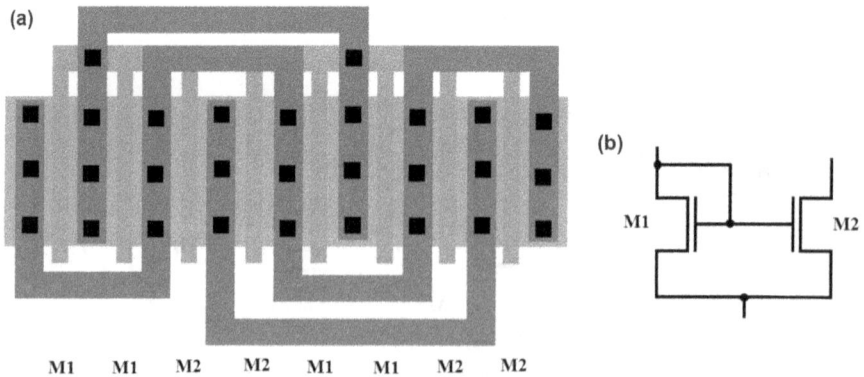

FIGURE 2.23 (a) Current mirror circuit: (a) layout using interdigitation and (b) equivalent circuit [23]

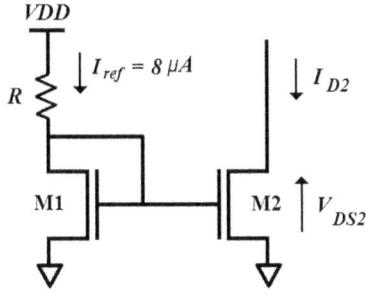

FIGURE 2.24   Current mirror based on NMOS

From Table 2.1, the resistor value is calculated as follows:

$$R = \frac{V_{DD} - V_{GS1}}{I_{REF}} = \frac{5 - 1.05}{8\mu A} = 493K\Omega$$

Figure 2.25 shows the SPICE simulation of the I-V characteristics of the NMOS current mirror. Current reference is approximately 8 μA. See that below, $V_{DSsat}=140\ mV$. The point where $V_{DS1}=V_{GS1}$ is where $I_{D2}=I_{REF}$. Finally, we conclude that I_{REF} and V_{GS1} are independent of V_{DS2}.

2.5.2   INVERTER

The CMOS inverter is a basic building block for digital circuit design [31, 32]. As Figure 2.26 shows, the inverter is based on CMOS technology. If the input is fixed to

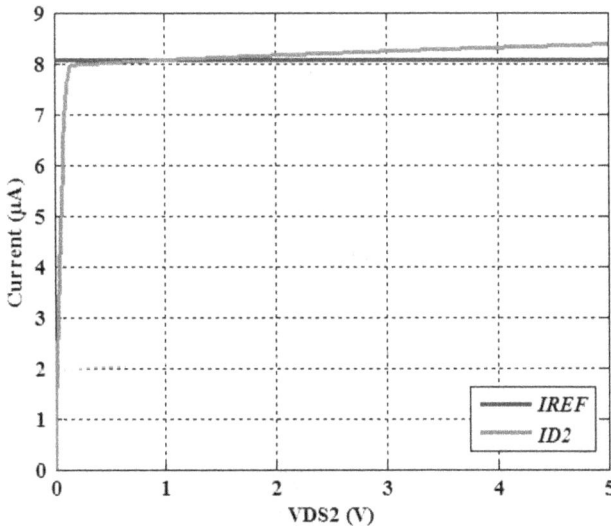

FIGURE 2.25   Output current of current mirror

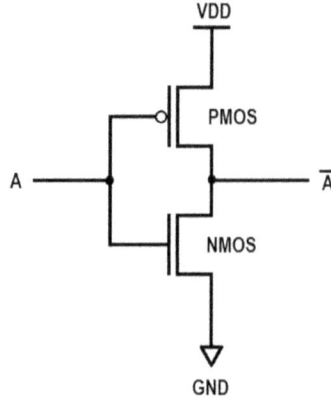

FIGURE 2.26 Inverter schematic

ground, the output is pulled to V_{DD} across the PMOS transistor. Whereas the input is fixed to V_{DD}, the output is pulled to ground across the NMOS transistor.

The operational point corresponds to the point on the curve when the input voltage is equal to the output voltage. At this point, the input voltage is called the inverter switching point voltage, V_{sp}, and both MOSFETs in the inverter are in the saturation region. Since the drain current in each MOSFET must be equal, the following is true

$$\frac{g_n}{2}\left(V_{SP}-V_{THN}\right)^2 = \frac{g_p}{2}\left(V_{SP}-V_{THP}\right)^2 \tag{17}$$

Solving for V_{sp} gives

$$V_{SP} = \frac{\sqrt{\dfrac{g_n}{g_p}}V_{THN}+\left(V_{DD}-V_{THP}\right)}{1+\sqrt{\dfrac{g_n}{g_p}}} \tag{18}$$

Figure 2.27 presents the mask layout of the inverter. We suppose that the design of the inverter is done with minimum-size transistors. The active area width is then fixed by the minimum diffusion contact size and the minimum separation between diffusion contact and active area boundaries. The width of the polysilicon line over the active area (which is the gate of the transistor) is typically taken as the minimum poly width. The PMOS transistor has to be located in an n-well region, and the minimum size of the n-well is dictated by the PMOS active area and the minimum n-well overlap over n+. The polysilicon gates of the NMOS and the PMOS transistors are usually aligned. The final step in the mask layout is the local interconnections in metal for the output node and for the V_{DD} and GND contacts. Notice that in order to be biased properly, the n-well region must also have a V_{DD} contact.

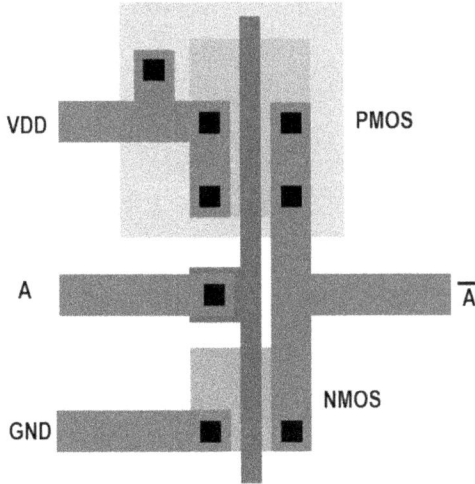

FIGURE 2.27   Mask layout of the inverter

Using SPICE simulation, the inverter voltage transfer curves for long- and short-channel CMOS processes are presented in Figure 2.28. Notice that the V_{DD} applied for the long-channel process is $5V$, while the V_{DD} applied for the short-channel process is $1V$. The output high voltage, V_{OH}, is V_{DD}, and the output low voltage, V_{OL}, is ground. V_{IL} is approximately $2.5V$ for the long-channel inverter, while it is 500 mV for the short-channel inverter.

In Figure 2.29, we also plotted the power consumption of the inverter for the long- and short-channel CMOS processes. It is noted that power consumption becomes important in the transition region.

Figure 2.30 shows the simulated low-to-high delay for different input patterns. For the case where the input transition A goes from low to high ($A=0{\rightarrow}1$), the time delay is 2.03 ns. On the other hand, for the case where $A=1$ transitions from $1{\rightarrow}0$,

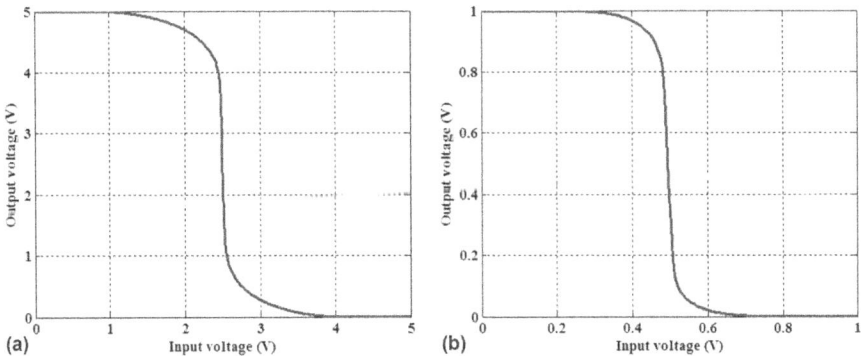

FIGURE 2.28   Characteristics of CMOS inverter: (a) long channel and (b) short channel

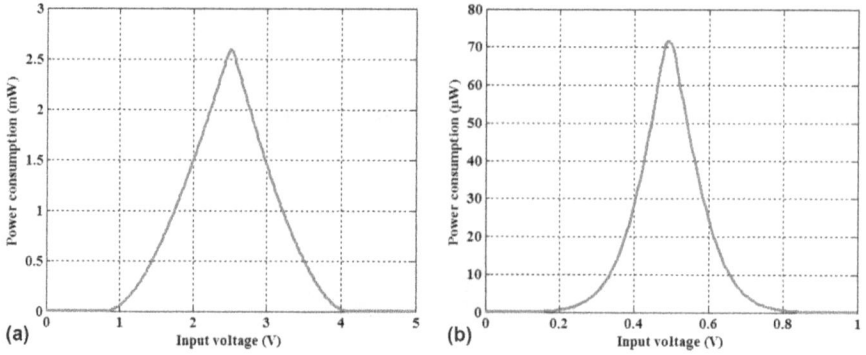

FIGURE 2.29   Power consumption of CMOS inverter: (a) long and (b) short channel

the time delay is 1.36 ns. The calculated average power in this dynamic simulation is 68.098 µW.

Increasing switching speed (reducing time delay) means increasing the (W/L) ratios of all MOSFETs in the inverter. This later, increases the gate, source, and drain areas and, as a result, the appearance of the parasitic capacitances charging the logic gates.

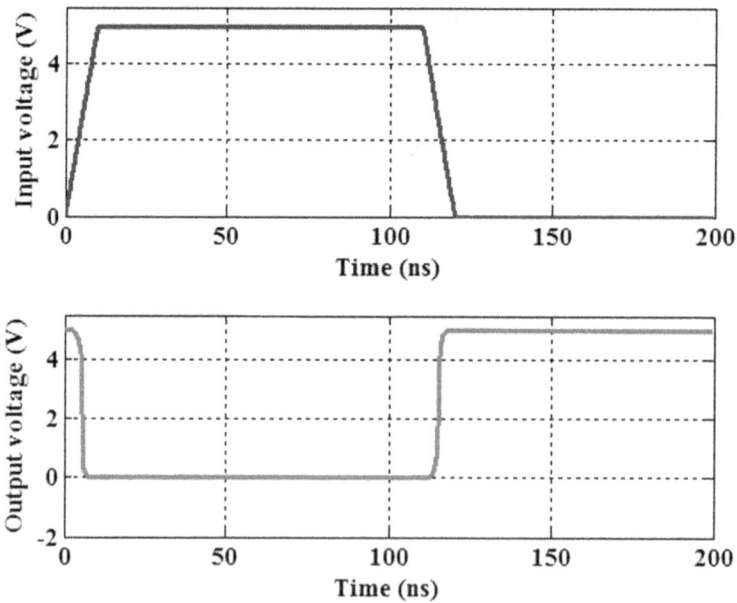

FIGURE 2.30   Input and output signals versus time of the inverter for a long-channel process

2.5.3 Nand and nor gates

Two input NAND and NOR gates are presented in Figure 2.31. The two logic gates need both inputs to be high until the output switches to low.

To explain the switching point voltage, we notice that the two parallel PMOS devices in Figure 2.31a are similar to one MOSFET; then we can write

$$W_3 + W_4 = 2W_p \tag{19}$$

We start our study by calculating the voltage transfer characteristic of the NAND gate with the NMOS and PMOS devices with the widths, W_n and W_p, and lengths, L_n and L_p respectively [33].

Supposing that all PMOS transistors have the same size. The transconductance parameter of a single PMOS can be written as

$$g_3 + g_4 = 2g_p \tag{20}$$

The two NMOS transistors are in series, then

$$L_1 + L_2 = 2L_n \tag{21}$$

and the transconductance is given by

$$g_1 + g_2 = \frac{g_n}{2} \tag{22}$$

Then the NAND gate can be modeled as an inverter in NMOS and PMOS devices with the ratios Wn/ 2Ln and 2Wp/Lp respectively; in this case, the transconductance of the NAND gate is given by

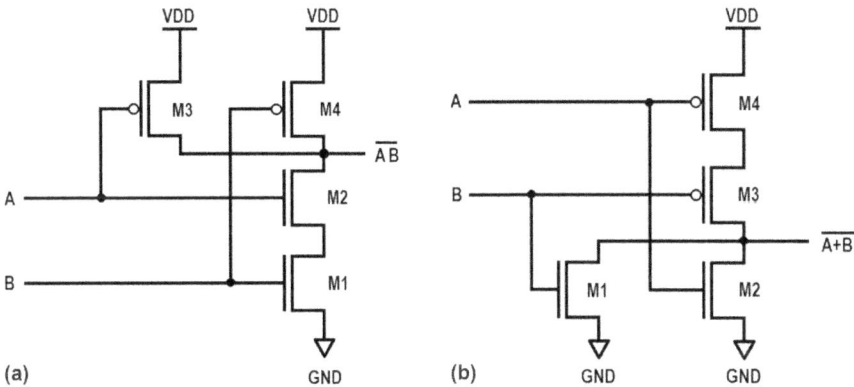

FIGURE 2.31 Logic gate schematic: (a) NAND gate and (b) NOR gate

$$g_{NAND} = \frac{g_n}{4g_p} \tag{23}$$

The expression of the switching point voltage is

$$V_{SP} = \frac{\sqrt{g_{NAND}}V_{THN} + (V_{DD} - V_{THP})}{1 + \sqrt{g_{NAND}}} \tag{24}$$

A similar analysis is done for the 2-input NOR gate (see Figure 2.31b).

Here, the switching point expression is given by

$$V_{SP} = \frac{\sqrt{g_{NOR}}V_{THN} + (V_{DD} - V_{THP})}{1 + \sqrt{g_{NOR}}} \tag{25}$$

The transconductance is given by

$$g_{NOR} = \frac{4g_n}{g_p} \tag{26}$$

Figure 2.32 shows the sample layouts of a 2-input NOR gate and a 2-input NAND gate, using one layer of polysilicon and one layer of metal. Here, the p-type and n-type diffusion areas for the PMOS and NMOS transistors are organized in parallel to have vertical gate polysilicon lines. Moreover, the two mask layouts present a high symmetry because the NOR and NAND gates have a symmetrical circuit topology.

The proposed NAND circuit simulation has been performed in SPICE simulation in long-channel technology (1 μm). The W/L ratio of NMOS transistors used in all

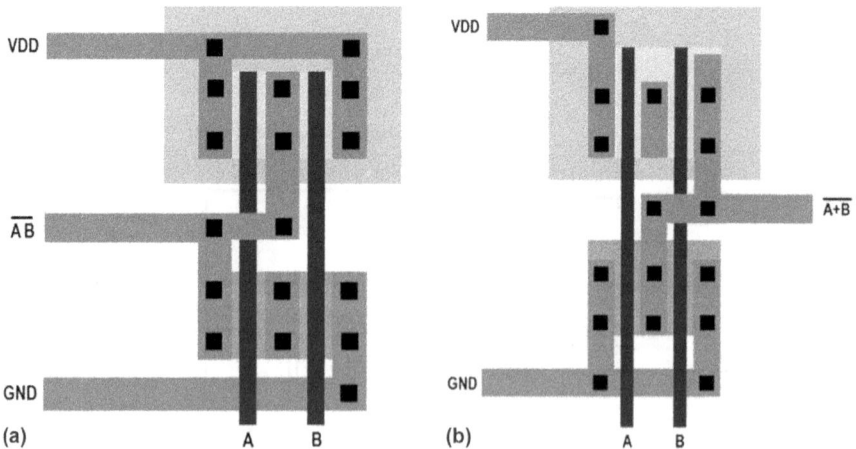

FIGURE 2.32 Layout representation: (a) 2-input NAND gate layout and (b) 2-input NOR gate layout

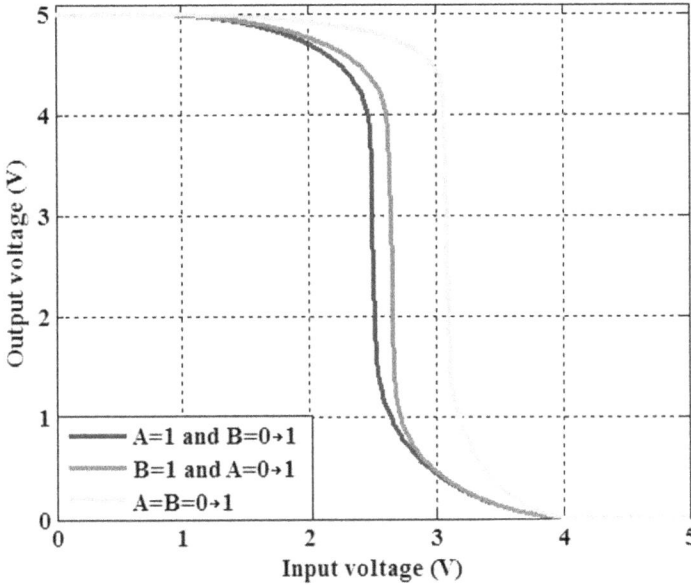

FIGURE 2.33 Voltage transfer characteristics of 2-input NAND in *1μm* CMOS process

circuits is preserved at its minimal value, which is 10 μ/2 and 30μ /2μ for PMOS transistors.

Three possible input combinations switch the output of the gate from high to low: A=B=0 →1; A=1, B=0→1; and B=1, A=0 → 1, which are performed using SPICE simulation. The resulting voltage transfer curves are depicted in Figure 2.33. The great change between case (a) and cases (b) and (c) is described by the fact that in the first case, both transistors are on at the same time. In the latter cases, only one of the pull up devices is on. The difference between (b) and (c) results mainly from the state of the inner node between the two NMOS devices.

The same simulations are done in the case of the static 2-input NOR gate. Three possible input combinations are simulated: (a) A=B=0 →1, (b) A=0, B=0 → 1, and (c) B=0, A=0→1. The resulting voltage transfer curves are shown in Figure 2.34.

We next consider the dynamic operation of the NAND and NOR gates in 1 *μm* CMOS technology.

The transient response of the NAND and NOR gates as calculated by SPICE are presented in Figure 2.35 and Figure 2.36 respectively. A transient analysis is requested to be performed over a *400ns* interval using a *0.2ns* time step. Concerning the NAND gate simulation, we see that both the rise and fall time of the output voltage signal are quite similar in duration. With the aid of the Probe facility, we find that the *90% to 10%* fall time t_{THL} is *2.87ns* and the high-to-low input-to-output transition delay t_{PHL} is *2.26ns*.

Whereas from the NOR simulation the 90% to 10% fall time t_{THL} is *3.21ns* and the high-to-low input-to-output transition delay t_{PHL} is *2.78ns*.

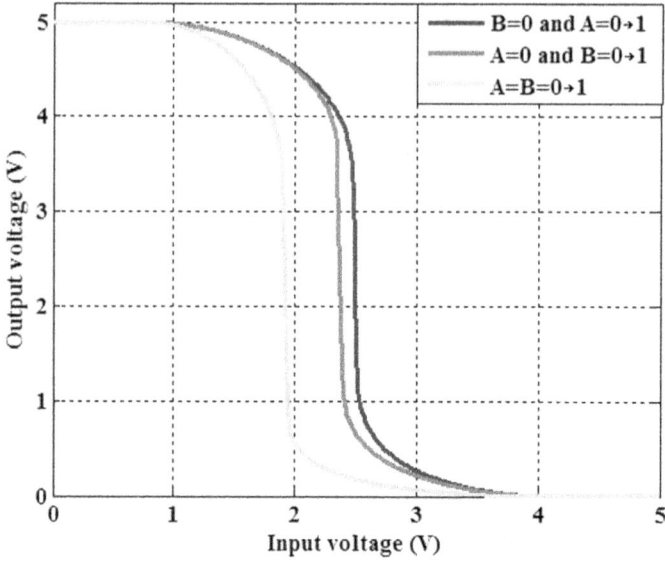

FIGURE 2.34   Voltage transfer characteristics of 2-input NOR in *1μm* CMOS process

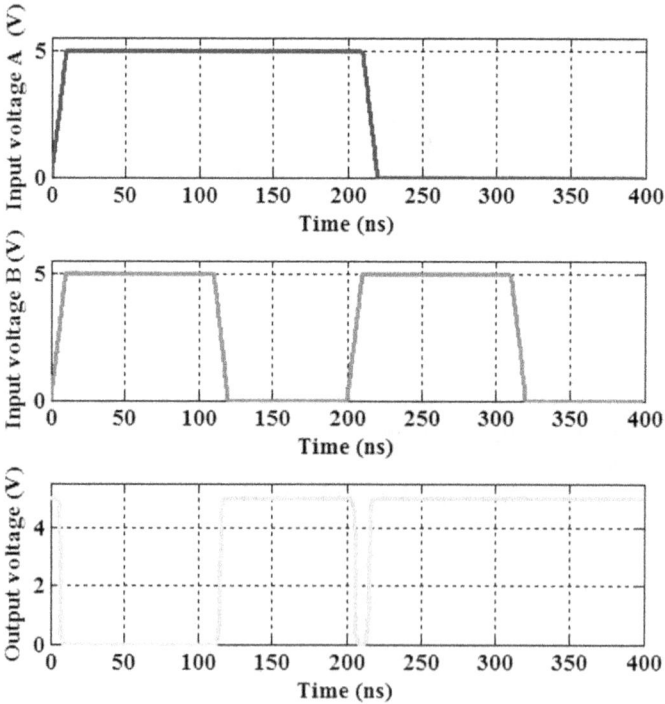

FIGURE 2.35   Simulated input and output waveforms of NAND gate in *1μm* CMOS technology

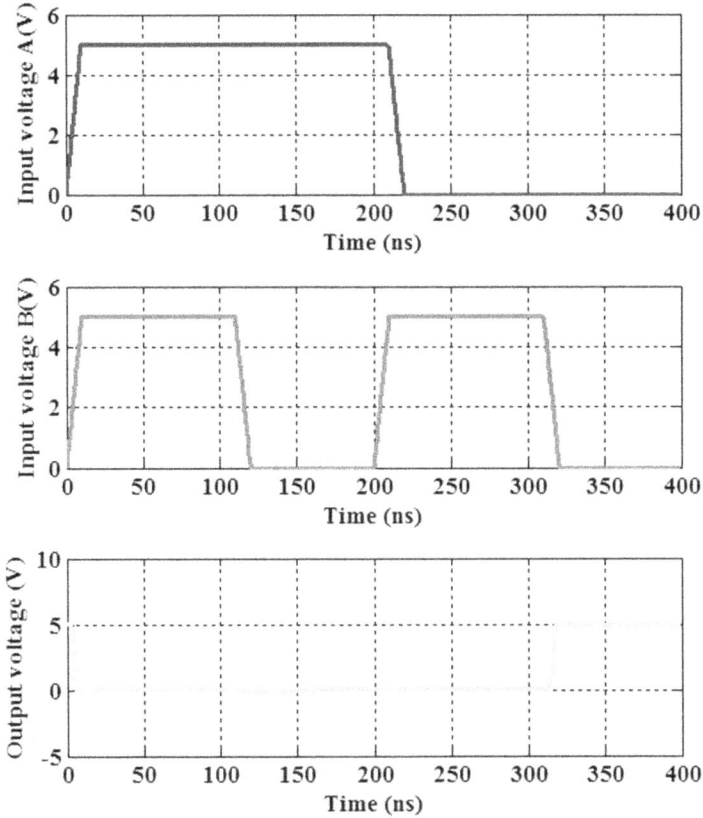

FIGURE 2.36 Simulated input and output waveforms of NOR gate in $1\mu m$ CMOS technology

2.5.4 FULL ADDER

In this part, we will design a one-bit binary full-adder circuit using long-channel CMOS technology [34–36]. Figure 2.37 shows the circuit diagram of the full adder based on CMOS technology. The circuit has three inputs A, B and C and two outputs sum and carry_out. The mask layout of the CMOS full-adder circuit is presented in Figure 2.38. Input and output pins have been ordered in vertical polysilicon lines. Moreover, the sum circuit and the carryout circuit were performed using one continued active area each.

Note that in this initial adder cell layout, all NMOS and PMOS transistors are placed in two parallel rows, between the horizontal power supply and the ground lines (metal). All polysilicon lines are laid out vertically. The area between the n-type and p-type diffusion regions is used for running local metal interconnections (routing). Also note that the diffusion regions of neighboring transistors have been merged as much as possible, in order to save chip area.

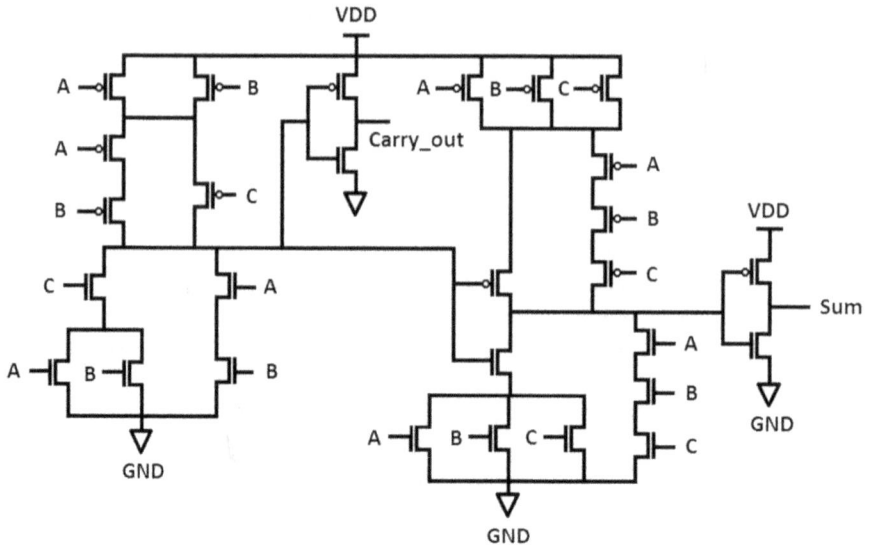

FIGURE 2.37 Circuit diagram of long-channel CMOS full adder

The regular gate-matrix layout style used in this example also has the inherent advantage of being easily adaptable to computer-aided design (CAD).

Figure 2.39 shows the simulated input and output waveforms using SPICE. Unfortunately, the simulation results show that the circuit does not meet all of the design specifications. The propagation delay times of the sum_out and carry_out signals are found to violate the timing constraints since the minimum-size transistors are not capable of properly driving the capacitive output loads.

2.6 CONCLUSION

This chapter examined the principles of designing a simple CMOS integrated circuit. The effect of parasitic effects such as storage capacitance, shift delay through the

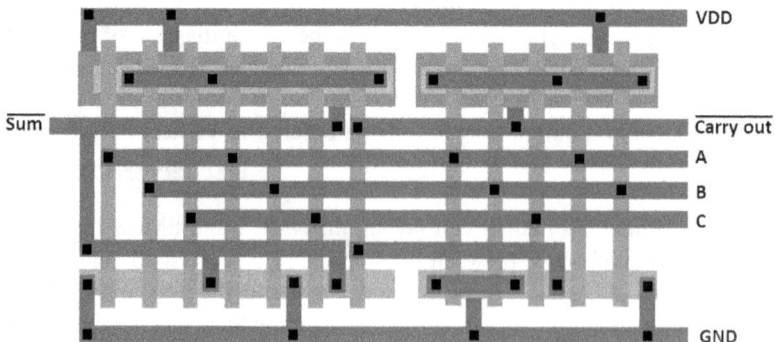

FIGURE 2.38 Mask layout of the CMOS full-adder circuit

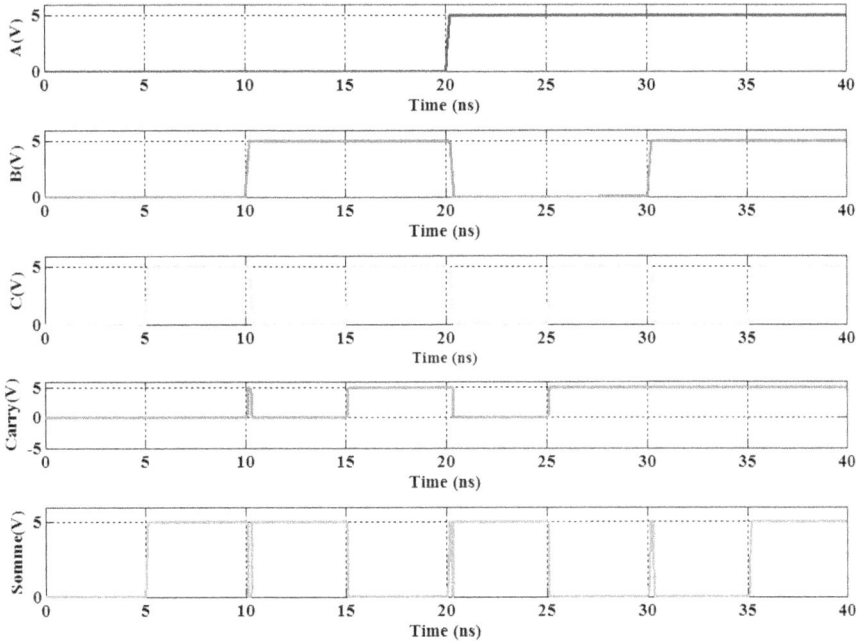

FIGURE 2.39 Simulated input and output waveforms of the full-adder circuit

well; and depletion capacitance have been simulated and investigated. We have given some information regarding the design rules for active area, polysilicon, and contact and metal interconnects.

We have introduced the main concepts for analog CMOS circuit design, and we have also performed a SPICE simulation of current mirrors. In addition, digital CMOS circuit designs such as inverter, NOR, NAND, and full-adder circuits have been designed and simulated. Based on the simulation results, it has been confirmed that the short-channel MOSFET model-based logic gates circuits have better output signal levels and consume less power compared to the long-channel MOSFET model at low supply voltage.

REFERENCES

1. D. Das. (2011). *VLSI Design* (1st edition), Oxford University Press.
2. K. N. Tu. (2003). "Recent advances on electromigration in very-large-scale-integration of interconnects", Journal of Applied Physics, 94(9), p. 5451. https://doi.org/10.1063/1.1611263.
3. S. K. Tolpygo, V. Bolkhovsky, T. J. Weir, A. Wynn, D. E. Oates, L. M. Johnson, M. A. Gouker. (2016). "Advanced fabrication processes for superconducting very large-scale integrated circuits", IEEE Transactions on Applied Superconductivity, 26. https://doi.org/10.1109/TASC.2016.2519388.

4. G. E. Moore. (1998). "Cramming more components onto integrated circuits", *Proceedings of the IEEE*, 86(1), pp. 82–85. https://doi.org/10.1109/N-SSC.2006.4785860.

5. J. M. Pimbley, M. Ghezzo, H. G. Parks, D. M. Brown. (1989). *Advanced CMOS Process Technology*, Academic Press, INC.

6. J. Lienig, J. Scheible. (2020). *Fundamentals of Layout Design for Electronic Circuits*, Springer Nature.

7. N. H. E. Weste, D. M. Harris. (2010). *CMOS VLSI Design a Circuits and Systems Perspective* (4th edition), Addison-Wesley.

8. J. F. Wakerly. (2001). *Digital Design: Principles and Practices* (3rd edition), Prentice-Hall.

9. BehzadRazavi. (2017). *Design of Analog CMOS Integrated Circuit* (2nd edition), McGraw-Hill education.

10. Y. Taur, T. H. Ning. (2009). *Fundamentals of Modern VLSI Devices* (2nd edition), Cambridge University Press.

11. S. A. Campbell. (2008). *Fabrication Engineering at the Micro- and Nanoscale* (3rd edition), Oxford University Press.

12. J. P. Uyemura. (2001). *CMOS Logic Circuit Design*, Kluwer Academic Publishers.

13. B. P. Wong, A. Mittal, Y. Cao, G. Starr. (2005). *Nano-CMOS Circuit and Physical Design*, Wiley-IEEE Press.

14. R. J. Baker. (2010). *CMOS Circuit Design, Layout, and Simulation* (3rd edition), John Wiley & Sons.

15. T. Hook, J. Brown, P. Cottrell, E. Adler, D. Hoyniak, J. Johnson, R. Mann. (2003). "Lateral ion implant straggle and mask proximity effect", *IEEE Transactions on Electron Devices*, 50(9), pp. 1946–1951. https://doi.org/10.1109/TED.2003.815371.

16. M. C. Schneider. (2010). *CMOS Analog Design Using All-Region MOSFET Modeling*, Cambridge University Press.

17. J. D. Plummer, M. D. Deal, P. B. Griffin. (2000). *Silicon VLSI Technology, Fundamentals, Practice, and Modeling*, Prentice-Hall Publishers.

18. R. S. Muller, T. I. Kamins, M. Chan. (2002). *Device Electronics for Integrated Circuits*, John Wiley and Sons Publishers.

19. R. J. Baker. (2019). *CMOS Circuit Design, Layout, and Simulation* (4th edition), IEEE Press.

20. H. Kaeslin, E. Zurich. (2008). *Digital Integrated Circuit Design from VLSI Architectures to CMOS Fabrication*, Cambridge University Press.

21. S. Labiod, S. Latreche, C. Gontrand. (2012). "Numerical modeling of MOS transistor with interconnections using lumped element-FDTD method", *Microelectronics Journal*, 43(12), pp. 955–1002. https://doi.org/10.1016/j.mejo.2012.10.005.

22. B. Smaani, S. B. Rahi, S. Labiod. (2022). "Analytical compact model of nanowire junctionless gate-all-around MOSFET implemented in Verilog-A for circuit simulation", *Silicon*. https://doi.org/10.1007/s12633-022-01847-9.

23. R. J. Baker. (2005). *CMOS Circuit Design, Layout, and Simulation* (2nd edition), IEEE Press and Wiley-Interscience.

24. S. Labiod, B. Smaani, S. Tayal, S. B. Rahi, H. Sedrati, S. Latreche. (2022). "Mixed-mode optical/electric simulation of silicon lateral PIN photodiode using FDTD method", *Silicon*. https://doi.org/10.1007/s12633-022-02081-z.

25. C. Mead, L. Conway. (1980). *Introduction to VLSI Systems*, Addison-Wesley.

26. N. H. E. Weste, D. M. Harris. (2010). *CMOS VLSI Design A Circuits and Systems Perspective* (4th edition), Addison-Wesley.

27. C. Pina. (2002). "Evolution of the MOSIS VLSI educational program", in *Proceedings Electronic Design, Test, and Applications Workshop*, pp. 187–191. https://doi.org/10.1109/DELTA.2002.994612.

28. S. M. Kang, Y. Leblebisi. (2003). *CMOS Digital Integrated Circuits: Analysis and Design*, (3rd edition), McGraw-Hill.

29. J. Ramírez-Angulo, R. G. Carvajal, A. Torralba. (2004). "Low-supply-voltage high-performance CMOS current mirror with low input and output voltage requirements", *IEEE Transactions on Circuits and Systems. Part II*, 51(3), pp. 124–129. https://doi.org/10.1109/TCSII.2003.822429.

30. A. Hastings. (2001). *The Art of Analog Layout*, Prentice Hall, Inc.

31. M. I. Elmasry. (1992). *Digital MOS Integrated Circuits II*, IEEE Press.

32. J. P. Uyemura. (2002). *Introduction to VLSI Circuits and Systems*, John Wiley and Sons Publishers.

33. J. P. Uyemura. (1992). *Circuit Design for Digital CMOS VLSI*, Kluwer Academic Publishers.

34. A. Yadav, R. Mehra. (2015). "Efficient layout design of 4-bit full adder using transmission gate", *International Journal of Computer Trends and Technology*, 23(3), pp. 116–119. http://doi.org/10.14445/22312803/IJCTT-V23P125.

35. S. Tayal, B. Smaani, S. B. Rahi, A. K. Upadhyay, S. Bhattacharya, J. Ajayan, B. Jena, I. Myeong, B. G. Park, Y. S. Song. (2022). "Incorporating bottom-up approach into device/circuit co-design for SRAM-based cache memory applications", *IEEE Transactions on Electron Devices*, 69(11), pp. 6127–6132. https://doi.org/10.1109/TED.2022.3210070.

36. M. A. Hernandez, M. L. Aranda. (2011). "CMOSFull-adders for energy-efficient arithmetic applications", *IEEE Transactions on Very Large Scale Integration (VLSI) Systems*, 19(4), pp. 718–721. https://doi.org/10.1109/TVLSI.2009.2038166.

3 Compact modeling of junctionless gate-all-around MOSFET for circuit simulation
Scope and challenges

Billel Smaani, Fares Nafa, Abhishek Kumar Upadhyay, Samir Labiod, Shiromani Balmukund Rahi, Mohamed Salah Benlatreche, Hamza Akroum, Maya Lakhdara, and Ramakant Yadav

CONTENTS

3.1   INTRODUCTION

In order to achieve more and higher performance of integrated circuits (ICs), complementary metal-oxide-semiconductor (CMOS) has been pushed to its physical and technical limits [1–2]. Therefore, several device architectures have been suggested [3]. Mainly, the junctionless gate-all-around (JLGAA) MOSFET has attracted much research attention [4–12]. Compared with the conventional inversion-mode devices [13, 14], the JLGAA MOSFET offers the best electrostatic control to reduce short-channel effects (SCE), higher I_{ON}/I_{OFF} ratios, good value of sub-threshold-slope (SS), lower gate-tunneling probability, and low-frequency noise (LFN) behavior [15–18].

DOI: 10.1201/9781003359234-3

In addition, the junctionless (JL) devices eliminate the source and drain junctions, making these types of transistors simple to fabricate and a key element for the down-scaling of CMOS technology [19–21]. Consequently, the junctionless gate-all-around MOSFET is one of the best emerging devices for future CMOS circuit implementation below 10nm of technology nodes [22–24].

Moreover, compact models of junctionless GAA MOSFET dedicated to circuit simulation are important for using these kinds of transistors in various integrated circuits. In fact, for possible implementation in different circuits' simulators, these types of models are usually accurate, simple, and have explicit analytical formulation [25] for possible implementation in different circuits simulators.

The main motivation for using compact models by circuit computer-aided design (CAD) in the industry of semiconductors is the efficient design optimization and cost-effectiveness of integrated-circuit products in the electronic design-automation (EDA) environment [26]. The interest of using compact models in circuit computer-aided design is the optimization of circuit performance for robust design integrated circuits chip. The task of improvement and optimization is complex due to the growing complexities of scaling down CMOS technology and devices. Indeed, targeting the scaling-down of CMOS technology to below10 nm has resulted in faster circuit speed, lower power dissipation, and several physical/quantum phenomena like short-channel effects, channel quantization, self-heating, band-to-band tunneling, non-quasi-static effects, and radio-frequency behaviors. These effects become important as the geometrical dimension of the device approaches its technology and physical limit [27]. It has been found that the performance evaluation and analysis of nanoscale and very large-scale integrated (VLSI) circuits with the prototype breadboard to characterize and build advanced integrated-circuit chips are expensive and time-consuming [28]. As a result, VLSI technology with reduced devices becomes more sensitive to process variability, which creates performance variability for both the circuits and the device [29]. To deal with this drawback, statistical circuit evaluation analysis is important for designing and developing VLSI chips. Actually, compact model FET designs are an excellent alternative for efficient design and cost-effective statistical device performance of VLSI circuits.

From the aforementioned background, a study about the compact modeling of junctionless (JL) GAA MOSFET for circuit simulation is presented in this chapter. Firstly, the specificities of compact models for FETs are exposed as a key element for circuits' simulation and design. Then, the interest in hardware description language (HDL) for circuit simulation is discussed. Furthermore, the architecture and physics of junctionless GAA MOSFET device operation are presented in a further section. Next, we respectively present the main approach and significant compact models of JLGAA MOSFET and the challenges of compact modeling of FET for future technology nodes. Finally, we wrap up the chapter with a conclusion.

3.2 SPECIFICITIES OF COMPACT MODELS

In the last two decades, the compact modeling (CM) of FETs has been organized as a conventional topic in semiconductor research and development (R&D) [30]. Indeed, the CM of FETs is directly related to the development of useful models for integrated semiconductor devices and for possible use in various circuit simulations [31]. These types of models are used to describe the transistor terminal behaviors with good accuracy, excellent computational efficiency, and acceptable simplicity for different circuits and system-level simulations for recent technology nodes [32]. In this context, the designer and users of CM are typically integrated-circuit designers. The semiconductor industry has therefore demonstrated reliance on fast and accurate compact model design in parallel with the increase of circuit frequencies, device scaling-down, the increase in the number of transistors (chips), and the analog content. Figure 3.1 shows the main interest of CM in the global research workflow for the development of industrial and new electronic applications. In fact, CM presents an efficient bridge between the transistor and the system-level outlooks. The first step illustrates the implementation of new emerging material. Next, the CM describes the electrical and physics behavior of the transistors. The final step is devoted to the design of different circuits and systems based on the developed CM [33].

A compact model for a specific transistor can be described as a set of mathematical formulations and accurate descriptions that evaluate the relationship between the device terminal characteristics and a variety of materials, as well as operational parameters such as the temperature and voltages [34], which are helpful for CAD and deep analysis of different integrated circuits [31, 35]. In addition, a good CM must take into consideration technology and physical problem that occur when scaling

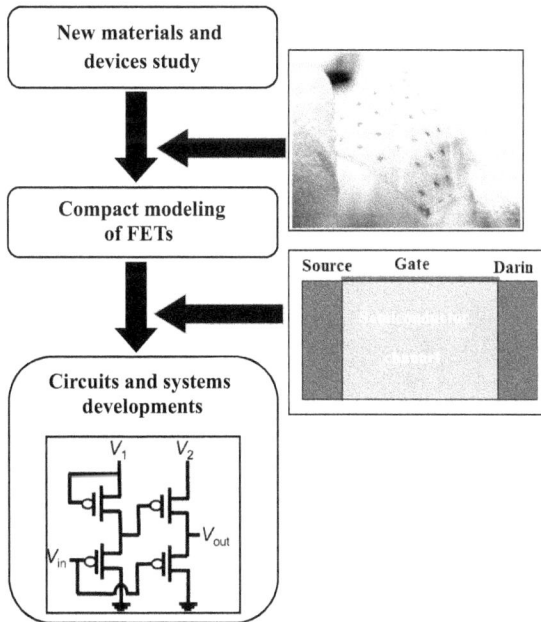

FIGURE 3.1   Global research workflow for the development of industrial and new electronic applications

down the CMOS circuits to below10 nm. This includes short-channel effects, reverse short-channel-effects, modulation of the channel length, DIBL (drain-induced barrier lowering), velocity overshoot, remote surface roughness scattering, impact ionization, degradation of the mobility, band-to-band tunneling, self-heating effect, quantization in the channel, polysilicon depletion, NQS effects, RF behaviors, and discrete dopants [36, 37].

The newly developed model is analyzed and validated using technology computer-aided design (TCAD) tools, like Silvaco [38, 39]. These TCAD tools helped to solve numerically the main semiconductor physics equations, such as Poisson's equation, Fermi–Dirac distribution, and continuity equations. Indeed, it provides a quantitative and comprehensive relationship between the transistor terminal characteristics with different materials and transistor input parameters. However, considering a numerical approach requires considerable computational load and consequently is usually not suitable for simulating circuits based on a large number of transistors. Nevertheless, the compact modeling approach simplifies this task by capturing the essential parts of leading mechanisms into simple analytical equations. Also, when a compact model for a new architecture is developed, the implementation of this model into a circuit simulator can be easily performed for possible system-level developments via the simulation, design, and prediction of practical circuits that include this new architecture. Table 3.1 illustrates the main required characteristics for useful and good compact models of FETs [33].

TABLE 3.1
Main required characteristics for good compact models of FETs

Characteristics	Description
Simplicity of equations	The simpler equations with a small number of parameters will be helpful for solving the currents and voltages within an integrated circuit, by improving the chance of reaching a convergence over a few periods (of time). In this context, descriptions with no numerical derivation or integration show big interest.
Wide range of applications	The difference between the operation regions is not important, and usually, a single equation can be devised in order to cover as many regions as possible. It will obviously minimize the number of equations and parameters for possible inclusion in the model library. If the compact model was validated by measuring a certain device, we should consider the extended derived models and applicability to other transistors (within the same kind of family) with, for instance, diverse sizes or/and at varying temperatures.
Transistor physics	Such a type of model should be based on physical theories. Simplification and also fitting capability are the most important issues of the compact model. Unfortunately, using too many empirical equations or parameters without or with less physical meaning can ultimately impact the full validity of the compact model.

3.3 INTEREST IN HARDWARE DESCRIPTION LANGUAGE

A hardware description language (or HDL) is a specialized powerful computer language frequently used to describe the design, structure, and operation for various types of electronic circuits, most frequently, digital circuits based on CMOS technology. These kinds of language are able to describe with good accuracy the formal description of any kind of circuit that allows automated simulation, analysis, and simulated testing of circuits with FETs. They also allow for the compilation of HDL programs into a lower-level specification of physics-based electronic devices, like the set of different masks habitually used to generate integrated circuits [40, 41].

In order to decrease the design complication of integrated circuits, there are important moves that raise the design's abstraction level using high-level synthesis with the help of HDL languages. The Verilog and VHDL languages are widely supported by various areas in the electronics industry. In addition, VHDL and Verilog share several characteristics, and both of them are widely suitable for analog and mixed-signal circuit simulation. The HDLs dedicated to analog circuit design are illustrated in Table 3.2 [42].

In addition, Verilog and VHDL form an important integral part of the electronic design-automation systems, especially for complex circuits, like microprocessors and VLSI systems.

3.4 DEVICE'S OPERATION AND PROPERTIES

As illustrated in Figure 3.2, the schematics of the 3-D gate-all-around MOSFET include the main geometry parameters, such as the dielectric thickness T_{ox}, the body radius R, and the channel length L. From the gate around the channel, the electrostatic potential in the body is ultimately improved. Therefore, the short-channel effects are more reduced than other architectures like FinFET or double-gate MOSFET [32].

Because we use the concept of "junctionless", the considered GAA MOSFET has no junctions or doping gradients type. This kind of device offers a low leakage current, less degradation of mobility, and less sensitivity to the thermal budget compared to the classical field-effect transistors. Furthermore, it presents a near ideal sub-threshold-slope (SS~ 60 mV decades at room temperature). Thus, the elimination

---

TABLE 3.2
Hardware description languages dedicated to analog circuit design

Name of HDL	Description
Verilog-AMS (Verilog for analog and mixed signal)	Open standard for extending Verilog language to analog and mixed analog-digital simulation as well.
VHDL-AMS (VHDL with Analog and Mixed-Signal extension)	A standard language dedicated to mixed analog and digital simulation.
Spectre HDL	A universal analog HDL.
HDL-A	An analog HDL language.

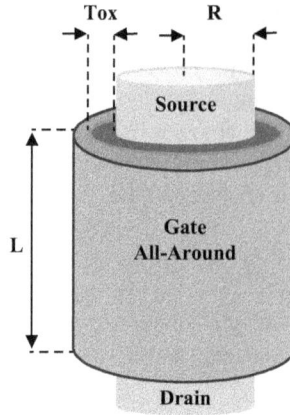

FIGURE 3.2 Architecture of 3-D junctionless GAA MOSFET

of source and drain junctions simplifies the fabrication process of CMOS technology to maintain the scaling-down [21].

Figure 3.3 shows the current-voltage characteristics of junctionless FETs describing the device's operation. It is clear that for $V_{GS}<V_{TH}$, the channel is fully depleted in electrons, while at the threshold voltage ($V_{GS}=V_{TH}$), a channel of neutral is formed and links drain and source. Where the gate voltage is greater than the threshold voltage ($V_{TH}<V_{GS}$), the formed neutral channel expands in thickness and width. The situation $V_{GS}>V_{FB}$ gives a fully accumulated channel.

Figures 3.4(a) and (b) show the transfer characteristics (I_{ds} vs. V_{GS}) of JL GAA MOSFET in linear and semilog scale, respectively. These characteristics are

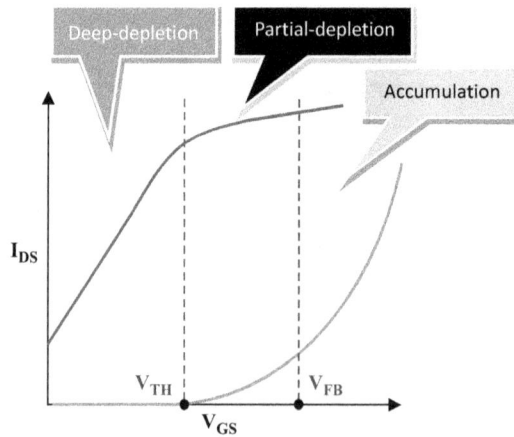

FIGURE 3.3 Transfer characteristics of junctionless FETs describing device operation, where I_{DS} is the drain current, V_{GS} is the gate voltage, V_{TH} is the threshold voltage, and V_{FB} is the flat-band voltage

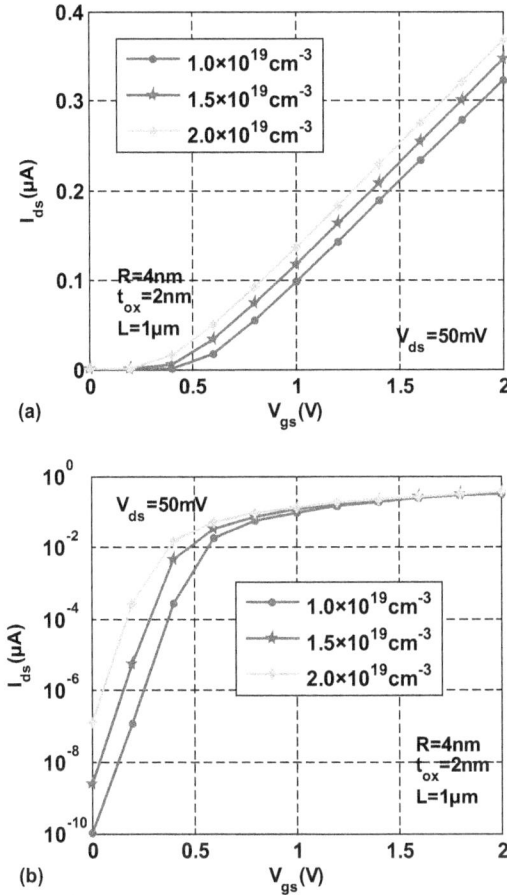

FIGURE 3.4   Transfer characteristics of 3-D JL GAA MOSFET in (a) linear scale, and (b) semilog scale, for different values of the doping concentration N_d=1.0×10^{19}cm^{-3}, 1.5×10^{19}cm^{-3}, and 2.0×10^{19}cm^{3}.

obtained from 3-D numerical simulation of JL GAA MOSFET with the help of Silvaco-TCAD Software. In this case, we consider different values of the doping concentration: N_d=1.0×10^{19}cm^{-3}, 1.5×10^{19}cm^{-3}, and 2.0×10^{19}cm^{3}. Moreover, the drain voltage V_{ds} is equal to 50mV, and the gate voltage V_{GSs} vary from 0 to 2V.

The transfer characteristics of JL GAA MOSFET are given in Figures 3.5(a) and (b) in linear and semilog scale, respectively, for different values of the channel radius R=4, 5, and 6nm.

At this stage of the study, we analyze the impact of varying the geometrical and technical parameters of the transistor, such as the body radius R and the body-doping N_d, on switching behaviors I_{ON}, I_{OFF}, and I_{ON}/I_{OFF} ratio. From Figures 3.4 and 3.5, it is clear that the current I_{ds} variation is strongly impacted by the channel radius R and the body-doping concentrations N_d, respectively. It is also apparent that increasing

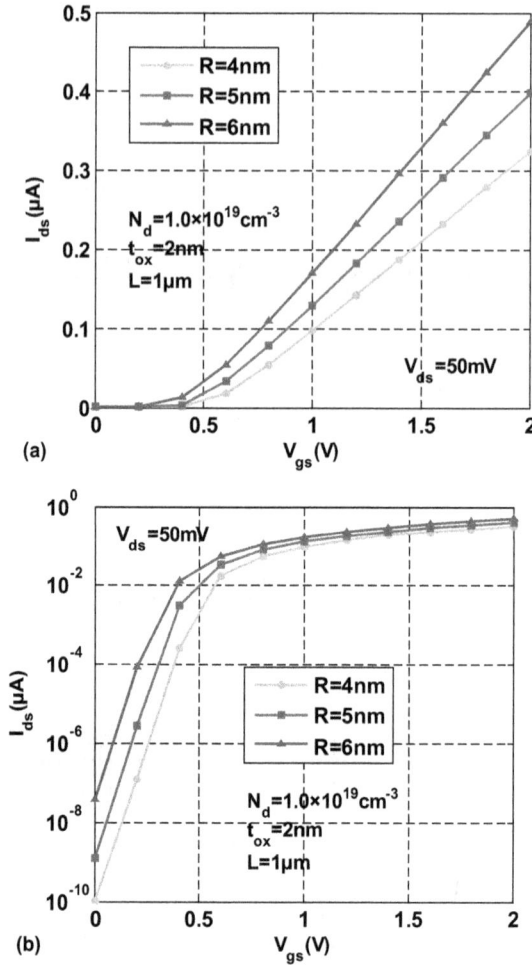

FIGURE 3.5   Transfer characteristics of 3-D JL GAA MOSFET in (a) linear scale, and (b) semilog scale, for different values of the channel radius R=4, 5, and 6nm

both N_d and R parameters create an important increase in the drain-current variation, particularly in the accumulation region.

However, the impact of varying the R parameter is more important than the N_d parameter as illustrated in Table 3.3 by means of OFF-current I_{OFF}, ON-current I_{ON}, and I_{ON}/I_{OFF} ratio [32].

3.5   MAIN APPROACH AND SIGNIFICANT COMPACT MODELS

The compact modeling of junctionless GAA MOSFET can be divided into three categories: the charge-based model, the surface-potential-based, and the

TABLE 3.3

Parameters of the JLN GAA MOSFET

Body-doping concentration (cm⁻³)	Channel radius (nm)	I_{OFF} (µA)	I_{ON} (µA)	I_{ON}/I_{OFF} ratio
1.0×10^{19}cm⁻³	4.0	1.07×10^{-10}	0.3337	3.11×10^{9}
1.5×10^{19}cm⁻³	4.0	2.57×10^{-9}	0.3572	1.38×10^{8}
2.0×10^{19}cm⁻³	4.0	1.26×10^{-7}	0.3794	3.01×10^{6}
1.0×10^{19}cm⁻³	5.0	1.23×10^{-9}	0.4196	3.41×10^{8}
1.0×10^{19}cm⁻³	6.0	3.91×10^{-8}	0.5087	1.30×10^{7}

threshold-voltage-based model (Figure 3.6). Next, we will explain the theoretical basis and main approach of the compact model developed from surface-potential-based and charge-based compact, which are widely used for compact modeling of these types of devices.

3.5.1 Charge-based Model

The charge-based approach is usually applied to the long-channel transistors [43]. Compared with the other compact models, the charge-based model is very helpful for developing a simple mathematical formula for the drain current in all regions of the transistor operation: fully depleted, partly depleted, and accumulated. This model is given by the Pao–Sah integral as [44–45]:

$$I_{ds} = \mu \frac{W}{L} \int_0^{V_{ds}} Q_m dV \tag{1}$$

where μ is the mobility, W is the channel width, L is the channel length, and V is the potential shift due to the electron's quasi-Fermi level, usually evaluated at the source side ($V=0$) and the drain side ($V_d=V_{ds}$), and Q_m is the mobile charge density.

These kinds of models are usually based on Boltzmann distribution and quasi-Fermi level description for the carrier densities. The drift-diffusion carrier model

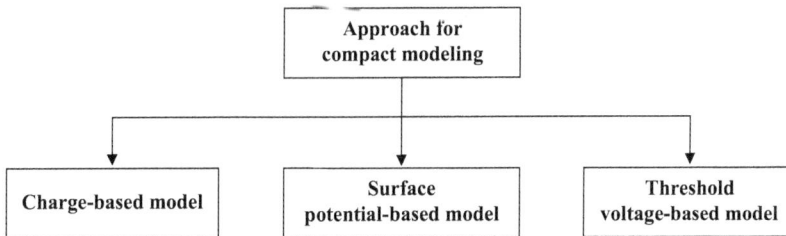

FIGURE 3.6 Main approaches for compact modeling of junctionless GAA MOSFET

is also considered for the current derivation. In the following, we will explain the theoretical basis of significantly developed charge-based compact models.

- **The approach based on Duarte et al. [44]**

In this case, a nonpiecewise analytical drain-current model has been developed for junctionless gate-all-around FET and for a long-channel case. The model begins with the parabolic-potential approximation [46]:

$$\Phi(r) = \frac{r^2}{R^2}(\Phi_S - \Phi_0) + \Phi_0 \tag{2}$$

where R is the body radius, and Φ_0 and Φ_S are the center and surface potentials in the body, respectively. The parameter r corresponds to the spatial distance in the radial direction.

Then, using Gauss' theorem and recalling Equation (2), an important expression can be developed regarding the mobile charge and voltages:

$$V_{gs} - V_{th} - V = \Phi_t \ln\left(-Q_m / 4\pi\varepsilon_{si}\Phi_t\right) - Q_m / C_{eff} + \Phi_t$$
$$\times\left[\left(1 + Q_m / qN_d\pi R^2\right) / \left(1 - \exp\left(-\left(Q_m + qN_d\pi R^2\right) / 4\pi\varepsilon_{si}\Phi_t\right)\right)\right] \tag{3}$$

where N_d represents the channel doping concentration, ε_{si} is the permittivity of silicon, q is the universal electronic charge of the electron, $\Phi_t(=KT/q)$ is the thermal voltage, ε_{si} is the permittivity of silicon, and V_{TH} is the device threshold voltage. The effective gate-capacitance C_{eff} controls the bulk charge in the semi-depleted region [47]. It is worth noticing that the charge-based model expressed by Equation (3) is valid in all regions of the transistor operation: (1) fully depleted, (2) semi-depleted, and (3) accumulated.

Accordingly, the mobile charge density Q_m is decoupled as $Q_m = Q_{dep} + Q_c$, where Q_{dep} corresponds to the mobile charge density in fully-depleted and semi-depleted modes (when the gate voltage V_{gs} is less than the flat-band voltage V_{fb}), and Q_c is a complementary-mobile charge density added to Q_{dep}.

This implies that Q_c is a main correction term to Q_{dep}, and both of these two terms are computed separately by the following set of equations:

$$V_{gs} - V_{th} - V = \Phi_t \ln\left(-Q_{dep} / 4\pi\varepsilon_{si}\Phi_t\right) - Q_{dep} / C_{eff} \tag{4}$$

$$V_{gs} - V_{fb} - V = \Phi_t \ln\left(-Q_c / 4\pi\varepsilon_{si}\Phi_t\right) - Q_c / C_c \tag{5}$$

with $Q_c = C_{ox} - C_{eff}$ corresponding to the complementary capacitance, and C_{ox} is the oxide capacitance.

In addition, the authors derived a drain-current expression using the Pao–Sah formula, as

$$I_{ds} = -\frac{\mu}{L}\left(\frac{(Q_{dep})^2}{2C_{eff}} - \Phi_t Q_{dep}\right)\Bigg|_{Qs_{dep}}^{Qd_{dep}} - \frac{\mu}{L}\left(\frac{(Q_c)^2}{2C_c} - \Phi_t Q_c\right)\Bigg|_{Qs_c}^{Qd_c} \tag{6}$$

The main benefit of this model is the nonpiecewise formulation and the simple analytical description of the surface and bulk current mechanisms in junctionless GAA FETs in all regions of device operation, with no fitting parameters. Moreover, the model meets all specifications of the numerical results in all regions of operation from deep depletion to accumulation.

- ***The approach based on Lime et al. [43] and Moldovan et al. [48]***

First, the authors developed a DC compact model considering the drift-diffusion equations. The final expression gives a simple analytical form:

$$I_{ds} = 2\pi\frac{R}{L}\mu\Phi_t\left(f\left(Q_m(0)\right) - f\left(Q_m(V_{ds})\right)\right) \tag{7}$$

Then, the mobile charge density is calculated using the following expression [43]:

$$f(Q) = \frac{Q^2}{2Q_{eq}} + 2Q - AQ\ln\left(1 + \exp\left(\frac{Q - Q_{dop}}{2AQ_{cp}}\right)\right) + Q_{dop}\ln\left(\frac{Q - Q_{dop}}{2Q_{cp}\left(\exp\left(\frac{Q - Q_{dop}}{2Q_{cp}}\right) - 1\right)}\right) \tag{8}$$

Where $Q_{dop}=qN_d(R/2)$, $Q_{cp}=2\epsilon_{si}\Phi_t/R$, and $A=1.425$.

The model proposed by Lime et al. has a simple explicit formulation and provides a continuous solution of the mobile charge density for all regions of operation, i.e., from depletion to accumulation mode.

Second, based on the analytical solution of the mobile charge density in [43], Moldovan et al. derived a quasi-static and continuous model to compute the intrinsic capacitances using the charge-conservation equation [48].

$$C_{SS} = C_{SG} + C_{SD} = C_{GS} + C_{DS} \tag{9}$$

$$C_{DD} = C_{DS} + C_{DG} = C_{SD} + C_{GD} \tag{10}$$

$$C_{GG} = C_{GS} + C_{GD} = C_{SG} + C_{DS} \tag{11}$$

where C_{gg} is the gate-to-gate capacitance, C_{dd} is the drain-to-drain capacitance, C_{ss} is source-to-source capacitance, C_{sg} is gate-to-source capacitance, C_{ds} is source-to-drain capacitance, C_{gs} is source-to-gate capacitance, C_{sd} is drain-to-source capacitance, C_{dg} is gate-to-drain capacitance, and C_{gd} is drain-to-gate capacitance.

In addition, the authors in [43, 48] proposed an explicit simple and continuous model for junctionless GAA MOSFET describing the DC and AC behavior of the device. Note that this model is the first complete model that has been implemented in CMOS with Verilog-A language.

- **The approach based on Gnani and Baccarani et al. [47]**

Based on the Bessel functions, the mobile charge in the depletion region is given by:

$$Q_m = -Q_0 \left\{ \frac{X^2}{4} - \frac{X}{2} \frac{\alpha^2 - 1}{\alpha} I_1 \left(\frac{X}{\alpha} \right) \right\} \quad for\ x_0 > X \tag{12}$$

and

$$Q_m = -Q_0 \left\{ \frac{(x_0)^2}{4} - \frac{x_0}{2} \frac{\alpha^2 - 1}{\alpha} I_1 \left(\frac{x_0}{\alpha} \right) \right\} \quad for\ x_0 < X \tag{13}$$

Where $Q_0 = 4\pi\epsilon_{si}\Phi_t R/\lambda$ corresponds to the normalized factor of the charge density, λ is the Debye length, $I_l(x)$ is the modified Bessel functions, $X = R/\lambda$ is the normalized body radius, $\alpha^2 = exp(v - u_c)$, $v = V/\Phi_t$ is the normalized potential shift due to the electron quasi-Fermi level, and u_c is the electrostatic potential at the transistor symmetry axis.

Also, the authors describe analytically the mobile charge in the accumulation region as

$$Q_m = -\frac{Q_0}{\sqrt{2}} \left(\exp\left(u_s - v\right) - \left(u_s - v\right) - 1 \right)^{1/2} - qN_d\pi R^2 \tag{14}$$

where u_s is the normalized electrostatic potential ($u_s = \Phi_s/\Phi_t$).

This model gives a very good physical explanation of the transistor behavior and provides an accurate description of the mobile charge density from sub-threshold to accumulation region assuming a cylindrical geometry. However, the mathematical formulation of the model is complicated due to the use of Bessel functions, which limits the implementation in circuit simulators.

- **The approach based on Jazaeri et al. [49]**

The authors adapted an analytical charge-based model derived for junctionless double-gate MOSFET to calculate the transcapacitance of JL GAA MOSFET [50]. The model describes the quasi-static behavior of JL GAA MOSFET as a set of the following equations:

$$C_{DG} = WL\alpha \left(-6\frac{\partial Q_{m,d}}{\partial V_{gs}} + \frac{\partial Q_{m,s}}{\partial V_{gs}} - 12\frac{\partial Q_{m,1}}{\partial V_{gs}} - 28\frac{\partial Q_{m,2}}{\partial V_{gs}} \right) \qquad (15)$$

$$C_{SG} = WL\alpha \left(\frac{\partial Q_{m,d}}{\partial V_{gs}} - 6\frac{\partial Q_{m,s}}{\partial V_{gs}} - 28\frac{\partial Q_{m,1}}{\partial V_{gs}} - 12\frac{\partial Q_{m,2}}{\partial V_{gs}} \right) \qquad (16)$$

$$C_{GG} = -5WL\alpha \left(\frac{\partial Q_{m,d}}{\partial V_{gs}} + \frac{\partial Q_{m,s}}{\partial V_{gs}} + 8\frac{\partial Q_{m,1}}{\partial V_{gs}} + 8\frac{\partial Q_{m,2}}{\partial V_{gs}} \right) \qquad (17)$$

$$C_{SD} = WL\alpha \left(\frac{\partial Q_{m,d}}{\partial V_{ds}} - 28\frac{\partial Q_{m,1}}{\partial V_{ds}} - 12\frac{\partial Q_{m,2}}{\partial V_{ds}} \right) \qquad (18)$$

$$C_{DD} = WL\alpha \left(-6\frac{\partial Q_{m,d}}{\partial V_{ds}} - 12\frac{\partial Q_{m,1}}{\partial V_{ds}} - 28\frac{\partial Q_{m,2}}{\partial V_{ds}} \right) \qquad (19)$$

$$C_{GD} = -5WL\alpha \left(\frac{\partial Q_{m,d}}{\partial V_{ds}} + 8\frac{\partial Q_{m,1}}{\partial V_{ds}} + 8\frac{\partial Q_{m,2}}{\partial V_{ds}} \right) \qquad (20)$$

where $Q_{m,1}$ and $Q_{m,2}$ are the internal mobile charge densities calculated at $y=0.25\times L$ and $0.75\times L$, respectively, $\alpha=0.111$. Next, to define the derivative of the local charge densities regarding the gate and drain voltages (V_{gs} and V_{ds}), the authors use the method published in [50].

Finally, the work of Jazaeri et al. is the first model describing the AC behavior of JL GAA MOSFET. However, compared to the model of Moldovan et al. [48], the model of Jazaeri et al. is not purely analytical and has not been implemented in circuit simulation.

3.5.2 SURFACE-POTENTIAL-BASED MODEL

The charge-based compact models of JL GAA MOSFETs described above can be implemented in Verilog-A for possible DC and AC simulations of circuits and transient properties. Nevertheless, for an accurate and qualified model for the EDA tool, these kinds of models should present high accuracy, powerful physical properties, and include small geometry effects, such as short-channel and quantum confinement effects. The surface potential-based approach is widely considered to achieve strong physical properties, high accuracy, and be easily simplified into threshold-voltage-based and charge-based models.

- *The approach based on Sorée et al. [51]*

A surface-potential-based and self-consistent quantum model has been developed to get the electronic structure. First, the authors derived a solution of the surface-potential Φ_S from the straight-forward integration of a 1-D Poisson equation in the abrupt-depletion approximation, and it can be written as

$$\Phi_S = V - \frac{qN_d}{4\varepsilon_{si}}\left(2(R-r_d)^2 \ln\left(\frac{R}{R-r_d}\right) - r_d(r_d - 2R)\right) \tag{21}$$

Where r_d corresponds to the depletion width.

Second, the authors developed an analytical expression for the electrostatic potential inside the oxide and the equation for the drain current proportional to both the body-doping density and the mobility.

Third, Sorée et al. carried out the self-consistent quantum-mechanical calculation of the charge density and the electrostatic-potential profile. The electronic eigen functions are derived from the universal Schrödinger equation.

It is important to mention that the model by Sorée et al. has the advantages of being analytically simple and including a quantum-mechanical effect. Despite these advantages, this model is only valid below the threshold-voltage region.

- **The approach based on Smaani et al. [32]**

By considering a regional approach [47], the authors developed an analytical-simple model without fittings parameters and with a physics-based concept. Furthermore, the model dedicated to junctionless GAA MOSFET has been also implemented in low-power circuits using Verilog-A language.

First, from the expression of surface-potential Φ_S given by Equation (21) and using adequate boundary conditions [32, 52], an important expression of the surface-potential Φ_S has been proposed in a deep-depletion regime so that

$$\Phi_S - V = \beta\left(V_{geff} - \Phi_S\right) + \left(2\beta\left(V_{geff} - \Phi_S\right) + \delta\right)$$

$$\times \left(\ln(R) - \frac{1}{2}\ln\left(R^2\left(1 + \frac{2}{\delta}\beta(V_{geff} - \Phi_S)\right)\right)\right) \tag{22}$$

where $V_{geff} = V_{gs} - V_{fb}$ is the effective gate voltage, $V_{fb} = \Phi_{ms} + \Phi_t ln(N_d/n_i)$ is the universal flat-band voltage, E_s refers to the surface electric field, $C_{ox} = \varepsilon_{ox}/Rln(1+t_{ox}/R)$ is the oxide capacitance, ε_{ox} is the oxide permittivity, Φ_{ms} is the work-function difference of the metal-gate and body-semiconductor, n_i corresponds to the intrinsic concentration, $\Phi_t(=KT/q)$ is the thermal voltage, T is the temperature, and K is the Boltzmann constant. $Q_{dep} = \pi q N_d R^2$ represents the fixed-charge density, $\beta = C_{ox}/(4\pi\varepsilon_{si})$, and $\delta = Q_{dep}/2\pi q\varepsilon_s$.

Second, the authors showed that the surface-potential in the accumulation region can be written as

$$\left(\Phi_S - 2V_{geff}\right)\Phi_S + V_{geff}^2 = \left(\exp\left(\frac{\Phi_S - V}{\Phi_t}\right) - 1\right)\eta \tag{23}$$

where $\eta = 2qNd\, Tsi\Phi t\, /Cox^2$.

Third, when the transistor is partly depleted, the authors show that the surface-potential can be written as

$$\xi = \beta\left(V_{geff} - \Phi_S\right) + \left(2\beta\left(V_{geff} - \Phi_S\right) + \delta\right) \times \left(\ln(R) - \frac{1}{2}\ln\left(R^2\left(1 + \frac{2}{\delta}\beta\left(V_{geff} - \Phi_S\right)\right)\right)\right)$$

(24)

with $\xi = -\Phi_t[\exp((\Phi_S - V)/\Phi_t) - ((\Phi_S - V)/\Phi_t) - 1]$ corresponding to the approximated solution of the surface electric field E_s in the accumulation region.

Consequently, the analytical model represents the drain current as a set of three equations:

$$I_{ds}^{sub} = 2\pi\Phi_t\mu\frac{R}{L}$$

$$\times \left\{ C_{ox}\left[i_c^p - \left(\frac{1}{2} + \left(\frac{1}{4} + \ln(R)\right)\beta\right)V_{geff}^2 + \frac{1}{4}\delta V_{geff} + \left(\frac{3}{4} + \ln(R) + \frac{1}{2\beta}\right)\frac{\delta^2}{4\beta}\right] + Q_{dep}V \right\}_S^D$$

(25)

with,

$$i_c^p = \left(\left(\frac{\Phi_S}{2} - V_{geff}\right)\Phi_S + \frac{V_{geff}^2}{2} - \frac{\delta^2}{8\beta^2}\right)\beta\ln\left(\frac{R^2}{\delta}\left(\left(V_{geff} - \Phi_S\right)2\beta + \delta\right)\right)$$

$$- \left(\left(\frac{1}{4} + \ln(R)\right)\beta + \frac{1}{2}\right)\Phi_S^2 + \left(\left(1 + \left(\frac{1}{2} + 2\ln(R)\right)\beta\right)V_{geff} - \frac{\delta}{4}\right)\Phi_S$$

$$I_{ds}^{par} = 2\pi\Phi_t\mu\frac{R}{L}$$

$$\times \left\{ C_{ox}\left[\left(\Phi_S - V_{geff} - \Phi_t\right)\Phi_t\exp\left(\frac{\Phi_S - V}{\Phi_t}\right) + i_c^p + \left(\delta - \beta V_{geff}\right)\frac{V_{geff}}{4} + \frac{3\delta^2}{16\beta}\right] + Q_{dep}V \right\}_S^D$$

(26)

$$I_{ds}^{acc} = 2\pi\Phi_t\mu\frac{R}{L}\left\{ C_{ox}\left[\left(V_{geff} + 2\Phi_t\right)\Phi_S - \frac{\Phi_S^2}{2} - 2\Phi_t\sqrt{\eta}\arctan\left(\frac{\Phi_S - V_{geff}}{\sqrt{\eta}}\right)\right] + Q_{dep}V \right\}_S^D$$

(27)

where Equations (25), (26), and (27) correspond to the drain current in deep-depletion, partly-depleted, and accumulation regimes, respectively. In the above equations,
S and D indicate the limits at the source side $\Phi_S(0)$ and the drain side $\Phi_S(L)$, respectively. In order to compute the current in the depletion regime $I_{ds}^{dep}=((I_{ds}^{sub})^2+(I_{ds}^{par})^2)^{1/2}$,
an interpolation function has been involved.

This is one of the most important contributions, where the authors provided
the analog and digital circuit simulation results of low-power circuits as well as
Verilog-A translation of this model.

3.5.3 THRESHOLD-VOLTAGE-BASED MODEL [53]

Compared with the charge-based and surface-based models, the threshold-voltage-
based approach is considered less often for compact modeling of JL GAA MOSFET,
and only a few works have been derived from the threshold-based model. In this context, Chiang et al. developed a quasi-2-D threshold-voltage-based model for short-
channel JL GAA MOSFET by considering the scaling equation.

Assuming a parabolic approximation of the electrostatic potential in the device's
body, the authors derived an analytical expression of the minimum central potential.
After that, by setting the minimum central potential to zero value and solving for the
gate voltage, the threshold voltage of short-channel JL GAA MOSFET can be written as:

$$V_{th} = \frac{2(\beta\gamma+\kappa\alpha)+\omega+\sqrt{(2(\beta\gamma+\kappa\alpha)+\omega)^2-(1-4\alpha\gamma)(\omega^2-4\beta\kappa)}}{1-4\alpha\gamma} \tag{28}$$

All the parameters of Equation (28) are well described by Chiang et al. in [53].

Although the approaches developed by Chiang et al. give good accuracy and a
simple formulation of the threshold voltage in JL GAA MOSFET, the proposed compact model covers only the sub-threshold region. Nevertheless, this model establishes
an important analytical formula for the threshold voltage, provides design guidance
for JL GAA MOSFET, and could be useful for device and circuit simulation.

3.6 CHALLENGES WITH COMPACT MODELING

In the aforementioned section, we presented the main approach and significant progress in the compact modeling of JL GAA MOSFET dedicated to circuit simulation.
However, there are important challenges, such as improving the robustness and practical applications of the models for circuit simulation. In the following section, we
introduce some open research fields that should be considered for future compact
modeling of JL GAA MOSFET.

- *Including short-channel effects*

To continue the aggressive device scaling in CMOS technology, analytical complete models incorporating short-channel effects (SCEs) in JL GAAMOSFET are

important concepts and crucial issues that should be addressed [53]. The SCEs are unwanted phenomena arising from the aggressive diminution of the device's channel length, especially, when the gate length becomes almost equal to the space-charge regions of the drain/source junctions with the substrate. The main effects are drain-induced barrier lowering (DIBL), threshold-voltage roll-off, velocity saturation, mobility degradation, and reverse leakage current rise [13, 54, 55]. In addition, an explicit and simple description of SCEs and their incorporation into a continuous compact model is highly desirable for fast and accurate computation in implemented CMOS circuits.

- *Model for gate-tunneling current*

The gate-tunneling current is a kind of unwanted negative phenomenon usually created from aggressive device scaling, and it is also essential for transistor design considerations [56, 57]. Moreover, for circuit simulator development [58, 59], an analytical compact model of JL GAA MOSFET for accurate calculation of gate-tunneling currents remains widely desirable and helpful. In this context, the confinement effect which is frequently caused by the electric field is significant for the very small diameter of the device's gate-all-around. Therefore, a compact model for the gate-tunneling current should incorporate structural and electrical quantum confinements.

- *Describing the threshold voltage*

The threshold voltage is a key concept in classical FETs. It allows for a simple and accurate calculation of the charge density based on its proportional relation with the gate overdrive voltage. In this context, less compact models have been focused on the threshold voltage [53], especially considering short-channel devices and including small geometry effects. In addition, solid fundamental research work on the threshold voltage of JL GAAMOSFET remains an important research subject.

- *Implementation in low-power circuits*

The implementation of compact models for JL GAA MOSFET in a low-power circuit is an important task that should be realized for future use in various integrated circuits. This task is usually performed through hardware description language and circuit simulator development. However, less compact models for JL GAA MOSFET have been implemented in low-power circuits [32, 48]. Therefore, more compact models for JL GAA MOSFET should be incorporated in both analog and digital low-power circuits in order to improve the practical applications in circuit simulation.

3.7 CONCLUSION

We have illustrated the interest and specificities of compact models for JL GAA MOSFET. In this context, we have shown that a compact model should present simplified mathematical equations and that it has a wide range of applications. We have also described the device's physics. We have introduced the significant compact models of JL GAA MOSFET and the main approaches that are considered for circuit simulation applications, such as charge-based, surface-potential-based, and threshold-voltage-based models. In this regard, most of the developed work regarding JL GAA MOSFET considered the charge-based and the surface-potential-based model. We found that compact models that are explicit, simple, and analytical-based are promising and helpful for the implementation of CMOS circuits through HDL. Furthermore, we have shown that the implementation of a compact model in low-power circuits, modeling the gate-tunneling current, including SCEs, and describing the threshold-voltage are the main issues that should be considered for future compact modeling in GAA MOSFET and for practical applications of these models in various circuit simulators.

REFERENCES

1. Mamaluy, D., Gao, X. The fundamental downscaling limit of field effect transistors, *Applied Physics Letters*, 106(19), 193503 (2015).
2. Noor, F. A., Bimo, C., Syuhada, I., et al. A compact model for gate tunneling currents in undoped cylindrical surrounding-gate metal-oxide-semiconductor field-effect transistors, *Microelectronic Engineering*, 216, 111086 (2019).
3. Tayal, S., Smaani, B., Rahi, S. B., et al. Incorporating bottom-up approach into device/circuit co-design for SRAM-based cache memory applications, *IEEE Transactions on Electron Devices*, 69(, 27–32 (2022).
4. Raut, P., Nanda, U. RF and linearity parameter analysis of junction-less gate all around (JLGAA) MOSFETs and their dependence on gate work function, Silicon, 68, 27–35 (2021).
5. Meriga, C., Ponnuri, R. T., Satyanarayana, B. V. V., et al. A novel teeth juncti on less gate all around FET for improving electrical characteristics, *Silicon*, 47, 79–84 (2021).
6. Wang, T., Lou, L., Lee, C. A junctionless gate-all-around silicon nanowire FET of high linearity and its potential applications, *IEEE Electron Device Letters*, 34(4), 478–480 (2013). https://ieeexplore.ieee.org/document/6471739.
7. Yamabe, K., Endoh, T. Ultimate vertical gate-all-around metal–oxide–semiconductor field-effect transistor and its three-dimensional integrated circuits, *Materials Science in Semiconductor Processing*, 134, 106046 (2021).
8. Sreenivasulu, V. B., Narendar, V. Characterization and optimization of junctionless gate-all-around vertically stacked nanowire FETs for sub-5 nm technology nodes, Microelectronics Journal, 116, 105214 (2021).
9. Djeffal, F., Ferhati, H., Bentrcia, T. Improved analog and RF performances of gate-all-around junctionless MOSFET with drain and source extensions, *Superlattices and Microstructures*, 90, 132–140 (2016).
10. Moon, D. I., Choi, S. J., Duarte, J. P., et al. Investigation of silicon nanowire gate-all-around junctionless transistors built on a bulk substrate, IEEE Transactions on Electron Devices, 60(4), 1355–1360 (2013).
11. Sharma, M., Gupta, M., Narang, R., et al. Investigation of gate all around junctionless nanowire transistor with arbitrary polygonal cross section. Paper presented at the 4th International Conference on Devices, Circuits and Systems (ICDCS), Coimbatore, India, 16–17 March (2018).

12. Thomas, S. Gate-all-around transistors stack up, *Nature Electronics*, 3(12), 728 (2020).
13. Smaani, B., Latreche, S., Iniguez, B. Compact drain-current model for undoped cylindrical surrounding-gate metal-oxide semiconductor field effect transistors including short channel effects, Journal of Applied Physics, 114(22), 224507 (2013).
14. Rahmana, I. K. M. R., Khan, Md. I., Khosru, Q. D. M. Analytical drain current and performance evaluation for inversion type InGaAs gate-all-around MOSFET, *AIP Advances*, 114, 065108 (2021).
15. Cao, W., Shen, C., Cheng, S. Q., et al. Gate tunneling in nanowire MOSFETs, IEEE Electron Device Letters, 32(4), 461–463 (2011).
16. Nowbahari, A., Roy, A., Marchetti, L. Junctionless transistors: State-of-the-art, *Electronics*, 9(7), 1174 (2020).
17. Talukdar, A., Raibaruah, A. K., Sarma, K. C. D. Dependence of electrical characteristics of junctionless FET on body material, *Procedia Computer Science*, 171, 1046–1053 (2020). https://doi.org/10.1016/j.procs.2020.04.112.
18. Jeon, C. H., Park, J. Y., Scol, M. L., et al. Joule heating to enhance the performance of a gate-all-around silicon nanowire transistor, IEEE Transactions on Electron Devices, 63(6), 2288–2292 (2016).
19. Lee, C.-W., Afzalian, A., Akhavan, N. D., et al. Junctionless multigate field-effect transistor, Applied Physics Letters, 94(5), 053511 (2009).
20. Lee, C.-W., Ferain, I., Afzalian, A., et al. Performance estimation of junctionless multigate transistors, *Solid-State Electronics*, 54(2), 97–103 (2010).
21. Colinge, J. P., Lee, C. W., Afzalian, A., et al. Nanowire transistors without junctions, *Nature Nanotechnology*, 5(3), 225–229 (2010).
22. Aditya, M., Rao, K. S., Balaji, B., et al. Comparison of drain current characteristics of advanced MOSFET structures - A review, *Silicon*, 61(14), 69–76 (2022).
23. Gupta, A., Rai, M. K., Pandey, A. K., et al. A novel approach to investigate analog and digital circuit applications of silicon junctionless-double-gate (JL-DG) MOSFETs, *Silicon*, 26, 77–84, (2021).
24. Talukdar, A., Raibaruah, A., Sarma, K. K. C. D. Dependence of electrical characteristics of junctionless FET on body material, Procedia Computer Science, 171, 1043–1056 (2020).
25. Jung, A., Bonnassieux, Y., Horowitz, G., et al. Advances in compact modeling of organic field-effect transistors. *IEEE Electron Devices Society*, 8, 1404–1415 (2020).
26. Saha, S. K. Modelling the effectiveness of computer-aided development projects in the semiconductor industry, *International Journal of Engineering Management and Economics*, 1(2/3), 162–178 (2010).
27. Paydavosi, N., Morsged, T. H., Lu, D. D., et al. *BSIM4v4.8.0 MOSFET model user's manual*, University of California, Berkeley, CA (2013).
28. Nagel, L. W. *SPICE2—A Computer Program to Simulate Semiconductor Circuits*, University of California, Berkeley, CA, Electronic Research Laboratory Memo ERL-M250 (1975).
29. Saha, S. K. Modeling process variability in scaled CMOS technology, *IEEE Design and Test of Computers*, 27(2), 8–16 (2010).
30. Gildenblat, G. S. *Compact Modeling: Principles, Techniques and Applications*, Springer, New York, 527, (2010).
31. McAndrew, C. C. Practical modeling for circuit simulation, *IEEE Journal of Solid-State Circuits*, 33(3), 439–448 (1998).
32. Smaani, B., Rahi, S. B., Labiod, S. Analytical compact model of nanowire junctionless gate-all-around MOSFET implemented in Verilog-A for circuit simulation, *Silicon*, 14, 67–76 (2022).

33. Jung, S., Bonnassieux, Y., Horowitz, G., et al. Advances in compact modeling of organic field-effect transistors. *IEEE Journal of the Electron Devices Society*, 8, 1404–1415 (2020), https://doi.org/10.1109/JEDS.2020.3020312.

34. Alvarado, J., Iñiguez, B., Estrada, M. Implementation of the symmetric doped double-gate MOSFET model in Verilog-A for circuit simulation, *International Journal of Numerical Modelling*, 23, 88–106 (2010).

35. Lundstrom, M. L., Antonidis, D. A. Compact models and the physics of nanoscale FETs, *IEEE Transactions on Electron Devices*, 61(2), 225–233 (2014).

36. Cheng, Y., Hu, C., *MOSFET Modeling and BSIM3 User's Guide*, Kluwer Academic Publishers, London (1999).

37. Paydavosi, N., Morsged, T. H., Lu, D. D. et al. *BSIM4v4.8.0 MOSFET Model User's Manual*, University of California, Berkeley, CA, (2013).

38. SILVACO international ATLAS user's manual (2007).

39. https://silvaco.com/.

40. Kaeslin, H. Top-down digital VLSI design from architectures to gate-level circuits and FPGAs, Elsevier Science, USA, 598, (2015).

41. Cobianu, O., Soffke, M. A., Glesner, A. Verilog-A model of an undoped symmetric dual-gate MOSFET, *Advances in Radio Science*, 4, 303–306 (2006).

42. Ciletti, M. D., *Advanced Digital Design with Verilog HDL*, Pearson Education India, Delhi, 1012, (2010).

43. Lime, F., Moldovan, O., Iñiguez, B. A compact explicit model for long-channel gate-all-around junctionless MOSFETs. Part I: DC characteristics, *IEEE Transactions on Electron Devices*, 61(9), 3036–3041 (2014).

44. Duarte, J. P., Choi, S.-J., Moon, D.-II. A nonpiecewise model for long-channel junctionless cylindrical nanowire FETs, *IEEE Electron Device Letters*, 33(2), 155–157 (2012).

45. Pao, H. C., Sah, C. T. Effects of diffusion current on characteristics of metal–oxide (insulator)–semiconductor transistors, *Solid-State Electronics*, 9(10), 927–937 (1966).

46. Auth, C. P., Plummer, J. D. Scaling theory for cylindrical, fullydepleted, surrounding-gate MOSFETs, *IEEE Electron Device Letters*, 18(2), 74–76 (1997).

47. Gnani, E., Gnudi, A., Reggiani, S., et al. Theory of the junctionless nanowire FET, *IEEE Transactions on Electron Devices*, 58(9), 2903–2910 (2012).

48. Moldovan, O., Lime, L., Iñiguez, B. A complete and Verilog-A compatible gate-all-around long-channel junctionless MOSFET model implemented in CMOS inverters, *Microelectronics Journal*, 46(11), 1069–1072 (2015).

49. Jazaeri, F., Barbut, L., Sallese, J.-M. Trans-capacitance modeling in junctionless gate-all-around nanowire FETs, *Solid-State Electronics*, 96, 34–37 (2014).

50. Sallese, J.-M., Jazaeri, F., Barbut, L., et al. A common core model for junctionless nanowires and symmetric double-gate FETs, *IEEE Transactions on Electron Devices*, 60(12), 4277–4280 (2013).

51. Sorée, B., Magnus, W., Pourtois, G. Analytical and self-consistent quantum–mechanical model for a surrounding gate MOS nanowire operated in JFET mode, Journal of Computational Electronics, 7(3), 380–383 (2008).

52. Smaani, B., Labiod, S., Nafa, F., et al. Analytical drain-current model and surface-potential calculation for junctionless cylindrical surrounding-gate MOSFETs, *International Journal of Circuits, Systems and Signal Processing*, 15, 1394–1399 (2021).

53. Chiang, T.-K. A new Quasi-2-D threshold voltage model for short-channel junctionless cylindrical surrounding gate (JLCSG) MOSFETs, *IEEE Transactions on Electron Devices*, 59(11), 3127–3129 (2022).

54. Moldovan, O., Lime, F., Barraud, B., et al. Experimentally verified drain-current model for variable barrier transistor, *Electronics Letters*, 51(17), 1364–1366 (2015).

55. Labiod, S., Smaani, B., Tayal, S., et al. Mixed-mode optical/electric simulation of silicon lateral PIN photodiode using FDTD method, *Silicon*, 15, 81–91(2022).
56. Noor, F. A., Bimo, C., Syuhada, I., et al. A compact model for gate tunneling currents in undoped cylindrical surrounding-gate metal-oxide-semiconductor field-effect transistors, *Microelectronic Engineering*, 216, 111086 (2019).
57. Cao, W., Shen, C., Cheng, S. Q., et al. Gate tunneling in nanowire MOSFETs, *IEEE Electron Device Letters*, 32(4), 461–463 (2011).
58. Roy, A. S., Sallese, J. M., Enz, C. C. A closed-form charge-based expression for drain current in symmetric and asymmetric double gate MOSFET, *Solid-State Electronics*, 50, 87–93 – (2006).
59. Labiod, S., Smaani, B., Latreche, S. Numerical modeling of electrical/optical combination for the simulation of PIN photodiode, in *2022 19th International Multi-Conference on Systems, Signals & Devices (SSD)*, 2022, pp. 1311–1317. https://doi.org/10.1109/SSD54932.2022.9955656.

4 Novel gate-overlap tunnel FETs for superior analog, digital, and ternary logic circuit applications

*Simhadri Hariprasad, Ramakant Yadav,
and Surya Shankar Dan*

CONTENTS

4.1 INTRODUCTION

The famous Moore's law given by Gordon Moore in 1960 states that every one to two years, the number of transistors in ICs doubles [1]. The problem that cripples this exponential growth is the increase in the power density generated due to the heating of the transistors. If the power density keeps scaling like this, the cost of cooling the

DOI: 10.1201/9781003359234-4

chips using contemporary methods will make the process economically unviable. Moreover, as the channel length keeps decreasing, the quantum tunneling effect starts to dominate the functioning of the CMOS technology. This happens because CMOS scaling leads to various short-channel effects like drain-induced barrier lowering (DIBL), mobility degradation, high leakage currents, and impact ionization. Hence the scaling of CMOS technology beyond the nanoscale range has caused various reliability issues [2]. It is well-known that the drift-diffusion transport in MOSFETs restricts the minimum *inverse sub-threshold slope SS* at around 60 mV/ decade change in the current level at room temperature. Recent research has highlighted the need for an alternative device providing better switching performance to counter the impending power crises at the nanoscale when the power consumption exceeds the limits of reliable device operation.

At the nanoscale dimensions, sub-threshold leakage becomes highly detrimental to the device operation [3, 4]. Hence, researchers worldwide have become interested in tunnel field-effect transistor (TFET) technology because of its lower *SS* and much smaller I_{off} than conventional MOSFETs. Therefore, TFET is suitable for ultra-low power [5, 6] applications. TFET technology replaces conventional diffusion-based minority carrier injection in the case of MOSFETs with *band-to-band* (BtB) tunneling-based minority carrier injection into the channel [7, 8]. The TFET basic structure is the gated PIN diode whose I_{on} arises from BtB generation [9]. This significantly improves *SS* and power consumption characteristics far beyond standard CMOS technology [10, 11]. SOI technology-based structures [12] have been reported to overcome the short-channel effects, lower *SS*, enhanced soft-error immunity, and improved electrostatics. If we want to replace the current MOSFET in VLSI circuits with TFETs as viable switches [13, 14], then the TFET-based circuits should work as fast as the MOSFETs and should have the same fan-out in the same circuit. Nevertheless, the major limitation of the conventional TFETs lies in the fact that the I_{on} is much lower than MOSFETs and the inherent ambipolar behavior of the standard TFETs. Unlike conventional TFETs, recent studies [15, 16, 17] have revealed that TFETs with higher I_{on} can be achieved with the gate stack overlapping the source region, using SiGe substrate and high-κ dielectric materials as the gate oxide. Various TFET structures with a large gate-source overlap, heavily doped source pocket, and gate-drain underlap regions have been reported to enhance the TFET performance [18–21]. For high I_{on} and low I_{off} currents, several device structures (double gate and triple gate) have been reported [22, 23].

4.2   GATE-OVERLAP TUNNEL FETS FOR DIGITAL APPLICATIONS

4.2.1   PROPOSED GOTFET STRUCTURES

Figure 4.1 shows the schematic of the GOTFET structure and its BtB generation. Device optimization has been done by optimizing the channel length, doping concentration, oxide thickness, and metalwork function to achieve higher I_{on} and lower I_{off}. Table 4.1 shows typical parameters used in the Synopsys TCAD simulator. Device simulation has been carried out using the drift-diffusion model, mobility

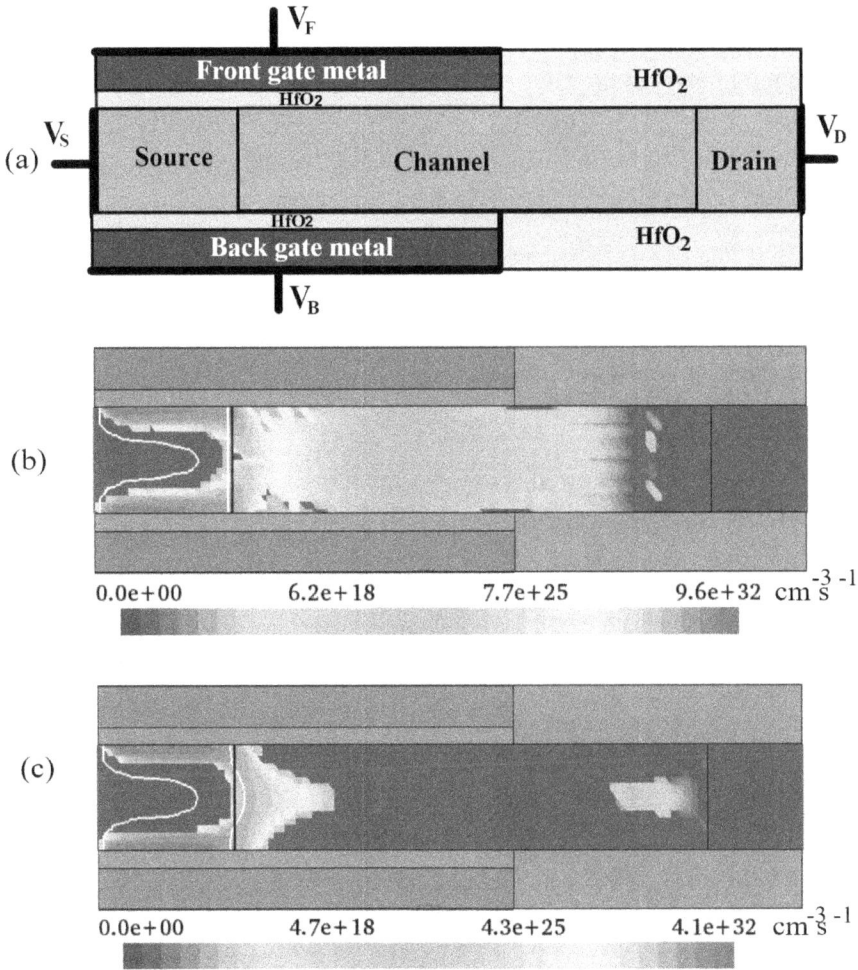

FIGURE 4.1   (a) Schematic of the proposed GOTFET device. (b) Electron and (c) hole BtB generation in nGOTFET and pGOTFET, respectively

model, high filed saturation models, Shenk TAT model, Auger recombination models, and dynamic non-local BtB model. Model and simulation parameters were extracted from experimental TFET reported by Kao (2011) to increase the validity and accuracy of the results. These extracted parameters have been incorporated in the simulation deck of the proposed GOTFETs to obtain accurate characteristics.

In this GOTFET device, the gate is overlapped on the source to enhance the I_{on} due to vertical BtB tunneling. BtB generation rate is achieved at 10^{32}/cm³s and results in an I_{on} more than MOSFET, while the I_{off} is one order of magnitude lower than equally sized MOSFET at the same 45 nm technology node. Due to the gate-overlapped source region in the proposed GOTFETs, BtB tunneling occurs from the

TABLE 4.1

Device parameters of the GOTFETs

| Region | nGOTFET | | pGOTFET | |
	Material	Doping cm^{-3}	Material	Doping cm^{-3}
Source	$Si_{0.1}Ge_{0.9}$	10^{20} p$^+$	$Si_{0.2}Ge_{0.8}$	10^{20} n$^+$
Channel	undoped Si	-	$Si_{0.2}Ge_{0.8}$	5×10^{17} p
Drain	Si	10^{20} n$^+$	$Si_{0.2}Ge_{0.8}$	5×10^{17} p
Gate	Al	-	Mo	-

bulk region of the p^+ source to the surface region of p^+ source region under the gate-stack overlapping source, as shown in Figure 4.1.

4.2.2 CHARACTERISTICS OF THE PROPOSED GOTFETS

Figure 4.2a shows I_D-V_{GS} characteristics of nGOTFET obtained using the dynamic non-local model with increasing V_{DS}. I_{on} = 903 μA/μm at $V_{GS} = V_{DS}$ = 1V, which is double that of MOSFET, while I_{off} = 0.3 pA/μm is one order of magnitude lower than the corresponding equally sized MOSFET at the same technology node. Threshold voltage extracted from the I_D-V_{GS} characteristics using the third derivative method has its first peak denoting as a V_{TL} = 0.4V for nGOTFET, as observed in Figure 4.2b.

Similarly, Figure 4.3a shows pGOTFET I_S-V_{SG} characteristics with increasing V_{SD}. At $V_{SG} = V_{SD}$ = 1 V, the on-current I_{on} = 559 μA/μm exceeds twice that of an equally sized MOSFET at the same 45 nm technology node. The leakage current I_{off} = 0.1 pA/μm at V_{SD} = 1 V is at least one order of magnitude lower than the equally sized MOSFET. $|V_{Tp}| \approx 0.36$ V obtained from optimized pGOTFET I_S-V_{SG} characteristics using the third derivative method, which is much lower than the equivalent pMOSFET, as shown in Figure 4.3b.

4.2.3 IMPLEMENTATION OF DIGITAL BASIC BUILDING BLOCKS

This section describes the implementation of digital basic building blocks for VLSI circuits design using the proposed GOTFET and its performance comparison with MOSFET in terms of speed and power consumption. The proposed GOTFET is a promising alternative to the MOSFET due to its lower inverse sub-threshold slope and low leakage currents for low-power circuits design. Higher I_{on}, low I_{off} and lower SS enable the CGOT-based digital circuits to operate faster and reduce static power consumption compared to the same circuit implemented with CMOS devices. The schematic of the CGOT-based inverter circuit and its delay characteristics are shown in Figure 4.4. The performance of the CGOT-based inverter has been benchmarked with an identical CMOS-based inverter with the same W/L ratio. The circuit has been simulated at a one GHz frequency with a load capacitance of 10 fF. The CGOT inverter operates 1.43 times faster than the corresponding CMOS inverter, as shown in Figure 4.4. Figure 4.5 compares the static power consumption of the CGOT inverter with the CMOS inverter.

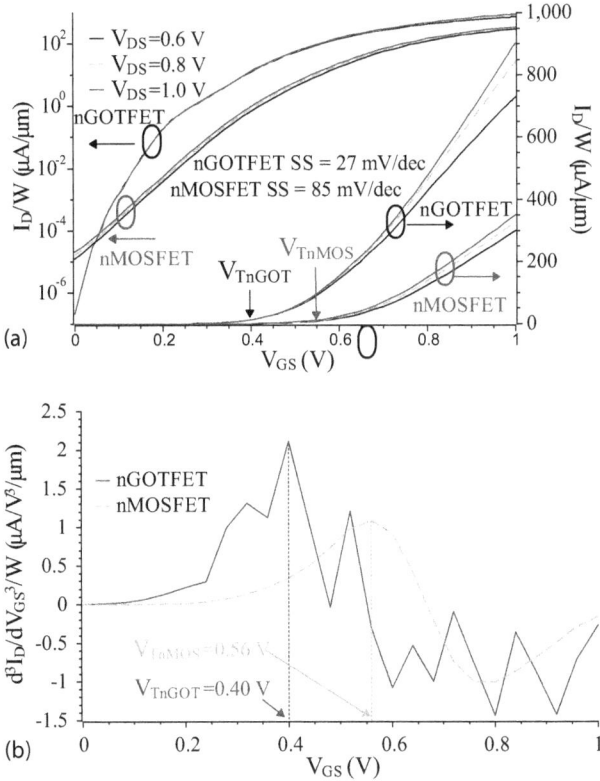

FIGURE 4.2 (a) I_D-V_{GS} characteristics of nGOTFET and nMOSFET for different values of V_{DS}. (b) Threshold voltage V_{tn} extracted from the characteristics of nGOTFET and nMOSFET using the third derivative method [27]

The CGOT inverter consumes 0.009 times the power consumed by the CMOS inverter. A total decrease of 99.45% of PDP can be achieved through CGOT-based inverter compared to CMOS-based inverter, as summarized in Table 4.2.

The CGOT-based 2-input NAND and NOR gates schematic is shown in Figure 4.6. The delay and static power characteristics comparison for NAND and NOR gates are shown in Figures 4.7 and 4.8, respectively. Table 4.2 benchmarks the performance parameters of the inverter, NAND, and NOR gates implemented with CGOT and CMOS technologies at 10 fF load capacitance.

4.3 GATE-OVERLAP TUNNEL FETS FOR TERNARY APPLICATIONS

4.3.1 PROPOSED GOTFET STRUCTURES

In VLSI applications, ternary logic circuits have recently gained considerable popularity over binary circuits. Their superiority over binary logic in digital design

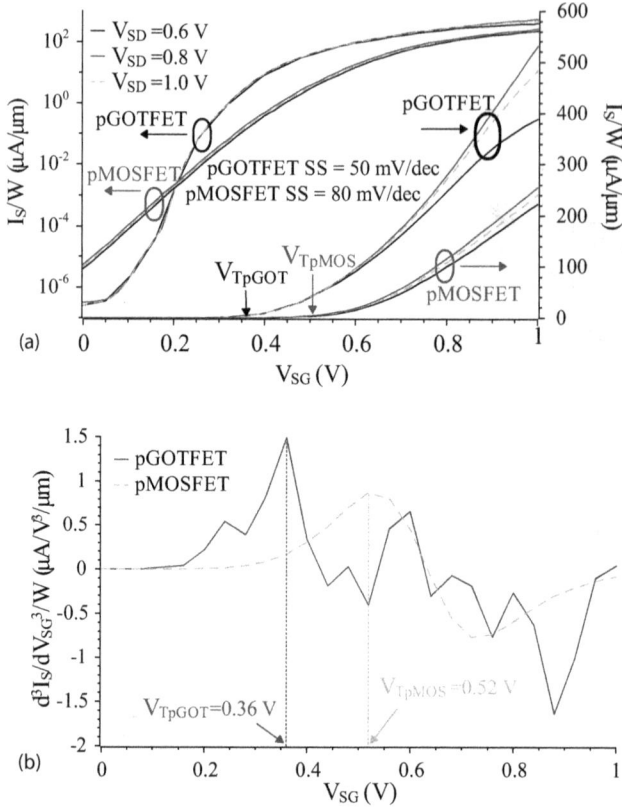

FIGURE 4.3 (a) I_S-V_{SG} characteristics of pGOTFET and pMOSFET for different values of V_{SD}. (b) Threshold voltage extracted from the characteristics of pGOTFET and pMOSFET using the third derivative method [27].

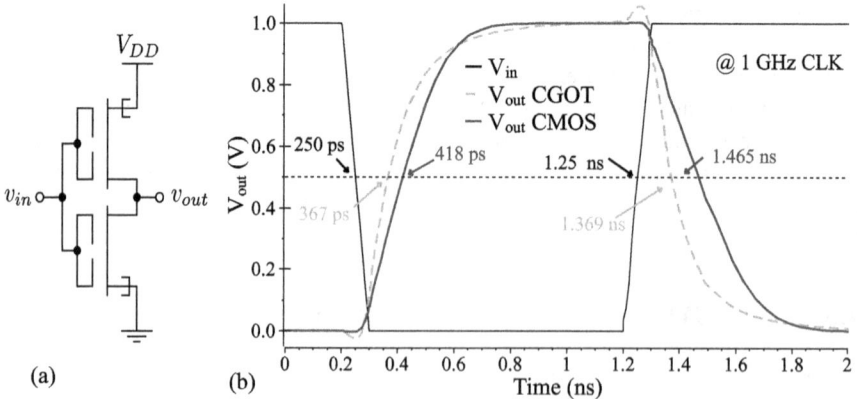

FIGURE 4.4 (a) Schematic GOTFET inverter. (b) Delay characteristics of GOTFET vs. MOSFET inverter

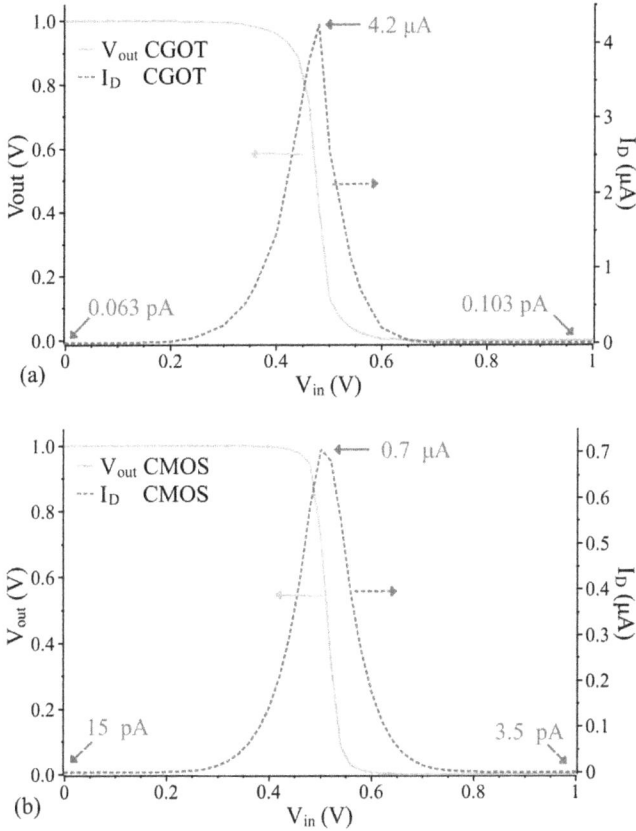

FIGURE 4.5 Comparison of static currents. (a) CGOT (b) CMOS

TABLE 4.2
Benchmarking the performance parameters of inverter, NAND, and NOR gates implemented with CGOT and CMOS technologies at 10 fF load capacitance

Circuit Parameter		Inverter		2-i/p NAND		2-i/p NOR	
	Units	CMOS	CGOT	CMOS	CGOT	CMOS	CGOT
Bias V_{DD}	V	1	1	1	1	1	1
Delay t_{pd} LH	ps	168	117	184	106	201	146
Delay t_{pd} HL	ps	215	119	254	127	337	157
Average delay	ps	191.5	118	219	127	269	151.5
Static I_{high}	pA	15	0.063	19	0.125	51	0.125
Static I_{low}	pA	3.5	0.103	9.4	0.206	5.4	0.165
Average P_{static}	pW	9.25	0.083	14.2	0.166	28.2	0.145
PDP ($*10^{-23}$)	J	177.1	0.98	310.98	2.1	758.6	2.2
Decrease in PDP		99.45 %		99.32 %		99.71 %	

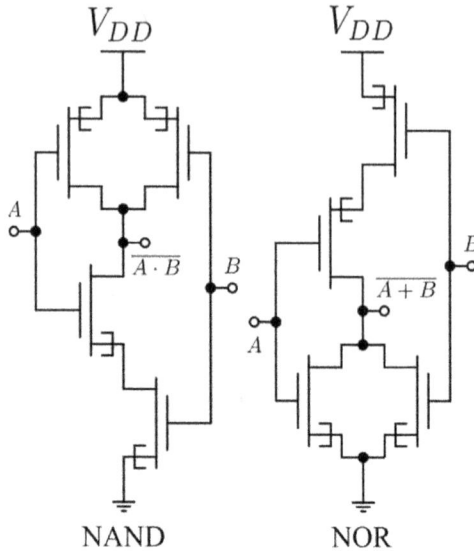

FIGURE 4.6   Schematic of CGOT digital circuits: 2-input (a) NAND and (b) NOR gates

systems is due to smaller chip sizes, fewer interconnects, and faster-operating speeds. Primarily, ternary logic requires that the device have two threshold voltages: low threshold voltage V_{TL} and high-threshold voltage V_{TH}. Therefore, we have modified the GOTFET structure proposed in Figure 4.1 for this application by changing the device parameters listed in Tables 4.3 and 4.4. The performance of the modified DG GOTFET structure is superior to that of the MOSFET at the same technology node. These devices are designed by changing the doping concentration and gate materials such that the low and high-threshold voltages (LVT and HVT) are $V_{DD}/3$ and $2V_{DD}/3$, with the ranges 0 to $V_{DD}/3$, $V_{DD}/3$ to $2V_{DD}/3$, and $2V_{DD}/3$ to V_{DD} representing the three logic states 0, 1, and 2 accordingly. In this work, we proposed the dual-threshold GOTFETs in the same device by changing the body terminal connections instead of device devices. As explained in the previous section, we have also included similar physical models in the ternary GOTFETs simulation deck. Devices are simulated at 1 V following the ITRS regulation on maximum bias limit at the 45 nm technology node.

4.3.2   CHARACTERISTICS OF THE PROPOSED GOTFETS

The LVT and HVT n-GOTFET device characteristics are obtained by changing the doping concentration and gate materials listed in Table 4.3 with $V_{TL}=V_{DD}/3$ and $V_{TH}=2V_{DD}/3$, respectively. We found that Al and TiSi$_2$ are the best gate materials to get LVT and HVT nGOTFETs characteristics, respectively. Figure 4.9 illustrates the transfer characteristics of LVT and HVT nGOTFETs as determined by non-local BtB generation in TCAD with increasing V_{DS}. As shown in Figure 4.9, the

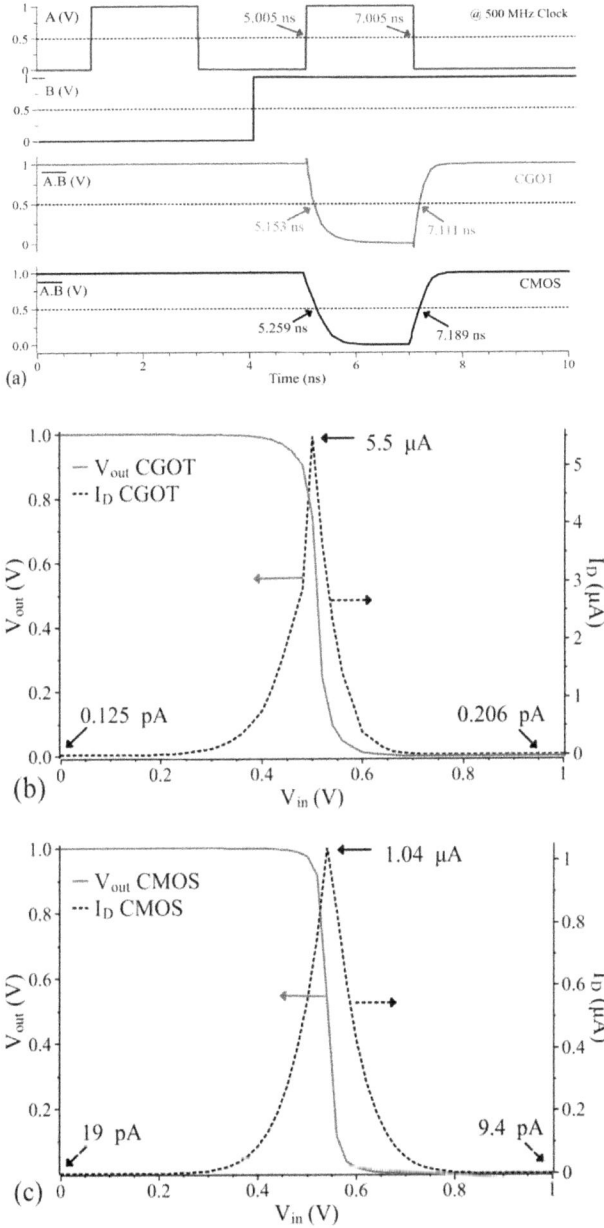

FIGURE 4.7 (a) Delay comparison CGOT vs. CMOS NAND gates. Comparison of static currents (b) CGOT vs. (c) CMOS NAND gates

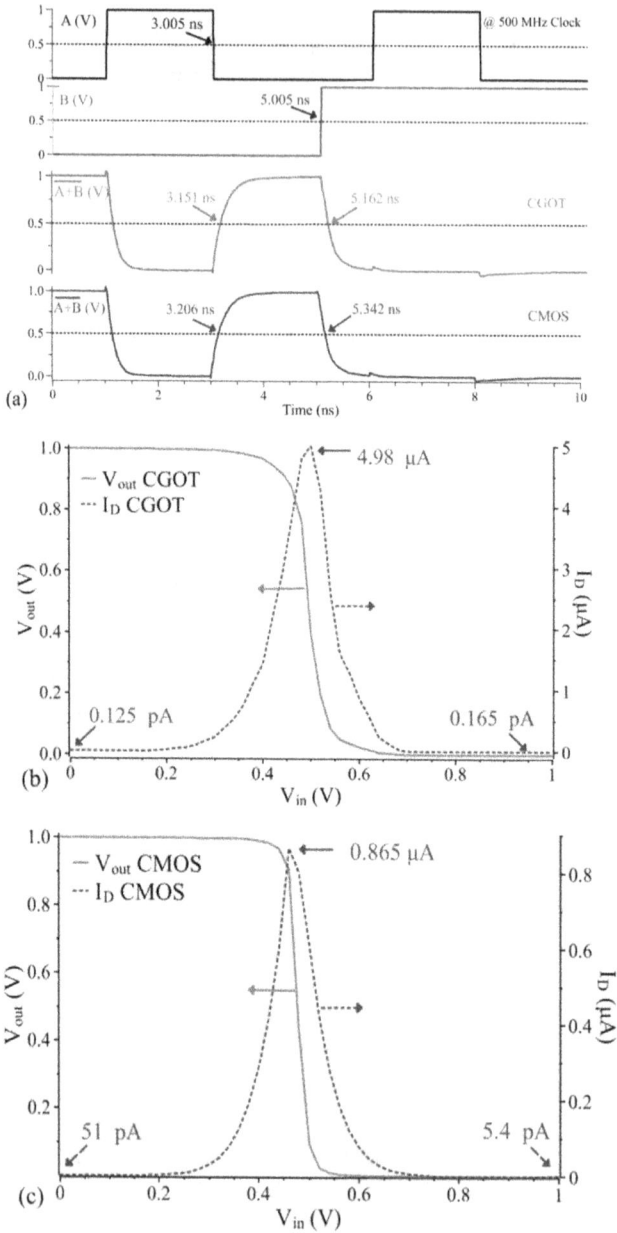

FIGURE 4.8 (a) Delay comparison CGOT vs. CMOS NOR gates. Comparison of static currents (b) CGOT vs. (c) CMOS NOR gates

TABLE 4.3

Device parameters of the LVT and HVT nGOTFET

	LVT nGOTFET		HVT nGOTFET	
Region	**Material**	**Doping cm^{-3}**	**Material**	**Doping cm^{-3}**
Source	$Si_{0.1}Ge_{0.9}$	1020 p+	$Si_{0.2}Ge_{0.8}$	1020 n+
Channel	undoped Si	-	$Si_{0.2}Ge_{0.8}$	5×10^{17} p
Drain	Si	10^{20} n$^+$	$Si_{0.2}Ge_{0.8}$	5×10^{17} p
Gate	Al	-	Mo	-

TABLE 4.4

Device parameters of the LVT and HVT pGOTFET

	LVT nGOTFET		HVT nGOTFET	
Region	**Material**	**Doping cm^{-3}**	**Material**	**Doping cm^{-3}**
Source	$Si_{0.26}Ge_{0.74}$	9×10^{20} n	$Si_{0.2}Ge_{0.8}$	10^{20} n$^+$
Channel	$Si_{0.26}Ge_{0.74}$	5×10^{16} p	$Si_{0.2}Ge_{0.8}$	5×10^{17} p
Drain	$Si_{0.26}Ge_{0.74}$	5×10^{18} p	$Si_{0.2}Ge_{0.8}$	10^{17} p
Gate	TiN	-	TiSi$_2$	-

I_D-V_{GS} characteristics of the proposed LVT nGOTFET exhibit an I_{on} that is almost double that of the LVT nMOSFET, while the I_{off} remains one order of magnitude lower. Furthermore, I_{on} is greater in an optimized HVT nGOTFET than in an HVT nMOSFET, although I_{off} is at least an order of magnitude lower. The lower threshold voltage V_{tnl} and higher threshold voltage V_{tnh} were extracted using the third derivative method [24]. The first peaks at V_{GS}=0.33 V and V_{GS}=0.66 V define the V_{tnl} and V_{tnh} of the nGOTFET, respectively, in Figure 4.9c. The reported LVT and HVT nGOTFETs have SS of 25 mV/dec and 20 mV/dec, respectively, which are much lower than the SS of most TFETs reported in the literature.

Similarly, LVT and HVT pGOTFET have been designed with the parameters listed in Table 4.4. TiN and TiSi2 were found to have the best DC characteristics for LVT and HVT pGOTFET, respectively, out of all the available gate materials in synopsys® TCAD tools. As shown in Figure 4.10, the I_{on} of the proposed LVT pGOTFET is about twice as high as the LVT pMOSFET, while I_{off} is still at least an order of magnitude lower than the I_{off} of the LVT pMOSFET. Also, I_{on} is higher in an optimized HVT pGOTFET than in a pMOSFET, but I_{off} is still at least an order of magnitude lower than in an HVT pMOSFET at the same technology node. The third-order derivative method was used to get the LVT and HVT pGOTFET threshold voltages of V_{tlp} =0.36V and V_{thp} =0.62 V, respectively. The SS of the proposed LVT and HVT pGOTFETs is 50 mV/dec and 33 mV/dec, respectively, as shown in Figure 4.10.

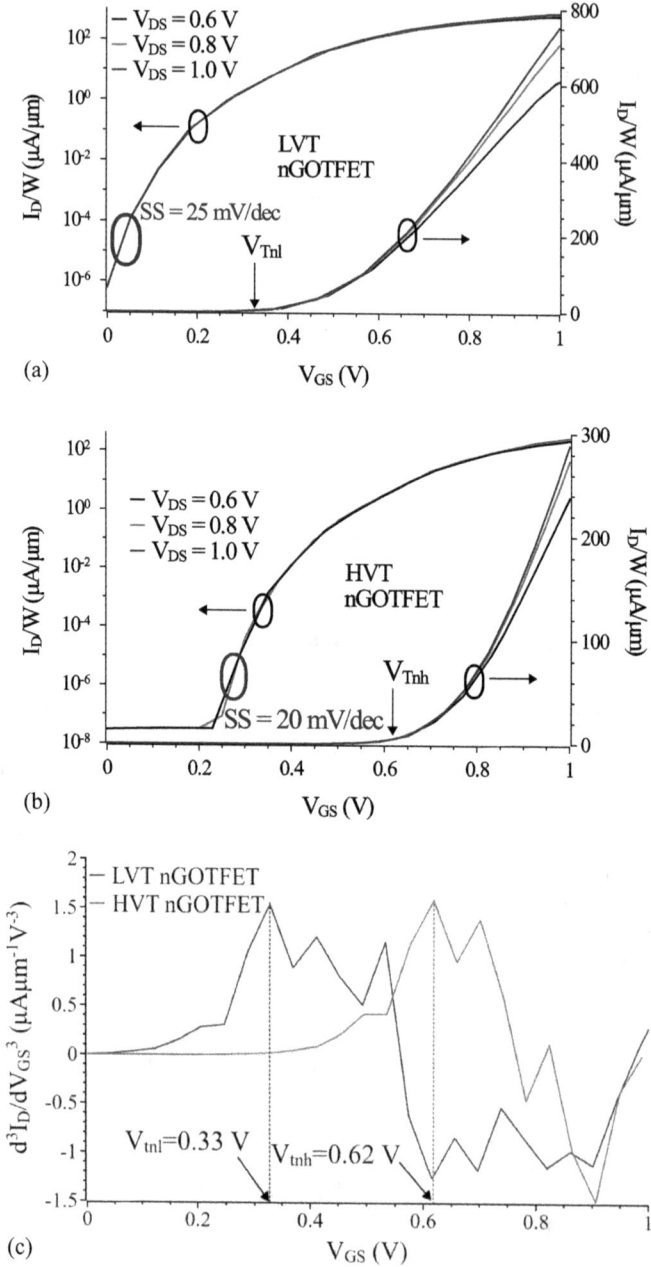

(a)

(b)

(c)

FIGURE 4.9 I_D-V_{GS} characteristics of (A) LVT (B) HVT nGOTFETs for different values of V_{DS}. (C) V_{tnl} and V_{tnh} extracted from the LVT and HVT characteristics of the nGOTFETs

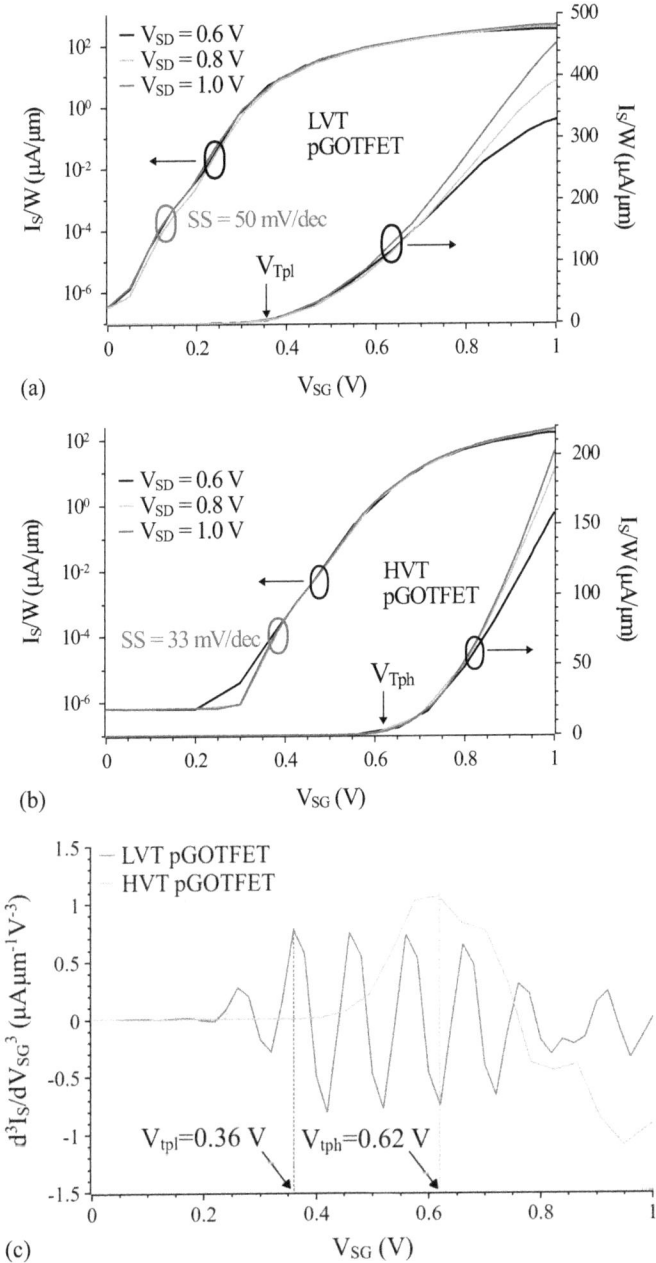

FIGURE 4.10 I_S-V_{SG} characteristics of (A) LVT (B) HVT pGOTFETs for different values of V_{SD}. (C) Vtpl and Vtph extracted from the LVT and HVT characteristics of the pGOTFETs

TABLE 4.5

Parameters of dual-threshold nGOTFETs and p GOTFETs

Region	LVT and HVT nGOTFET		LVT and HVT pGOTFET	
	Material	Doping cm^{-3}	Material	Doping cm^{-3}
Source	Si$_{0.8}$Ge$_{0.92}$	10^{20} P+	Si$_{.28}$Ge$_{.72}$	10^{20} P+
Channel	Si	10^{15} n	Si$_{.28}$Ge$_{.72}$	-
Drain	Si	10^{20} n+	Si$_{.28}$Ge$_{.72}$	10^{18} N
Front Gate	TiSi2	-	TiSi$_2$	-
Back Gate	Al	-	TiN	-

4.3.3 PROPOSED DUAL-THRESHOLD GOTFETS IN THE SAME DEVICE

The most exciting feature of the proposed GOTFET is that, by changing the material and doping parameters as listed in Table 4.5, we can get the optimal performance of LVT and HVT pGOTFETs in the same device instead of dedicated devices. These structures have been optimized such that LVT characteristics obtained $V_{TL} = V_{DD}/3$ by providing the front and back terminals with higher bias ($V_{FS} = V_{BS} = V_{GS} = 1$ V). At the same time, HVT characteristics were obtained by $V_{TH} = 2V_{DD}/3$, with the front gate having a higher bias and the back gate connected to the source terminals. Figure 4.11a depicts the I_D-V_{GS} properties of LVT and HVT nGOTFETs acquired from a non-local BtB generation model with an increasing V_{DS}. For the proposed nGOTFETs, the lower $V_{tln} = 0.36$ V and the higher $V_{thn} = 0.6$ V were derived using the third derivative approach [5], as shown in Figure 4.11b. Figure 4.11c shows how the I_S-V_{SG} characteristics of LVT and HVT pGOTFETs change as the V_{SD}. The proposed pGOTFET has a lower V_{tlp} of about 0.32 V and a higher V_{thp} of about 0.6 V, as shown in Figure 4.11d.

4.3.4 IMPLEMENTATION OF NTI, PTI, AND STI TERNARY LOGIC CELLS

The primary logic cells in ternary logic applications are *negative ternary inverter* (NTI), *positive ternary inverter* (PTI), and *standard ternary inverter* (STI). This subsection benchmarks the characteristics of complementary GOTFET (CGOT) based NTI, PTI, and STI cells with CMOS-based cells at a 45 nm technology node. Figure 4.12 shows the schematics of the NTI, PTI and STI logic cells. The performance of the CGOT-based NTI logic cell has been benchmarked with a CMOS NTI cell at 45 nm technology using the industry-standard cadence EDA tool [32]. Figure 4.13 shows the NTI logic cell simulation results using the cadence EDA tool.

The average static power in CGOT NTI is 0.174 pW, which is significantly lower than the CMOS NTI, which consumes 73.7 pW, as highlighted in Table 4.6. Similar to the NTI cells, the performance of a CGOT vs. CMOS PTI logic cell has been shown in Figures 4.14a, 4.14b, and 4.14c at the 45 nm technology node. The average

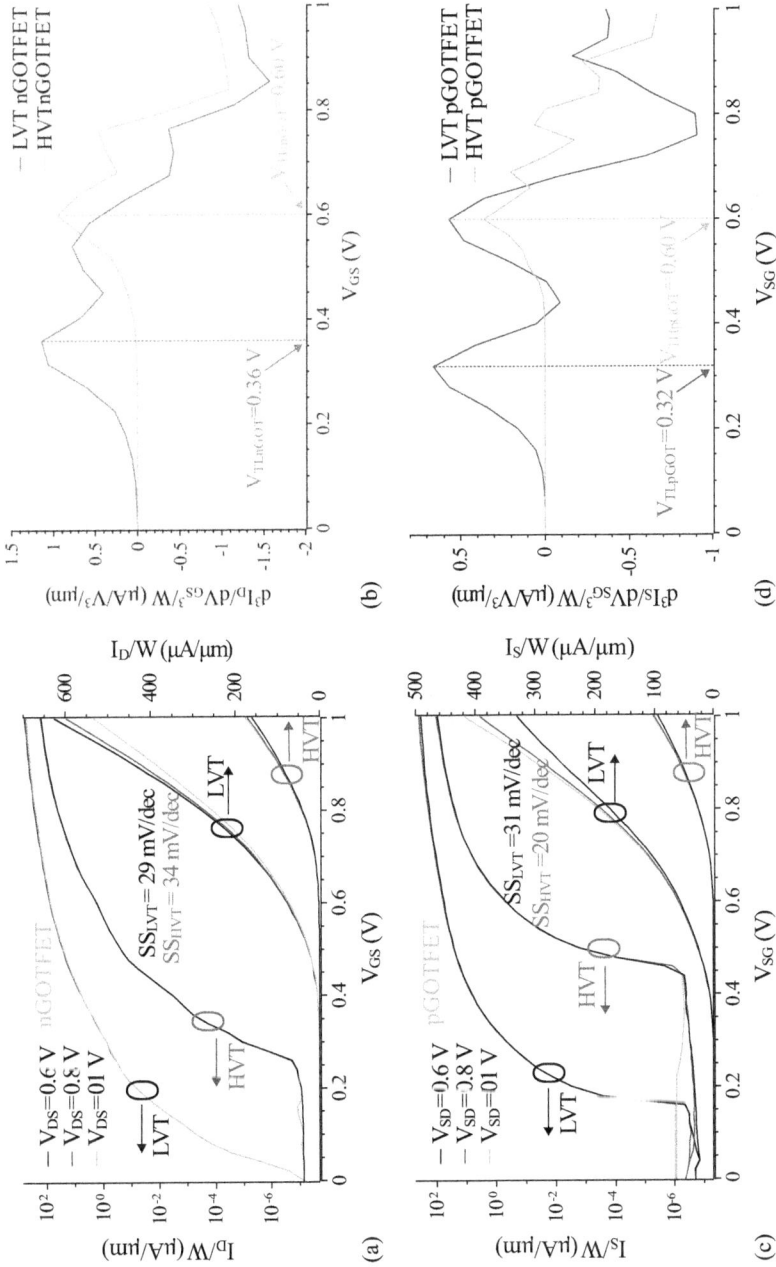

FIGURE 4.11 (A) I_D-V_{GS} characteristics of dual-threshold nGOTFETs for different values of V_{DS} (B) V_{tln} and V_{thn} extracted from the nGOTFET characteristics using the third derivative method. (C) I_S-V_{SG} characteristics of dual-threshold pGOTFETs for different drain biases values V_{SD}. (D) V_{tlp} and V_{thp} extracted from the pGOTFET characteristics using the third derivative method.

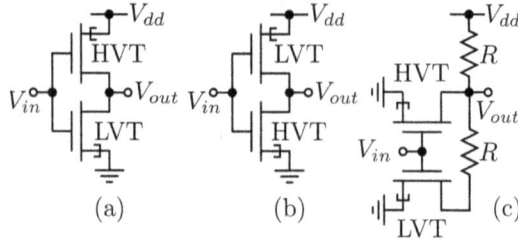

FIGURE 4.12 Schematics of the LVT and HVT CGOT (a) NTI, (b) PTI, and (c) STI cells with $100 \text{ k}\Omega \leq R \leq 100 \text{ M}\Omega$

static power consumed in CGOT PTI is 0.041 pW, significantly lower than the CMOS PTI, which consumes 22.60 pW, highlighted in Table 4.7.

The PDP of the CGOT PTI is 0.31×10−23 J, which is only 0.11% of the PDP of standard 45 nm CMOS PTI cells (293.8×10−23 J). The overall decrement in PDP owing to the proposed CGOT PTI logic cell is 99.89%. The average delay in CGOT STI at the 45 nm technology node is 0.037 ns, which is significantly lower than the CMOS STI cell, which is 0.114 ns, shown in Figure 4.15 and highlighted in Table 4.8. The PDP of CGOT STI is 9.7×10−24 J, which is only 0.00014% of the PDP of standard 45 nm CMOS STI cells (7.15×10−18 J). The overall decrement in PDP owing to the proposed CGOT STI logic cell is 99.9999%.

4.4 DOUBLE GATE LINE-TUNNELING FETS (DGLTFET) FOR ANALOG APPLICATIONS

4.4.1 PROPOSED DGLTFET STRUCTURES

Earlier, GOTFET was proposed for digital applications, which are unsuitable for analog applications due to poor drain saturation characteristics affecting the output resistance r_o. A new device, DGLTFET, has been proposed to improve the saturation characteristics, whose schematic is shown in Figure 4.16. This device has an ep-layer sandwiched between the oxide layer and the source region. This leads to flatter saturation characteristics, resulting in increased r_o that leads to improved $A_{vo}=g_m r_o$, crucial for analog operation. Device parameters, doping concentrations, and gate materials of n and p DGLTFET devices are listed in Table 4.9. We have calibrated the simulation deck of the DFLTFET with a previously published experimental work. Figure 4.17 shows the I_D-V_{GS} characteristics of the simulated device with experimental work [25].

4.4.2 CHARACTERISTICS OF THE PROPOSED GOTFETS

Figure 4.18 shows the I_D-V_{GS} characteristics of the nDGLTFET at $V_{DS} = 1$ V using the non-local BtB tunneling model. At $V_{GS} = V_{DS} = 1$ V, $I_{on}=1090$ $\mu A/\mu m$, which is more than three times the I_{on} of the MOSFET at the 45 nm technology node. In these

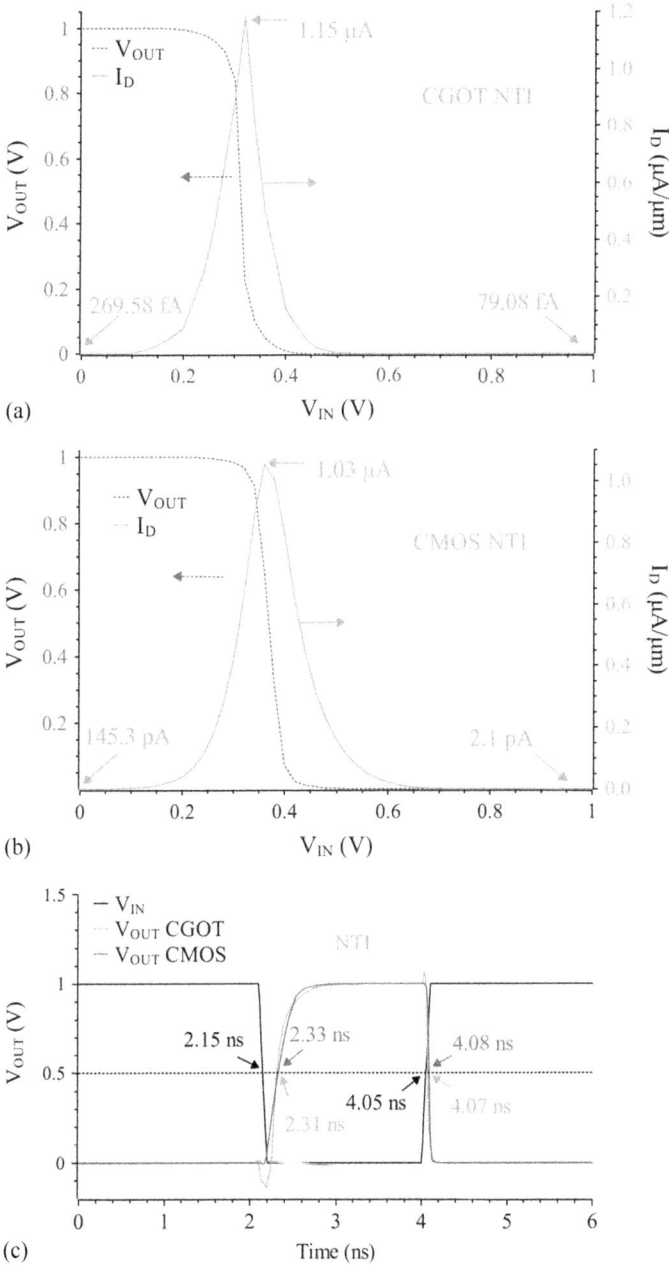

FIGURE 4.13 Static power consumption of (a) CGOT and (b) CMOS NTI cells. (c) Comparison of the delay characteristics of CGOT vs. CMOS NTI cell

TABLE 4.6

GOTFET and CMOS-based NTI cell delay and static power consumption comparison

Circuit parameter	Units	GOTFET	CMOS
Powersupply V_{DD}	V	1	1
Average Delay	ns	0.09	0.11
Average static power P_{static}	pW	0.174	73.7
PDP ($*10^{-23}$)	J	1.57	810.7
Decrease in PDP		99.81%	

devices, point tunneling occurs at lower voltages from the source to the channel region; at higher bias, line tunneling dominates from the source region to the epi-layer region. Point tunneling and line tunneling result in higher I_{on}, which is 2.5 times higher than MOSFETs, as shown in Figure 4.18. As gate-overlapping length on the source side increases (L_{ov}), vertical BtB tunneling increases, which improves I_{on} and g_m; however, this affects the rise of C_{GS} (decreases BW), so we need to optimize the L_{ov} according to the applications. Figure 4.19a shows the output characteristics of the nDGLTFET device for different gate biases. Saturation characteristics are explained by electron density on the surface of the epi-layer. For a given gate bias, as drain bias increases, electron density initially decreases; after a particular drain bias, it is constant, which we call $V_{DS}(sat)$ [31].

Analog performance has improved with epi-layer doping (n_{ep}) and thickness (t_{ep}). The tunneling width between the source's valence band and the epi-layer conduction band decreases with reducing the t_{ep}, which improves the g_m. As changing the t_{ep}, tunneling width remains constant between the epi-layer and source region as drain bias changes, so r_o remains constant, as shown in Figure 4.19a.

As n_{ep} increases, the electric field increases, which enhances the BtB tunneling and improves the g_m. As n_{ep} increases, the onset of saturation increases with drain bias, reducing the r_o, as shown in Figure 4.20 [30]. Quantum confinement severely affects lower technology nodes. We observed that threshold voltage shifts in 150 mv in DGLTFET structure with and without field-induced quantum confinement (FIQC) as observed in the I_D-V_{GS} characteristics in Figure 4.21 [28].

Figure 4.22 depicts an analysis of the output resistances for various $|V_{GS}|$ values for the n and p DGLTFET and MOSFET. In extreme saturation, the nDGLTFET's r_o is at least two orders of magnitude more than the MOSFET. Due to the larger g_m and improved saturation characteristics, the intrinsic gain $g_m r_o$ in DGLTFET devices is almost two orders of magnitude more than in MOSFET devices. Due to the greater g_m at lower gate biases (V_{GS}), f_T is higher in DGLTFET. However, gate capacitance dominates at higher gate biases V_{GS}, resulting in f_T being lower in DGLTFET relative to MOSFET. The gate capacitances C_{GS} and C_{GD} are extracted by the AC analysis in TCAD. Fs 4.23a and 23b illustrate the gate capacitance C_{GS} and C_{GD} for various values of V_{GS}.

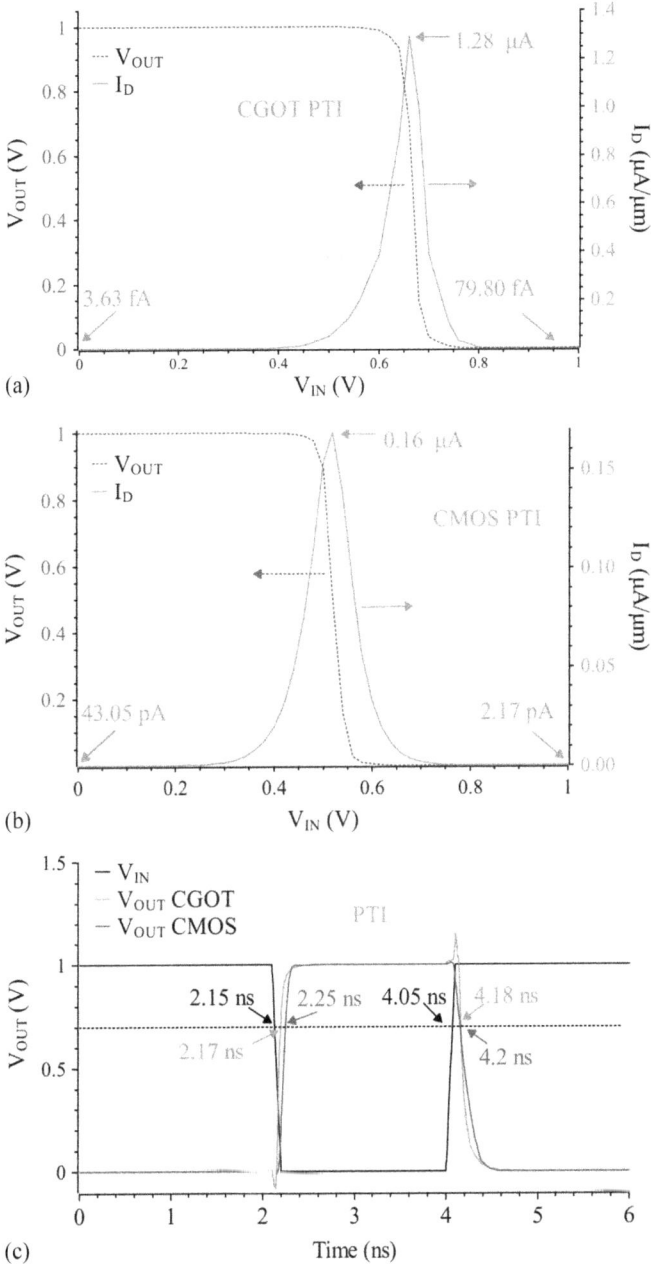

FIGURE 4.14 Static power consumption of (a) CGOT and (b) CMOS PTI cells. (c) Comparison of the delay characteristics of CGOT vs. CMOS PTI cell

TABLE 4.7

GOTFET and CMOS-based PTI cell delay and static power consumption comparison

Circuit parameter	Units	GOTFET	CMOS
Powersupply V_{DD}	V	1	1
Average Delay	ns	0.075	0.13
Average static power P_{static}	pW	0.041	22.6
PDP ($*10^{-23}$)	J	0.31	293.8
Decrease in PDP		99.89%	

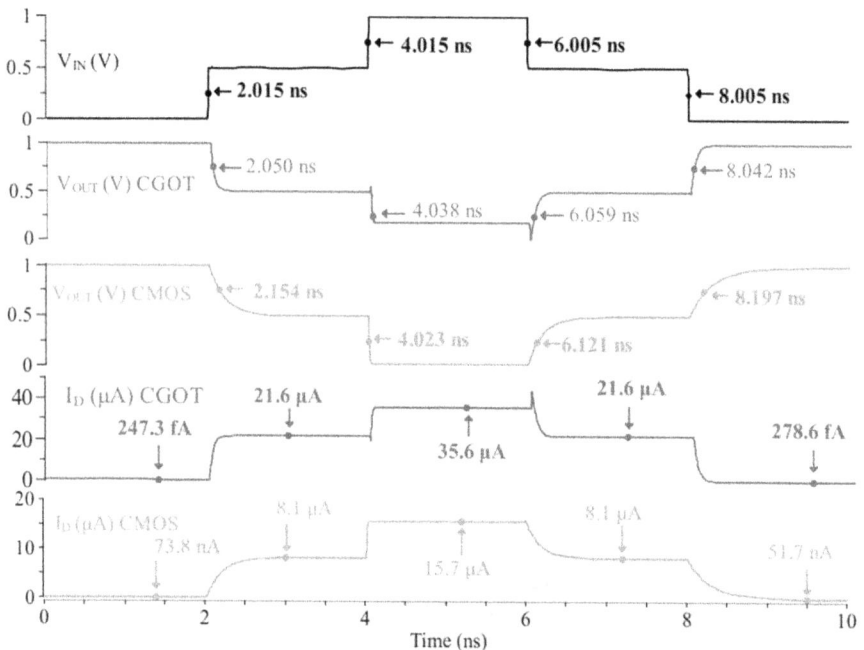

FIGURE 4.15 CGOT vs. CMOS STI cells delay and power characteristics comparison

4.4.3 ANALOG APPLICATIONS OF THE LINE-TUNNELING TFETS

Earlier sections explained the device characteristics of both n and p DGLTFET devices. This section also discusses the performance of the analog circuits designed using these devices. Figure 4.24 shows the schematics of resistive load and cascade configuration of the CS amplifier-based DGLTFET device. Figure 4.25 depicts the voltage transfer characteristics (VTC) of V_{out} vs. V_{in} and the differential gain dV_{out}/dV_{in} vs. V_{in} on the same graph. It demonstrates that the transition slope of the DGLTFET-based CS amplifier is much steeper than that of the CMOS-based CS

TABLE 4.8

GOTFET and CMOS-based STI cell delay and static power consumption comparison

Circuit parameter	Units	GOTFET	CMOS
Powersupply V_{DD}	V	1	1
Average Delay	ns	0.037	0.114
Average static power P_{static}	pW	0.262	62750
PDP ($*10^{-23}$)	J	0.97	715350
Decrease in PDP		99.99%	

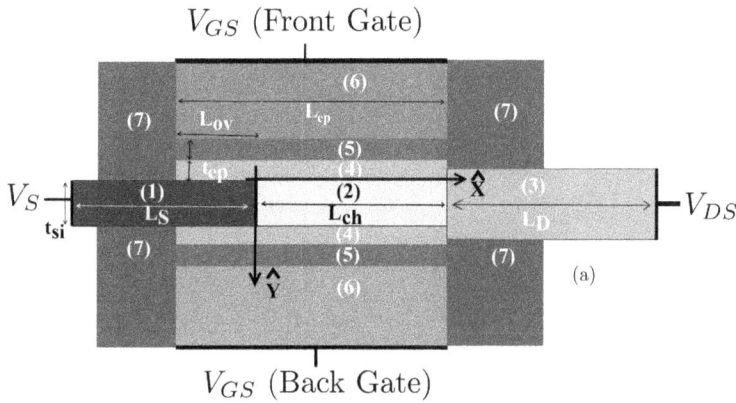

FIGURE 4.16   (a) Schematic cross-sectional view of the proposed DGLTFET device. (1) p+ Si0.5Ge0.5 source (2) p– Si channel (3) n+ Si drain (4) n+ Si epitaxial layer (5) HfO2 gate oxide (6) Al metal gate (7) SiO2 spacer oxide

amplifier. Due to higher r_o and g_m, the gain of the DGLTFET-based CS amplifier is three times as high as the gain of the CMOS-based CS amplifier, which can be observed in Figure 4.26.

A comparison of the circuit performance of both CS amplifier configurations is indicated in Table 4.10. The unity-gain BW of the DGLTFET-based CS amplifier with resistive load is 806 GHz and 16.98 GHz without load and with a capacitive load CL of 10 fF, respectively, as compared to the unity-gain BW of 210.6 GHz (without load) and 8.1 GHz (with load) of MOSFET-based CS amplifier with resistive load. Furthermore, the unity-gain BW of the DGLTFET-based cascode load CS amplifier is 71 GHz without load and 15 GHz with capacitive load C_L =10 fF, compared to the unity-gain BW of 23 GHz (without load) and 10 GHz (with the same 10-fF capacitive load) of the MOSFET-based CS amplifier [36].

Figure 4.27 depicts the circuitry of the DGLTFET-based single-stage and cascode current mirrors. The circuits were co-simulated using identical circuits constructed

TABLE 4.9

Parameters of the DGLTFET devices

Parameter	Units	nDGLTFET	pDGLTFET
Channel length	nm	45	45
Source length	nm	50	50
Drain length	nm	40	40
Gate length	nm	65	65
Gate oxide thickness	nm	3	3
Channel thickness	nm	10	10
Epi-layer thickness	nm	3	3
Gate-source overlap	nm	20	20
Spacer length	nm	30	30
Source doping	/cm3	2×10^{20} p$^+$ SiGe	2×10^{20} n$^+$ Si
Channel doping	/cm3	10^{16} p Si	10^{16} n Si
Drain doping	/cm3	2×10^{20} n$^+$ Si	5×10^{19} p$^+$ Si
Epi-layer doping	/cm3	2×10^{20} n$^+$ Si	2×10^{18} p$^+$ SiGe
Gate material	-	Al	Poly Zn

using 45-nm CMOS transistors. Reference current I_{ref} has been used in DGLTFET and MOSFET devices using R_{ref} = 5 kΩ and 2.5 kΩ, respectively. Similarly, R_{ref} = 5 kΩ and 2.5 kΩ have been used for cascode current mirrors for biasing the same I_{ref} In conventional MOSFETs, the channel length modulation effect is high at lower technology nodes which significantly degrades the R_{out} of current mirror circuits. However, In DGLTFET devices, its effect is minimal. We observed three times higher R_{out} in DGLTFET-based simple current mirror circuits compared to the CMOS technology node.

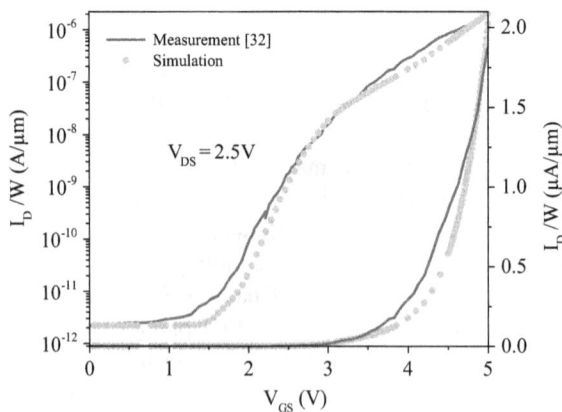

FIGURE 4.17 Calibration of the simulation deck with the I_D-V_{GS} characteristics measured from a prefabricated device at V_{DS}=2.5 V [25]

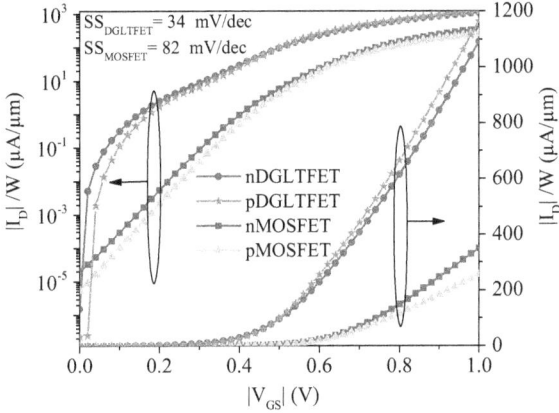

FIGURE 4.18 $|I_D|$-$|V_{GS}|$ characteristics of DGLTFET and MOSFET (both p- and n-channel) for $|V_{DS}|=1$ V plotted on linear and logarithmic scales

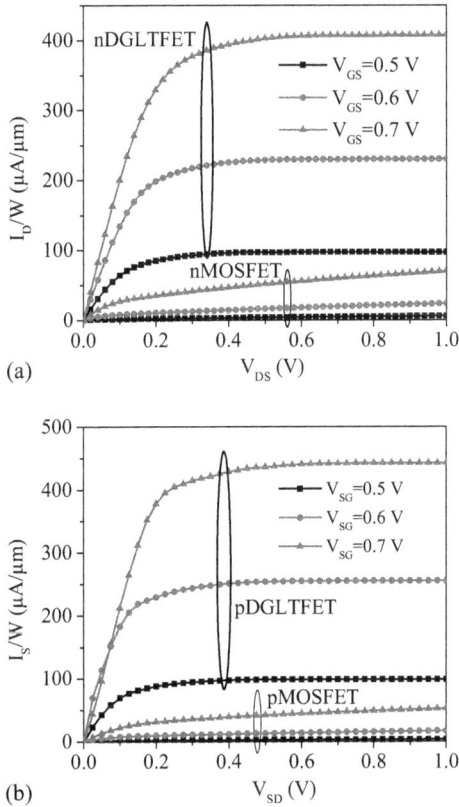

FIGURE 4.19 I_D-V_{DS} characteristics of nDGLTFET and nMOSFET for different values of V_{GS} and (b) I_S-V_{SD} characteristics of pDGLTFET and pMOSFET for different values of V_{SG}

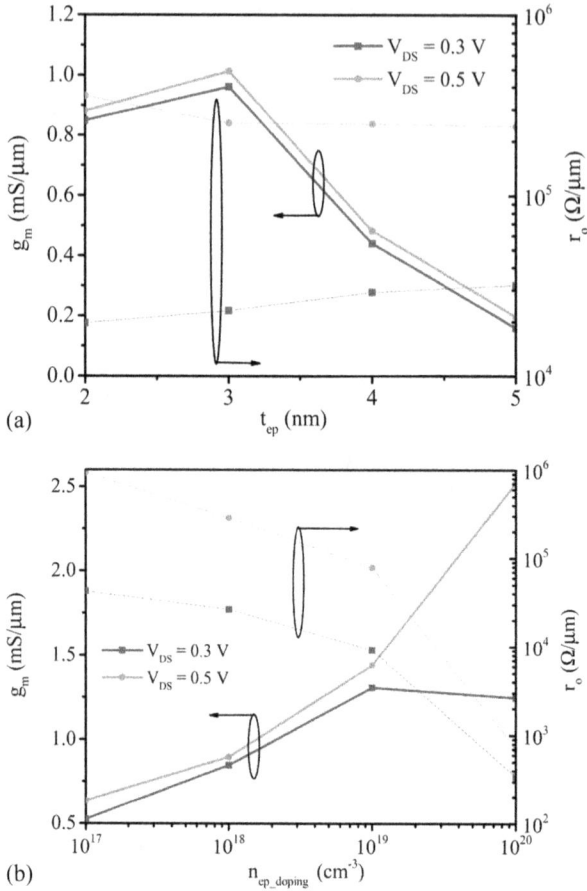

FIGURE 4.20 Effect of g_m and r_o with epitaxial parameters (a) t_{ep} and (b) n_{ep} in the nDG-LTFET device

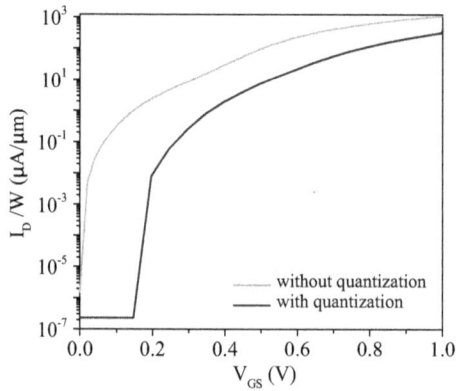

FIGURE 4.21 Impact of field-induced quantum confinement (FIQC) effects on the I_D-V_{GS} characteristics of DGLTFET

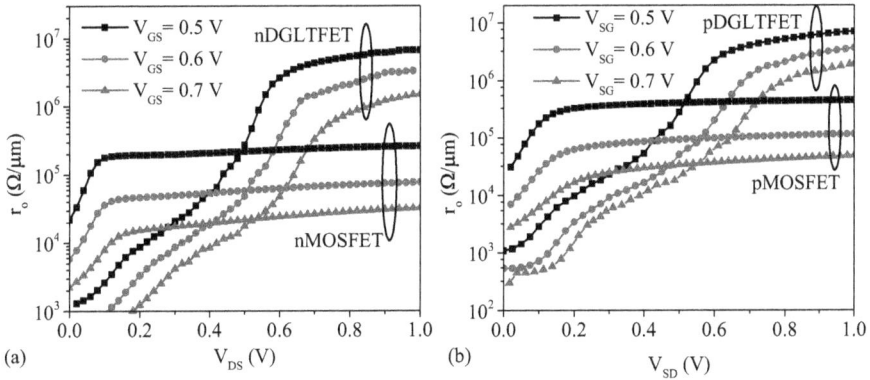

FIGURE 4.22 The output resistance r_o characteristics for n- and p-type DGLTFET and MOSFET for different gate biases V_{GS} (V_{SG})

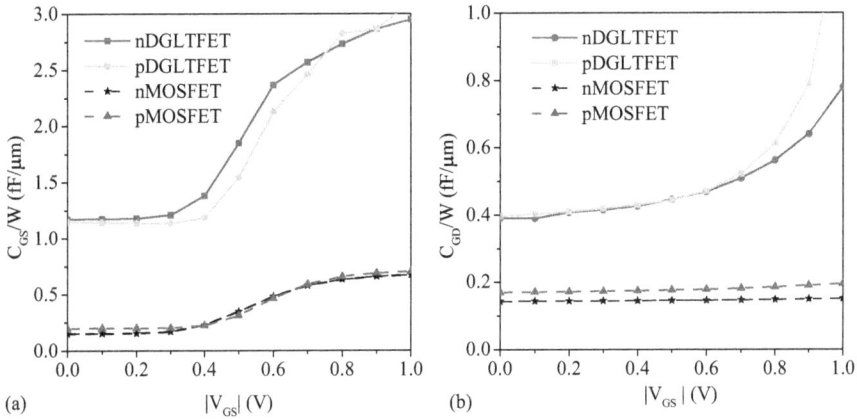

FIGURE 4.23 Comparison of the (a) C_{GS}-$|V_{GS}|$ (b) C_{GD}-$|V_{GS}|$ characteristics of n- and p-type DGLTFET, MOSFET at $|V_{DS}|=1$ V

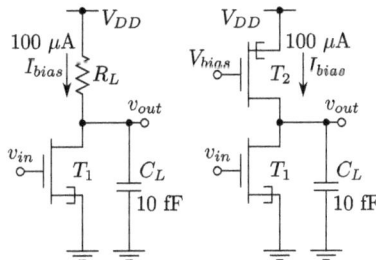

FIGURE 4.24 Schematics of the CS amplifier in resistive load and cascode variants

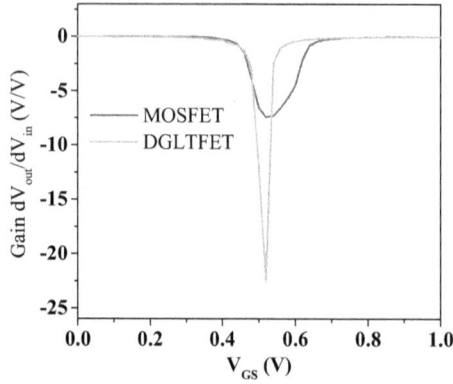

FIGURE 4.25 The gain dV_{out}/dV_{in} as a function of V_{in} of cascade CS amplifier under $V_{bias}=0.5$ V

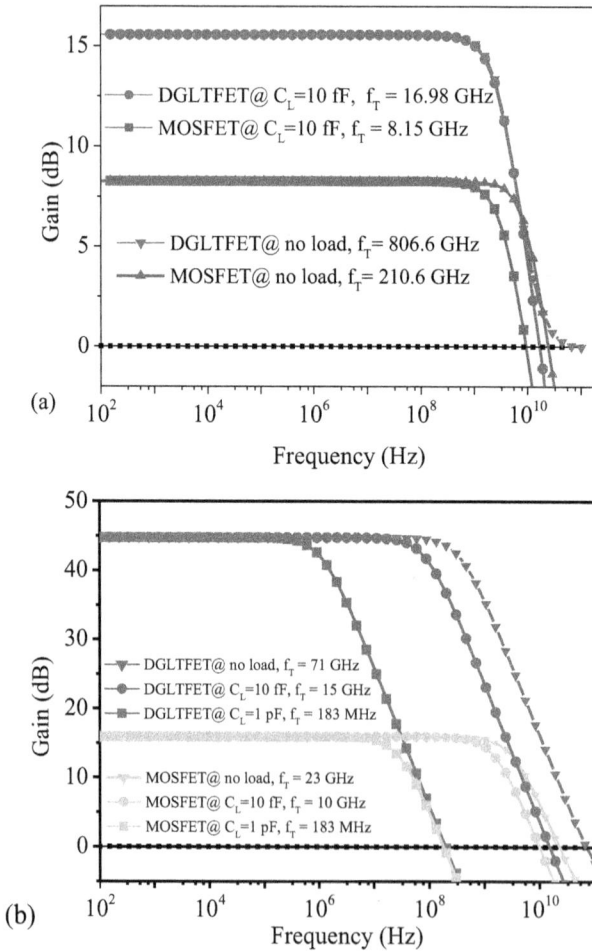

FIGURE 4.26 AC analyses of (a) resistive load and (b) cascode CS amplifier configuration under different C_L values

TABLE 4.10

Comparison of the DGLTFET with the equivalent MOSFET for the resistive load and cascode CS amplifiers

Parameter	Units	nDGLTFET	MOSFET
I_{bias}	μA	100	100
channel length L	nm	45	45
In voltage Vin	uV	100	100
In voltage Vin	uV	4	4
Unity-gain frequency f_T	MHz	5	5
Resistive load CS amplifier			
Width W	um	1	1
Resistive load R_L	kOhm	5	5
Output voltage V_{out}	mV	1	0.506
Intrinsic gain A_0	dB	15.56	8.254
Unity-gain frequency f_T	GHz	16.98	8.15
3dB bandwidth f_{3dB}	kHz	7.7	120
Cascode load CS amplifier			
Width Wp	um	1	1
Width Wn	um	1	1
Cascode bias V_{bias}	V	0.5	0.4
Output voltage V_{out}	mV	17.5	0.623
Gain A_0	dB	44.7	15.95
Unity-gain frequency Ft	GHz	15	10

As observed from Figure 4.28, I_{out} is more dependent on V_{out} in MOSFET-based current mirror circuits, but DGLTFET-based current mirror circuit I_{out} is more independent of V_{out}. The R_{out} of the DGLTFET-based cascade current mirror circuit is two orders of improvement than that of simple current mirror circuits and five orders higher than the MOSFET-based cascade current mirror circuit. As a result of DGLTFET's lower V_t, the smaller value of V_{out} at which the current saturates is substantially lower in DGLTFET-based current mirror circuits than in CMOS.

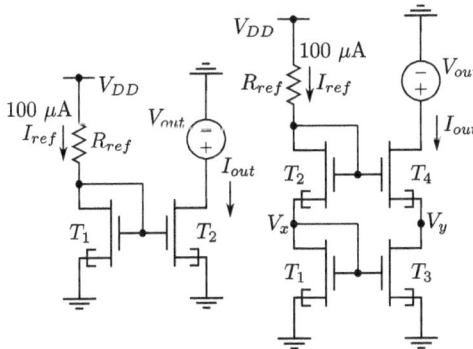

FIGURE 4.27 Schematics of the (a) single-stage and (b) cascode current mirror configurations

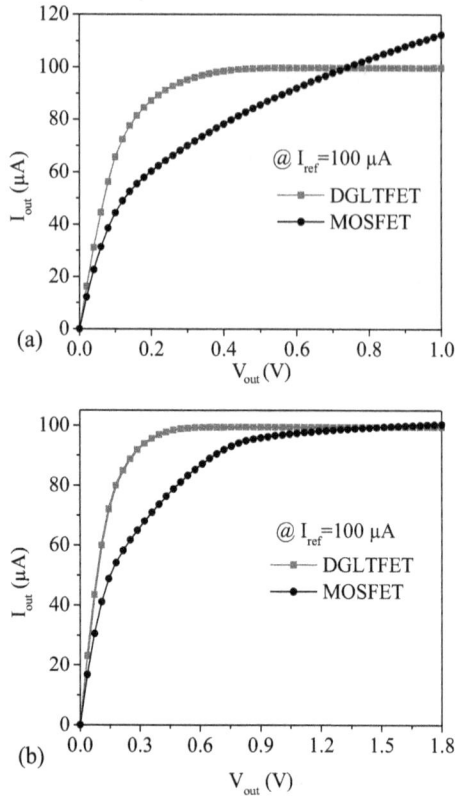

FIGURE 4.28 I_{out} vs. V_{out} (a) the single stage (b) the cascode current mirror

As a further extension, we have designed complex circuits like op-amp for unity-gain frequency f_T =5 MHz at V_{DD}=1 V. The circuit performance is benchmarked with the same circuit designed using CMOS technology at a 45 nm technology node. In Figure 4.29, the nDGLTFETs T_1 and T_2 of the first stage act as input transistors for reaching the greater g_m, while the pDGLTFETs T_3 and T_4 function as current mirror loads. T_7 and T_8 are DGLTFETs that make up the second-stage CS amplifier. Table 4.11 demonstrates that the gain of the DGLTFET-based two-stage op-amp is 26 dB more than that of the MOSFET-based op-amp for the same f_T. Due to the better saturation zone attributes of the p-type DGLTFET-based current mirror load and the larger g_m of the input n-type DGLTFET transistors, as seen in Figure 4.30, the differential gain of the DGLTFET op-amp is more than six times that of the MOSFET-based op-amp. In addition, the common-mode gain of the DGLTFET is roughly 6 dB less than that of the MOSFET, owing to the greater output resistance of its current mirror-based tail current source. The common-mode rejection ratio (CMRR) of the op-amp observed with DGLTFET devices is 23.5 dB (15 times) greater than the equivalent CMOS circuit, principally due to the larger differential gain and lower common-mode gain of the first stage.

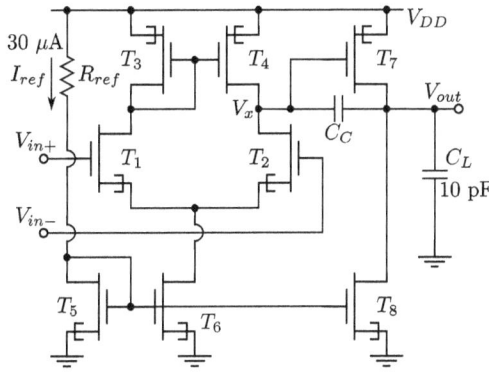

FIGURE 4.29 Schematic of a two-stage DGLTFET-based operational amplifier

TABLE 4.11

Comparison of two-stage op-amp designed with DGLTFET and the equivalent MOSFET for f_T =5MHz

Parameter	Units	nDGLTFET	MOSFET
Bias current I_{ref}	μA	30	30
Channel length L	nm	45	45
Load capacitance C_L	pF	10	10
Comp capacitance C_C	pF	4	4
Unity-gain frequency f_T	MHz	5	5
Gain V_{out}/V_d	dB	57	31
Differential gain V_x/V_d	dB	30	13.5
Common-mode gain	dB	−27	−20
CMRR	dB	51	33.5
Phase margin (PM)	degree	76	62
3dB bandwidth f_{3dB}	kHz	7.7	120

4.4.4 VERTICAL LTFET DEVICES FOR ANALOG CIRCUIT APPLICATIONS

Vertically grown TFETs are preferable because they allow more TFETs to be placed on a single chip, which increases the number of devices on the chip [26, 33, 34]. In this section, the VLTFET device has been proposed for analog circuit applications. The influence of gate-to-source overlapping length L_{ov} on the device and circuit performance has been studied and explained using physics.

Figure 4.31 shows the n-type VLTFET device structure and BtB tunneling rate schematic. An epi-layer of n-type Si is inserted to sharpen the band profile between the source and pocket [26], and its material parameters are shown in Table 4.12. Regarding the device's reliability, the source pocket with an intermediate mole fraction, especially in contrast to the source and channel regions, makes the band profiles

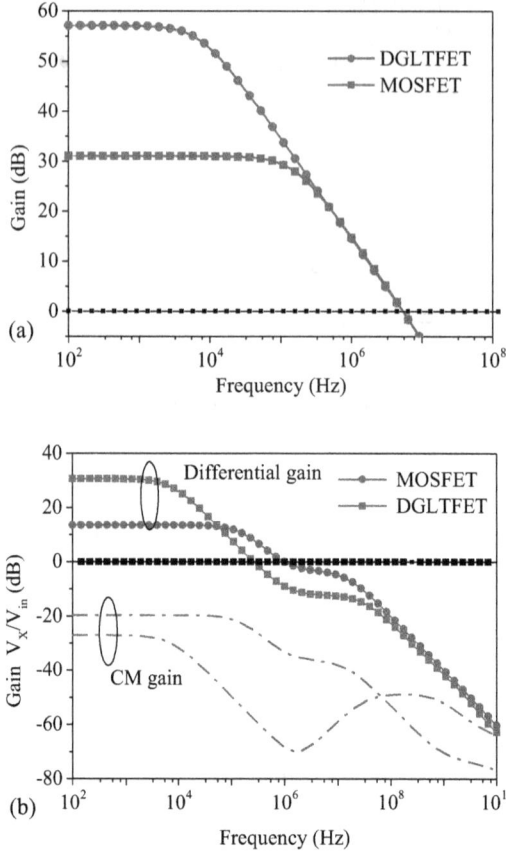

FIGURE 4.30 (a) Total gain A_v of op-amp for $f_T = 5$ MHz. (b) The differential A_d and common-mode (CM) gains A_{cm} of the op-amp differential stage for the overall $f_T = 5$ MHz

sharper without the risk of junction breakdown. The SiGe source and Si epi-layer improve device performance. The dynamic non-local path tunneling model was used to capture the BtB generation at all the interfaces in the 2D numerical simulations of the VLTFET device structure. In the early stages, the device was simulated with the TCAD simulator's default settings [26, 35]. Then, the parameters were adjusted so that the final characteristics matched the experimental work published by [26]. Figure 4.32 shows how the results of the calibrated simulation deck match up with the experimental work.

Figure 4.33 shows the transfer characteristics when L_{ov}=30 nm. I_D is found to be I_{on} = 2.4 µA/µm for V_{GS}= 1 V and I_{off} = 5 pA/µm for V_{GS} = 0 V when V_{DS}=1 V. The third derivative method is used to find the threshold voltage. The first peak of $\partial\,^3I_D/\partial V^3{}_{GS}$ for V_{DS} = 1 V gives $V_{tn}\approx0.4$ V. In these devices, vertical tunneling dominates, which is electrons are tunneling from VB of p⁺ source to CB of n⁺ epi-layer due to a high electric field along F_y. Soft saturation is part of the I_D-V_{DS} characteristics between

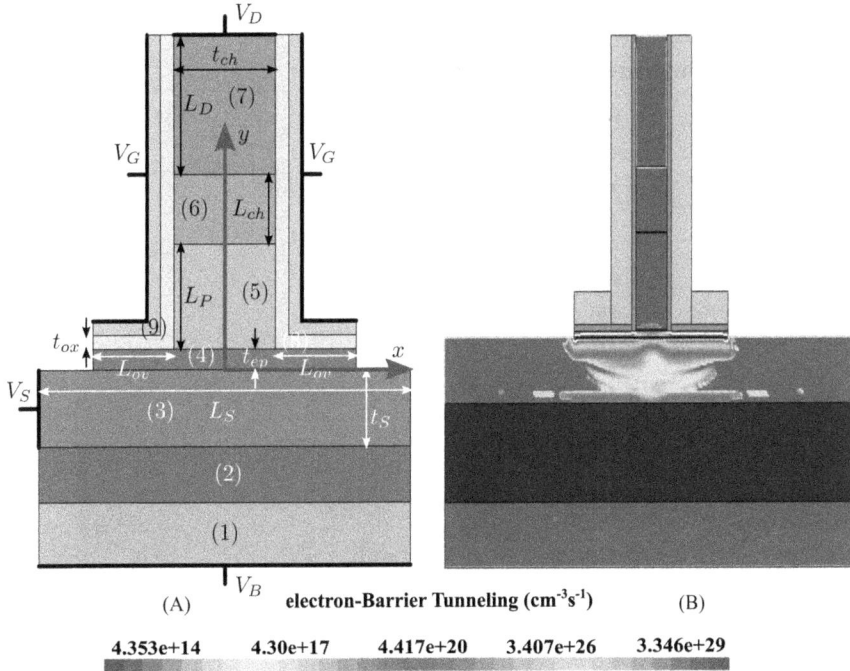

FIGURE 4.31 (a) Schematic cross-sectional view (1) Si substrate, (2) SOI, (3) p+ SiGe source, (4) n+ Si epitaxial layer, (5) i-SiGe pocket, (6) i-Si channel, (7) n+ Si drain, (8) HfO2 gate oxide, and (9) TiN metal gate. (b) Electron BtB tunneling rate of nVLTFET device

$0 < (V_{GS}-V_t) \leq V_{DS}$. Deep saturation happens between $V_t < V_{GS} \leq V_{DS}$, where V_{DS} loses control of the carrier density, as shown in Figure 4.34 [30, 31]. As a result, the carrier density remains the same. At first, as V_{DS} increases for a given value of V_{GS}, the electron density in the epi-layer over the source region decreases. As shown in Figure 4.35a, increasing V_{DS} has no effect on the electron density after a certain V_{DS}. Figure 4.35b shows that V_{DS} does not affect the surface potential near the source region after saturation.

Figure 4.36 shows the effect of temperature on these devices. As observed, an increase in temperature reduces the bandgap, affecting tunneling probability. Therefore, BtB generation increases, leading to increasing currents. Due to FIQC observing that shift in the V_t, which affects reduction in I_{on}.

After saturation, tunneling length W_{tun} (Figure 4.38a) stays constant at its minimum. Because V_{DS} has less influence on I_D after saturation, r_o is on the order of 100 MΩ/μm, as illustrated in Figure 4.38b for various V_{GS} biases. Increasing the L_{ov} increases the vertical BtB tunneling owing to the increase in tunneling cross-section, resulting in a rise in I_{on}, as seen in Figure 4.39 [29]. Increased I_{on} significantly affects the g_m, as observed in Figure 4.40a. The drain bias has an insignificant influence on the I_{on} with L_{ov}. Moreover, the r_o is very high and almost constant, as shown in Figure

TABLE 4.12

Parameters of the n and p VLTFET devices

Parameter	Units	nDGLTFET	pDGLTFET
Channel length	nm	20	20
Source length	nm	1000	1000
Drain length	nm	30	30
Gate length	nm	20	20
Pocket length	nm	30	30
Gate oxide thickness	nm	2	2
Channel thickness	nm	15	15
Epi-layer thickness	nm	3	3
Gate-source overlap	nm	30	30
Pocket length	nm	20	20
Source doping	/cm3	2×10^{20} p$^+$ Si$_{0.5}$Ge$_{0.5}$	2×10^{20} n$^+$ Si
Pocket doping	/cm3	10^{16} iSi$_{0.75}$Ge$_{0.25}$	10^{16} iSi$_{0.75}$Ge$_{0.25}$
Channel doping	/cm3	10^{16} Si	10^{16} Si
Drain doping	/cm3	5×10^{19} n$^+$ Si	5×10^{19} p$^+$ Si
Epi-layer doping	/cm3	5×10^{18} n$^+$ Si	5×10^{18} p$^+$ Si$_{0.5}$Ge$_{0.5}$
Gate material	-	TiN	Ag

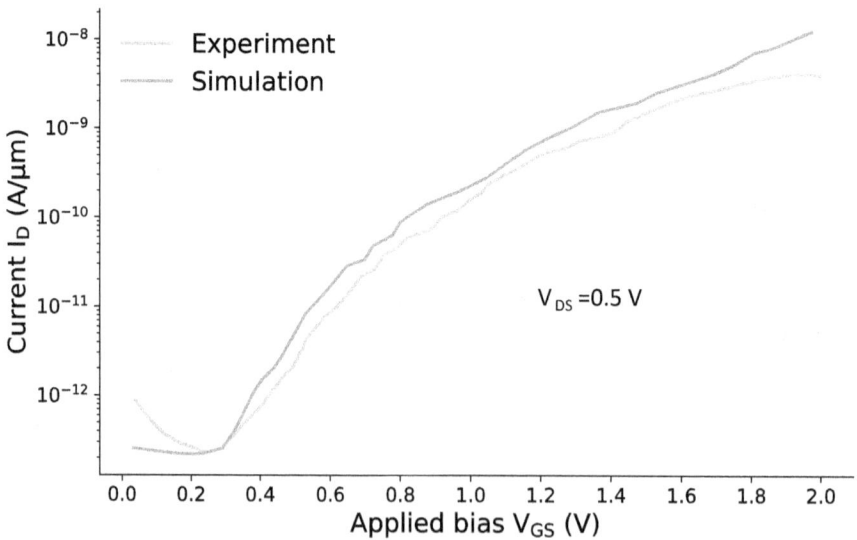

FIGURE 4.32 Simulation model calibration using a prefabricated device at V_{DS}=0.5 V [26]

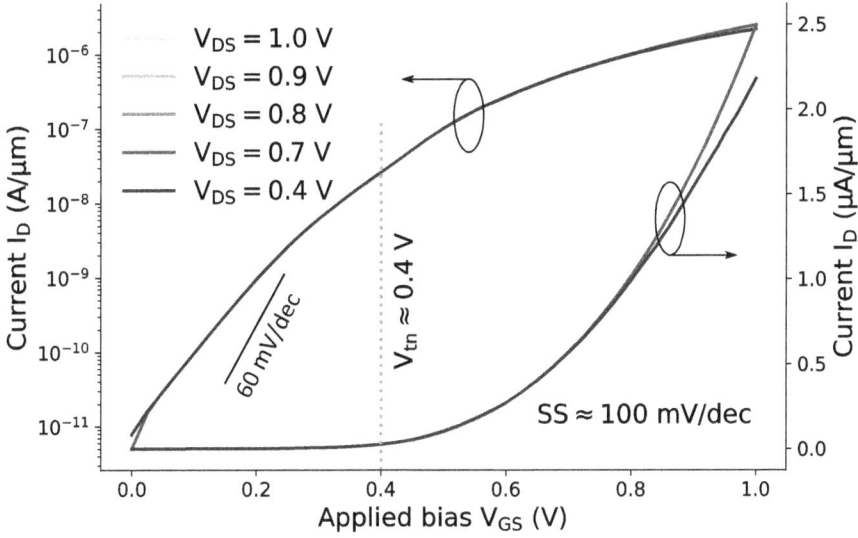

FIGURE 4.33 *ID-VGS* characteristics of nVLTFET for different *VDS*

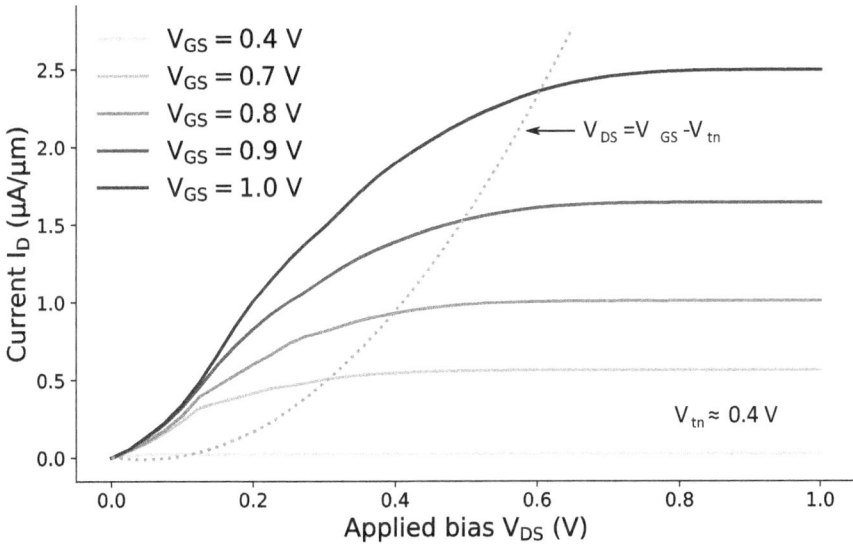

FIGURE 4.34 *ID-VDS* characteristics of nVLTFET for different VGS

4.40b. A_{vo} can be enhanced in these devices by increasing L_{ov} since g_m improves with L_{ov} and r_o is constant, as shown in Figure 4.41.

4.4.5 ANALOG CIRCUIT DESIGN USING VERTICAL LTFET DEVICES

In Table 4.13, the impact of L_{ov} on the analog performance characteristics of VLTFETs is summarized and compared to those of MOSFETs with gate length L_G. The earlier

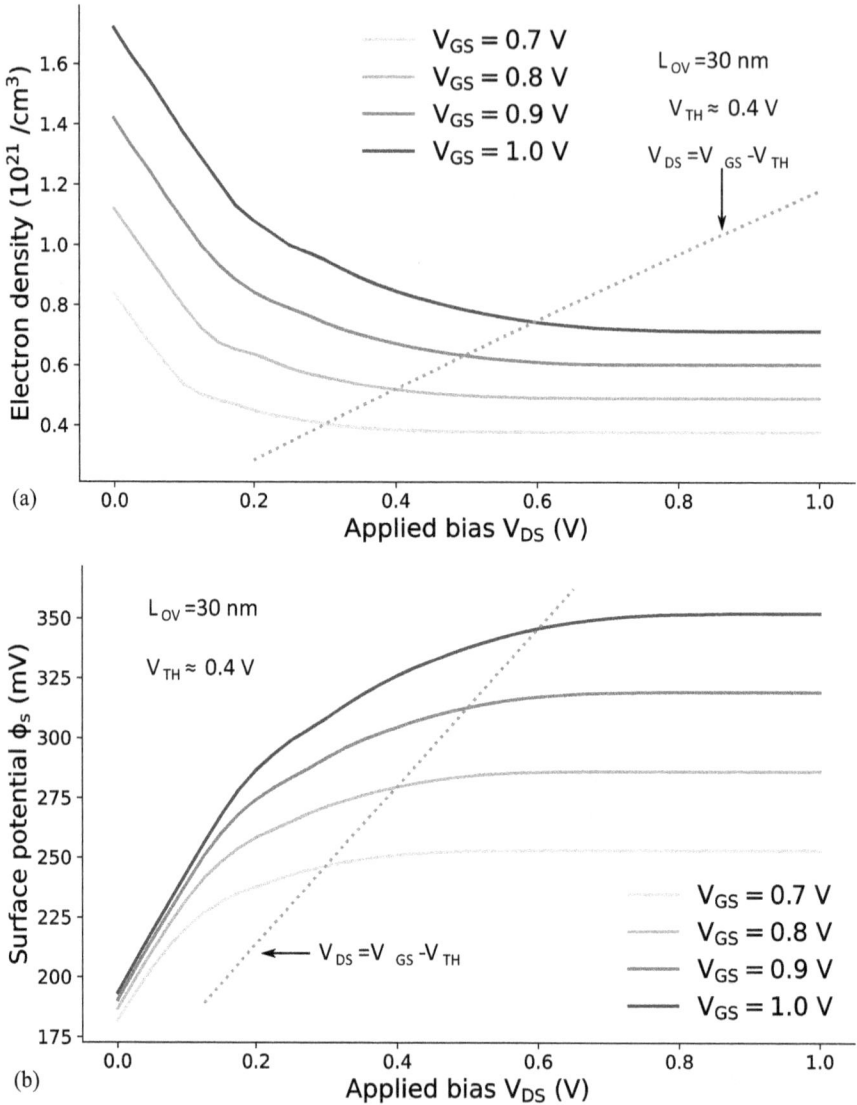

(a)

(b)

FIGURE 4.35 Influence of (a) density (b) surface potential in the epi-layer area for different drain biases

section explained the effect of L_{ov} on analog performance in VLTFET devices. This section explains the impact of L_{ov} on analog circuits. P-channel VLTFET device used as a current source load with fixed V_{bias}. The pVLTFET has equal, opposite doping concentrations with comparable dimensions to the nVLTFET, except for the materials and properties summarized in Table 4.1 and its characteristics in Figure 4.37 [37].

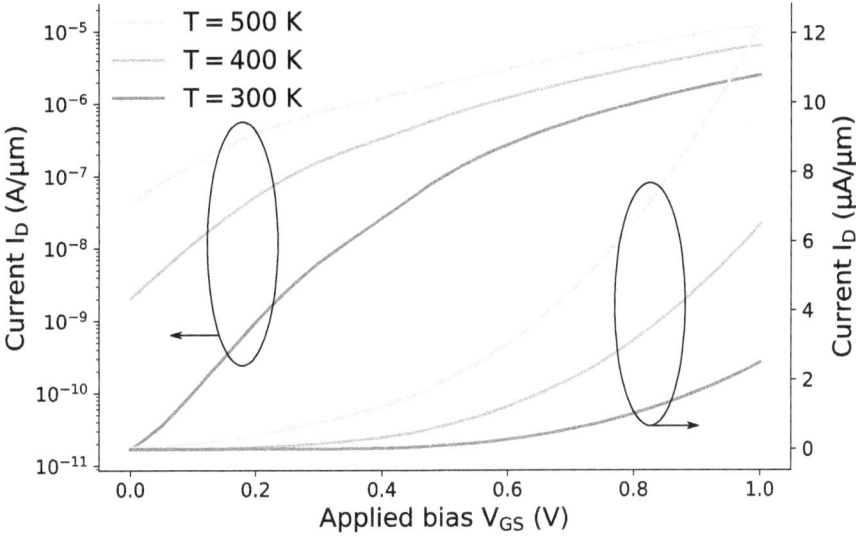

FIGURE 4.36 Effect of temperature on the VLTFET characteristics

First, we design a CS amplifier with L_{ov}=30 nm with n and p VLTFET devices. For all the analyses, the DC output voltage (operating voltage) of the CS amplifier is set to $V_{DD}/2$. Under no-load circumstances, the voltage gains A_V and unity-gain bandwidth f_T of the CS amplifier are measured to be 45 dB and 34 MHz, respectively. Subsequently, the VLTFET's L_{ov} is raised to 100 nm without affecting other parameters. This yields A_v = 52 dB and f_T = 81 MHz, as g_m rises by three times, as observed in Figure 4.42. In CMOS circuits, the gain has been improved with the width (W) parameter owing to the g_m increase, however, which affects the r_o. However, in VLTFET devices, the gain has been improved with the L_{ov} parameter without affecting the r_o. In addition to W, we can improve the gain with L_{ov} parameter in VLTFET devices. Figure 4.42 shows that the percentage increase in the gain of CS amplifier is high since increasing L_{ov}, which improves g_m without affecting r_o [37].

In addition, with the CS amplifier, we developed the cascode current mirror circuit with a reference current of I_{ref}=1 A. In typical CMOS systems, the channel length modulation effect is more pronounced at lower technology nodes, diminishing the R_{out}. The source-gate overlap predominantly affects I_D in VLTFETs. Therefore the impact of channel length modulation is insignificant. Figure 4.43 demonstrates that I_{out} is constant and irrespective of V_{out}. Consequently, as illustrated in Figure 4.44, R_{out} transcends several MΩ and achieves a maximum of 10^{11} Ω, leading to the perfect current mirror/source functioning. In addition, extending the L_{ov} from 30 nm to 100 nm enhances the R_{out} by a factor of ten for the same V_{out} [37].

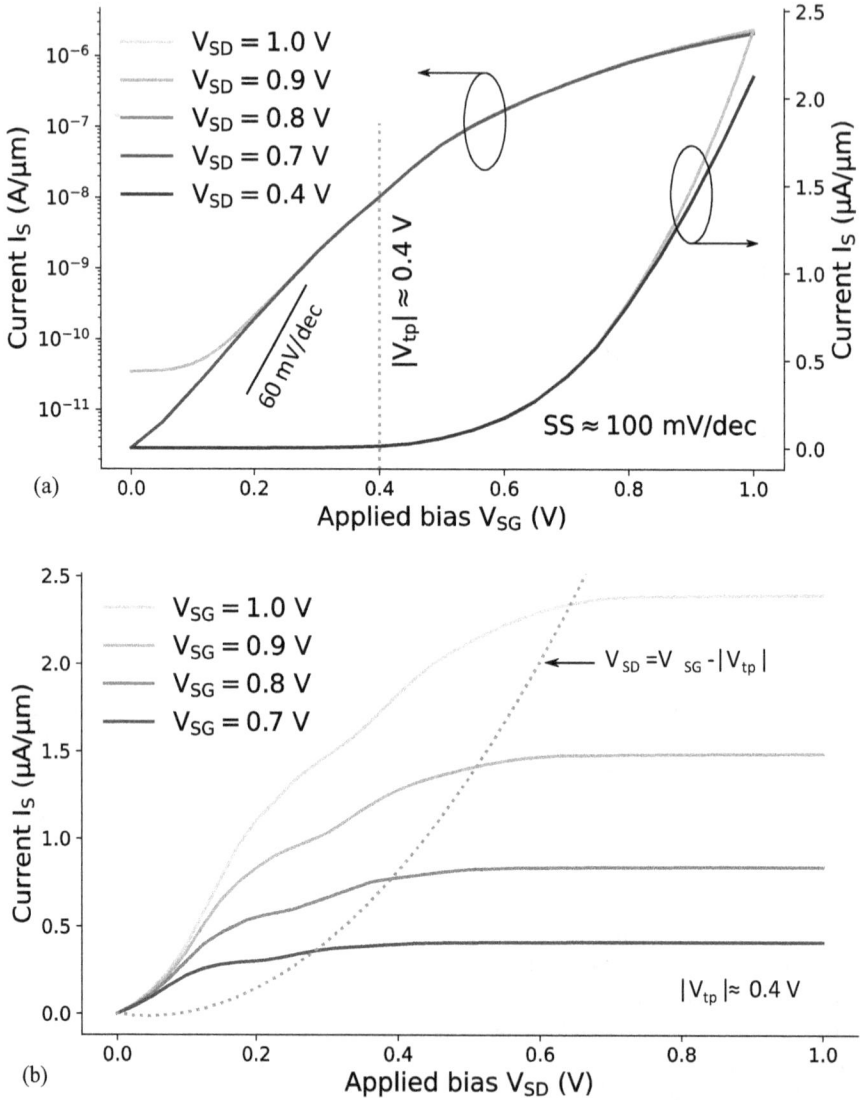

FIGURE 4.37   (a) I_S-V_{SG} characteristics of pVLTFET for various V_{SD} and (b) I_S-V_{SD} characteristics of pVLTFET for various V_{SG}

4.5   SUMMARY

This chapter explains the various gate-overlap tunnel field-effect transistors (GOTFETs) for digital, ternary, and analog circuits. Their performance has been benchmarked with industry-standard 45 nm CMOS technology. In this chapter, the I_{off} of the GOTFETs suggested for ultra-low-power circuits is at least one order of magnitude lower. In addition, I_{on} exceeds double that of a typical MOSFET of equal

(a)

$V_{TH} \approx 0.4$ V

$V_{DS} = V_{GS} - V_{TH}$

$V_{GS} = 0.7$ V
$V_{GS} = 0.8$ V
$V_{GS} = 0.9$ V
$V_{GS} = 1.0$ V

$L_{OV} = 30$ nm

Tunneling width W_{tun} (nm)

Applied bias V_{DS} (V)

(b)

$V_{GS} = 0.7$ V
$V_{GS} = 0.8$ V
$V_{GS} = 0.9$ V
$V_{GS} = 1.0$ V

$V_{TH} \approx 0.4$ V

$L_{OV} = 30$ nm

$V_{DS} = V_{GS} - V_{TH}$

Resistance r_o (Ω/μm)

Applied bias V_{DS} (V)

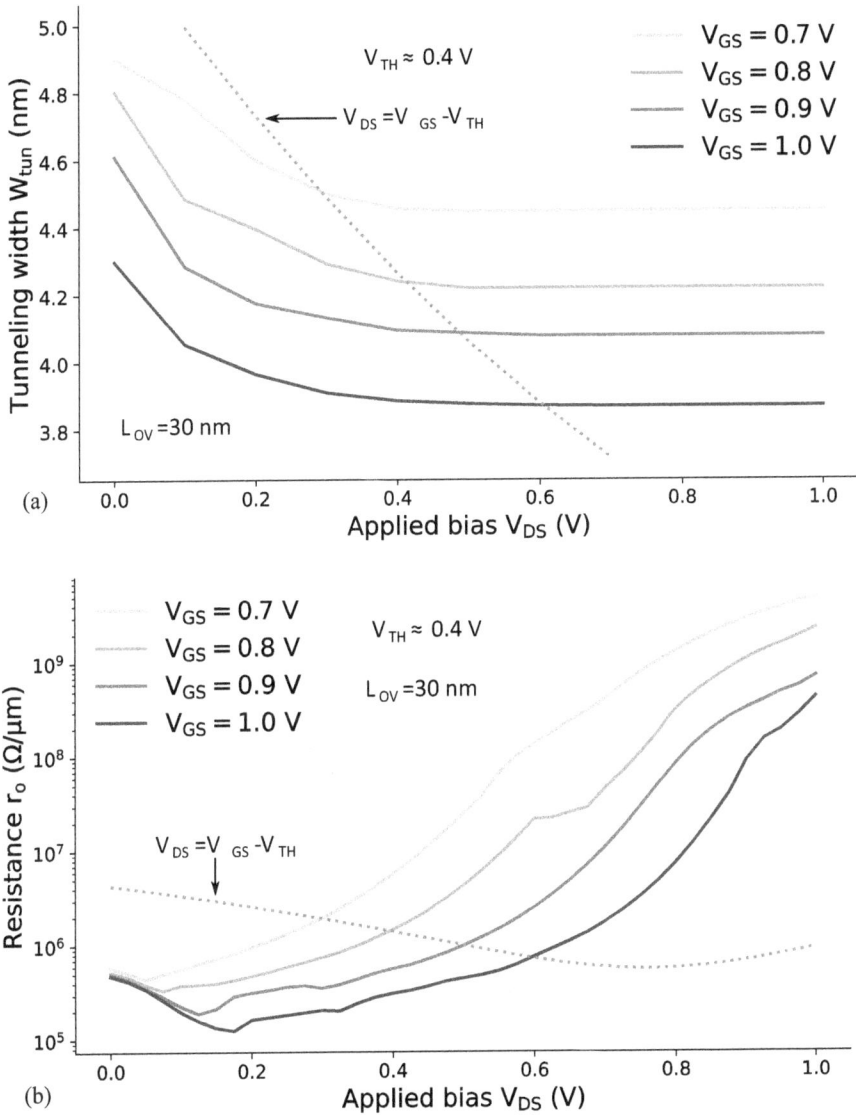

FIGURE 4.38 (a) Tunneling width W_{tun}.(b) Output resistance r_o variation with V_{DS} for distinct V_{GS} values

size at the same 45 nm technology node. Higher I_{on} makes the circuits more robust and improves performance, whereas a lower I_{off} significantly reduces static (leakage) power dissipation. This chapter also introduces LVT and HVT GOTFET devices for ultra-low-power ternary logic circuits since enhancing the GOTFETs' performance in digital circuits. To the best of our knowledge, for the first time, innovative low and high-threshold GOTFET devices have been reported for ternary logic applications.

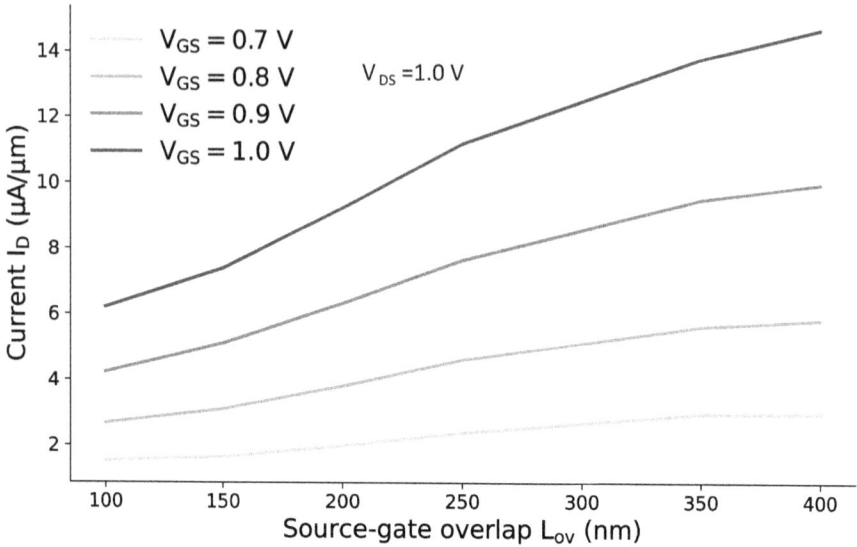

FIGURE 4.39 I_D fluctuation with L_{ov} for various V_{GS} values at V_{DS}=1 V

These devices are designed so that the low and high-threshold voltages (LVT and HVT) are $V_{DD}/3$ and $2V_{DD}/3$, respectively. The most exciting feature of the proposed GOTFET is that, in the same device structure, just by changing the material and doping parameters, we can get the optimal performance of LVT and HVT GOTFETs. The LVT and HVT GOTFET devices described in this chapter have I_{off} that is at least one order of magnitude less than that of the MOSFET, while I_{on} is nearly double that of the MOSFET at the same technology node. Higher I_{on} speeds up the operation of ternary logic circuits, whereas lower I_{off} significantly reduces static power consumption. This work also shows dual-threshold GOTFETs that can be both LVT and HVT by changing their terminals connections in the same device instead of using two separate devices.

This chapter extended this work to analog circuits. Due to the epi-layer between the source and oxide layer, the proposed LTFET exhibited excellent drain current saturation characteristics. The DGLTFET presented in this chapter for analog circuits has about three times the I_{on} and at least one order lower I_{off} than a standard MOSFET of the same size at the same technology node. Owing to its higher g_m and r_o, the intrinsic gain of a DGLTFET is much higher than a MOSFET or other traditional TFET. The design of multiple standard analog circuit designs, viz., CS amplifier, current mirror, and an op-amp using the proposed DGLTFET devices. For the same bias currents, the DGLTFET-based CS amplifier has three times the gain of a MOSFET-based amplifier. Because the DGLTFET does not have much channel length modulation effect, its current mirror circuits have better I_{out} saturation characteristics than those based on MOSFETs. Vertical FETs are preferred because significantly more transistors can be placed on a single chip. This chapter further described VLTFET devices at 20 technology nodes

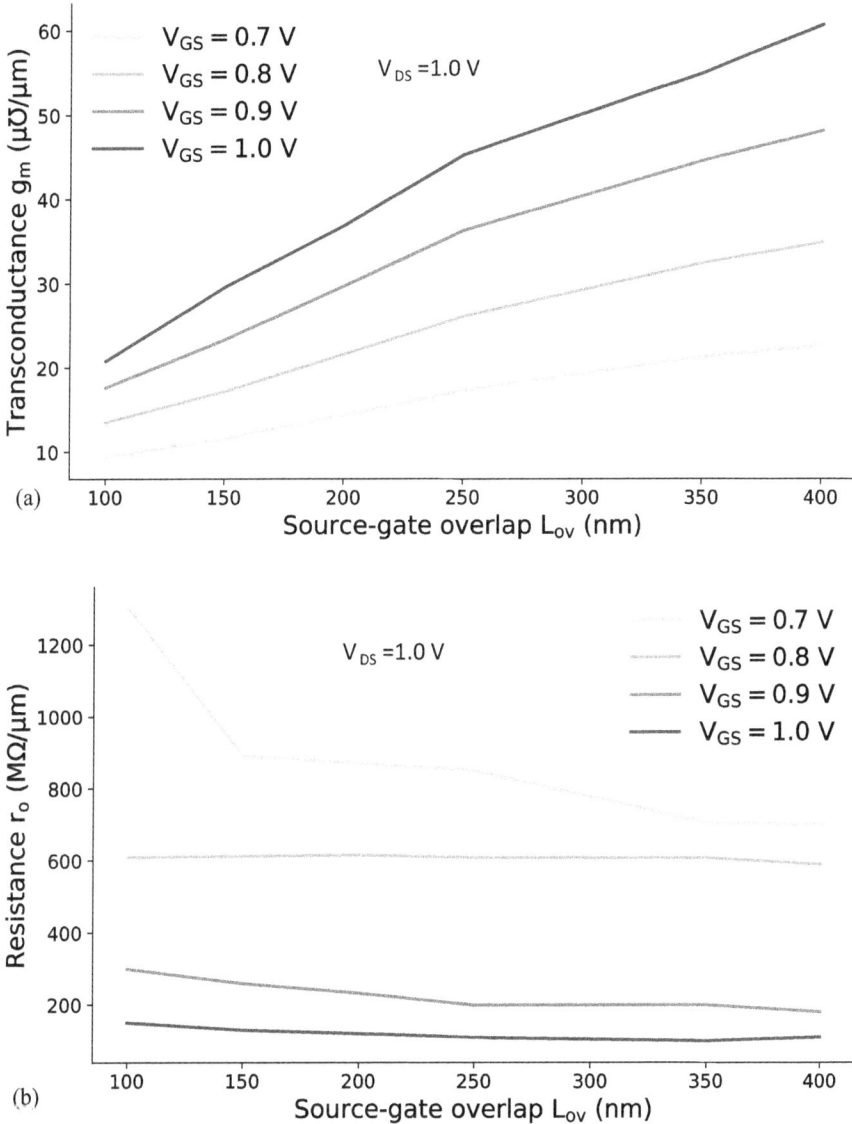

FIGURE 4.40 Variation in (a) transconductance g_m and (b) output resistance r_o with Lov for various V_{GS} values at V_{DS}=1 V

and explained L_{ov} impact on the analog device and circuit performance. As L_{ov} increased, I_{on} increased linearly, which improved g_m linearly. In addition, these devices are insignificant, affected by drain bias for BtB generation. Due to this, r_o was observed to be very high, significantly improving the gain. We observed a significant increase in gain in the CS amplifier by changing the L_{ov} parameter

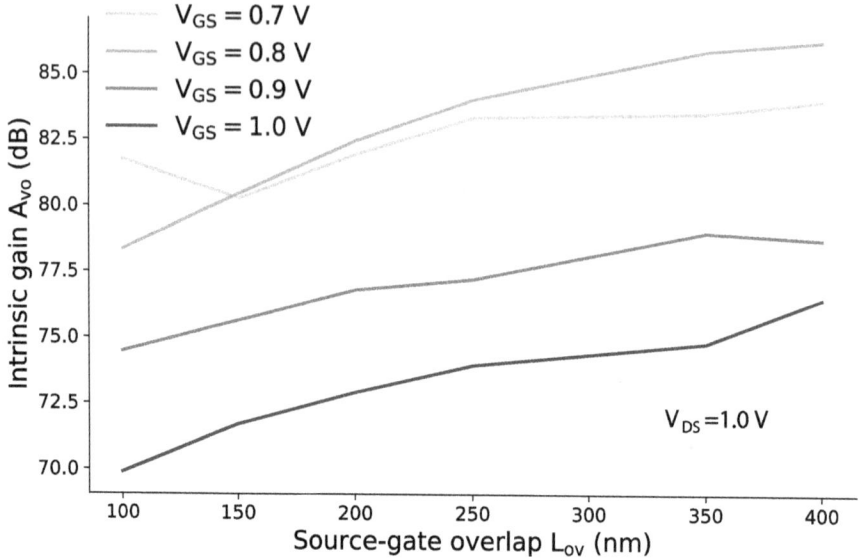

FIGURE 4.41   Variation of A_{vo} with L_{ov} for various V_{GS} values at V_{DS}=1 V

over the width. Also, we observed that *Rout* improved in the cascode current mirror circuit using these devices. We concluded that L_{ov} plays a significant role in addition to width in analog circuits.

TABLE 4.13

Analytical overview of the relationship between MOSFET and VLTFET analog performance on device parameters

Parameters	VLTFET	MOSFET
I_D	$W \times L_{OV}$	W/L_G
g_m	$W \times L_{OV}$	W/L_G
r_o	$1/W$	L_G^2/W
A_{v0}	L_{ov}	L_G

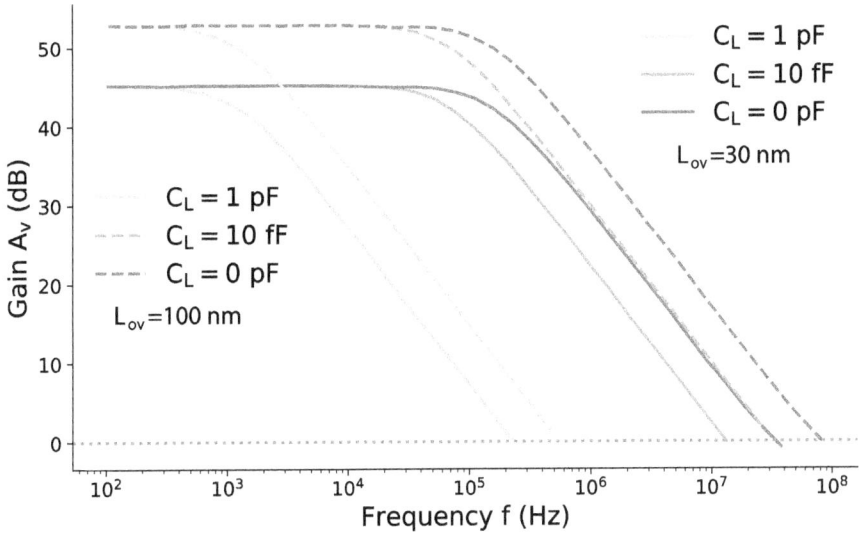

FIGURE 4.42 AC analyses of the CS amplifier under C_L=no load, 10 fF, and 1 pF

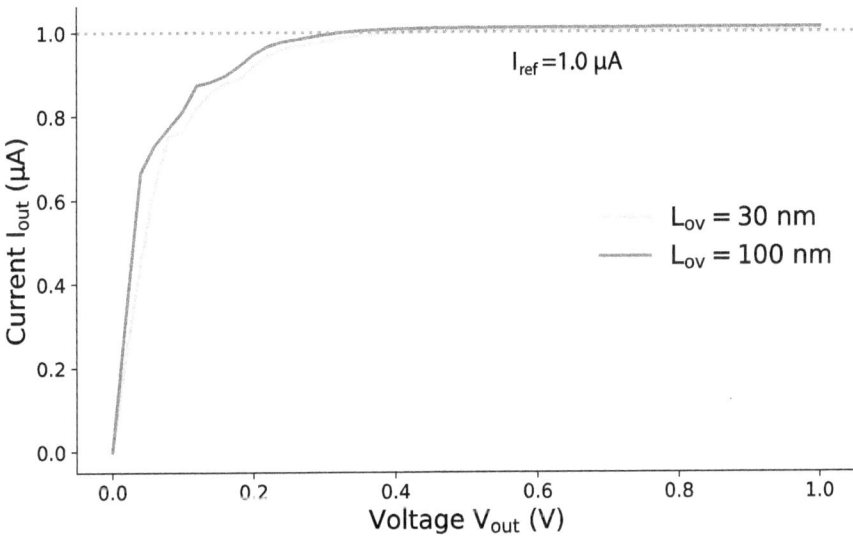

FIGURE 4.43 I_{out} vs. V_{out} of cascode current mirror under I_{ref} = 1 μA

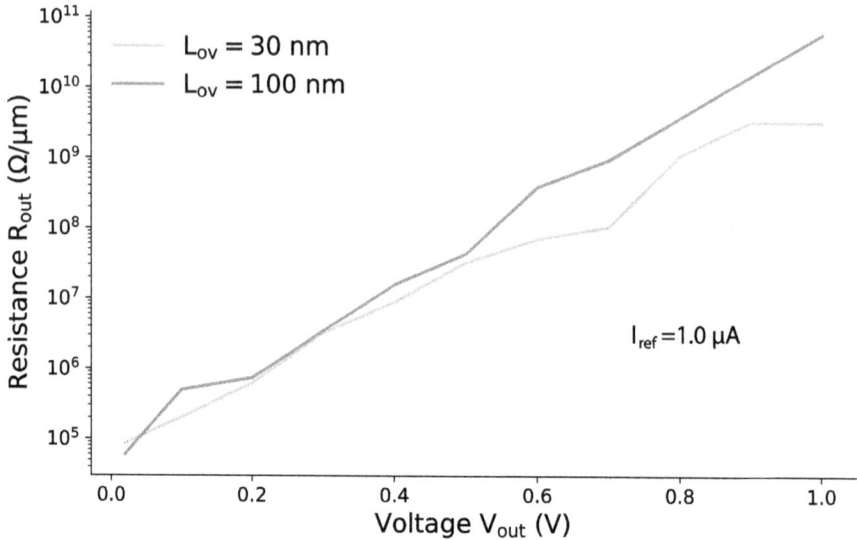

FIGURE 4.44 R_{out} vs. V_{out} of cascode current mirror under I_{ref}=1 µA

REFERENCES

1. E. Mollick. "Establishing Moore's Law". In: IEEE Annals of the History of Computing 28.3 (July 2006). Conference Name: IEEE Annals of the History of Computing, pp. 62–75. issn: 1934-1547. doi: 10.1109/MAHC.2006.4

2. K. Roy, S. Mukhopadhyay, and H. Mahmoodi-Meimand. "Leakage Current Mechanisms and Leakage Reduction Techniques in Deep-Submicrometer CMOS Circuits". In: *Proceedings of the IEEE* 91.2 (Feb. 2003), pp. 305–327. doi:10.1109/JPROC.2002.808156.

3. Y. Taur, and T. H. Ning. *Fundamentals of Modern VLSI Devices*. Cambridge University Press, 1998.

4. S. Saurabh, and M. J. Kumar. *Fundamentals of Tunnel Field-Effect Transistors*. CRC Press, Oct. 2016.

5. W. Choi et al. "Tunneling Field-Effect Transistors (TFETs) with Subthreshold Swing (SS) Less Than 60 mV/dec". In: *IEEE Electron Device Letters* 28.8 (Aug. 2007), pp. 743–745. doi:10.1109/LED.2007.901273.

6. N. Gupta et al. "Ultra-low-Power Compact TFET Flip-Flop Design for High-Performance Low-Voltage Applications". In: *2016 17th International Symposium on Quality Electronic Design (ISQED)* (Mar. 2016), pp. 107–112. doi:10.1109/ISQED.2016.7479184.

7. M. Kumar, and S. Jit. "Effects of Electrostatically Doped Source/Drain and Ferroelectric Gate Oxide on Subthreshold Swing and Impact Ionization Rate of Strained-Si-on-Insulator Tunnel Field-Effect Transistors". In: *IEEE Transactions on Nanotechnology* 14.4 (July 2015), pp. 597–599. doi:10.1109/TNANO.2015.2426316.

8. Y. Khatami, and K. Banerjee. "Steep Subthreshold Slope n- and p-Type Tunnel-FET Devices for Low-Power and Energy-Efficient Digital Circuits". In: *IEEE Transactions on Electron Devices* 56.11 (Nov. 2009), pp. 2752–2761. doi:10.1109/TED. 2009.2030831.

9. V. Nagavarapu, R. Jhaveri, and J. C. S. Woo. "The Tunnel Source (PNPN) n-MOSFET: A Novel High Performance Transistor". In: *IEEE Transactions on Electron Devices* 55.4 (Apr. 2008), pp. 1013–1019. doi:10.1109/TED.2008.916711.

10. B. Bhushan, K. Nayak, and V. R. Rao. "DC Compact Model for SOI Tunnel Field-Effect Transistors". In: *IEEE Transactions on Electron Devices* 59.10 (Oct. 2012), pp. 2635–2642. doi:10.1109/TED.2012.2209180.

11. S. S. Dan et al. "A Novel Extraction Method and Compact Model for the Steepness Estimation of FDSOI TFET Lateral Junction". In: *IEEE Electron Device Letters* 33.2 (Feb. 2012), pp. 140–142. doi:10.1109/LED.2011.2174027.

12. M. K. Anvarifard, and A. A. Orouji. "Proper Electrostatic Modulation of Electric Field in a Reliable Nano-SOI with a Developed Channel". In: *IEEE Transactions on Electron Devices* 65.4 (Apr. 2018), pp. 1653–1657. doi:10.1109/TED. 2018.2808687.

13. S. Strangio et al. "Assessment of InAs/AlGaSb Tunnel-FET Virtual Technology Platform for Low-Power Digital Circuits". In: *IEEE Transactions on Electron Devices* 63.7 (July 2016), pp. 2749–2756. doi:10.1109/TED.2016.2566614.

14. M. Alioto, and D. Esseni. "Tunnel FETs for Ultra-Low Voltage Digital VLSI Circuits: Part II–Evaluation at Circuit Level and Design Perspectives". In: *IEEE Transactions on Very Large Scale Integration (VLSI) Systems* 22.12 (Dec. 2014), pp. 2499–2512. doi:10.1109/TVLSI.2013.2293153.

15. A. Pal et al. "Insights Into the Design and Optimization of Tunnel-FET Devices and Circuits". In: *IEEE Transactions on Electron Devices* 58.4 (Apr. 2011), pp. 1045–1053. doi:10.1109/TED.2011.2109002.

16. K. Kao et al. "Optimization of Gate-on-Source-Only Tunnel FETs With Counter-Doped Pockets". In: *IEEE Transactions on Electron Devices* 59.8 (Aug. 2012), pp. 2070–2077. doi:10.1109/TED.2012.2200489.

17. C. Schulte-Braucks et al. "Fabrication, Characterization, and Analysis of Ge/GeSn Hetero-junction p-Type Tunnel Transistors". In: *IEEE Transactions on Electron Devices* 64.10 (Oct. 2017), pp. 4354–4362. doi:10.1109/TED.2017.2742957.

18. Ajay et al. "Analysis of Cylindrical Gate Junctionless Tunnel Field Effect Transistor (CG-JL-TFET)". In: *2015 Annual IEEE India Conference (INDICON)* (Dec. 2015), pp. 1–5. doi:10.1109/INDICON.2015.7443557.

19. M. Alioto, and D. Esseni. "Tunnel FETs for Ultra-Low Voltage Digital VLSI Circuits: Part II–Evaluation at Circuit Level and Design Perspectives". In: *IEEE Transactions on Very Large Scale Integration (VLSI) Systems* 22.12 (Dec. 2014), pp. 2499–2512. doi:10.1109/TVLSI.2013.2293153.

20. S. Chander, and S. Baishya. "A Two-Dimensional Gate Threshold Voltage Model for a Heterojunction SOI-Tunnel FET With Oxide/Source Overlap". In: *IEEE Electron Device Letters* 36.7 (2015), pp. 741–716. doi:10.1109/LED.2011.2174027.

21. M. K. Anvarifard, and A. A. Orouji. "Enhancement of a Nanoscale Novel Esaki Tunneling Diode Source TFET (ETDS-TFET) for Low-Voltage Operations". In: *Silicon* (Dec. 6 2018). doi:10.1007/s12633-018-0043-6.

22. M. W. Cheng-Yu. "Current Degradation by Carrier Recombination in a Poly-Si TFET with Gate-Drain Underlapping". In: *IEEE Transactions on Electron Devices* 64.3 (2017), pp. 1390–1393. doi:10.1109/LED.2011.2174027.

23. A. Chattopadhyay, and A. Mallik. "Impact of a Spacer Dielectric and a Gate Overlap/Underlap on the Device Performance of a Tunnel Field-Effect Transistor". In: *IEEE Transactions on Electron Devices* 58.3 (2011), pp. 677–683. doi: 10.1109/LED.2011.2174027.

24. A. Ortiz-Conde et al. "Threshold Voltage Extraction in Tunnel FETs". In: *Solid-State Electronics* 93 (Mar. 2014), pp. 49–55. doi:10.1016/j.sse.2013.12.010.

25. O. M. Nayfeh, J. L. Hoyt, and D. A. Antoniadis. "Strained $Si_{1-x}Ge_xSi$ Band-to-Band Tunneling Transistors: Impact of Tunnel-Junction Germanium Composition and Doping Concentration on Switching Behavior". In: *IEEE Transactions on Electron Devices* 56.10 (2009), pp. 2264–2269.

26. S. Glass et al. "Examination of a New SiGe/Si Heterostructure TFET Concept Based on Vertical Tunneling". In: *2017 Fifth Berkeley Symposium on Energy Efficient Electronic Systems & Steep Transistors Workshop (E3S)*. IEEE, 2017, pp. 1–3.

27. A. Ortiz-Conde et al. "Threshold Voltage Extraction in Tunnel FETs". In: *Solid-State Electronics* 93(2014), pp. 49–55.

28. W. G. Vandenberghe et al. "Impact of Field-Induced Quantum Confinement in Tunneling Field-Effect Devices." *Applied Physics Letters* 98.14 (2011): 143503.

29. A. Acharya et al. "Impact of Gate–Source Overlap on the Device/Circuit Analog Performance of Line TFETs." *IEEE Transactions on Electron Devices* 66.9 (2019): 4081–4086.

30. A. Acharya, A. B. Solanki, S. Dasgupta, and B. Anand. "Drain Current Saturation in Line Tunneling-Based TFETs: An Analog Design Perspective." *IEEE Transactions on Electron Devices* 65(1) (2017), pp. 322–330.

31. A. Acharya, S. Dasgupta, and B. Anand. "A Novel V_{Dsat} Extraction Method for Tunnel FETs and Its Implication on Analog Design." *IEEE Transactions on Electron Devices* 64(2) (2016), pp. 629–633.

32. *Virtuoso Spectre Circuit Simulator User Guide*. Cadence Design System Inc., 2011.

33. J. H. Kim, S. W. Kim, H. W. Kim, and B.-G. Park. "Vertical Type Double Gate Tunneling FETs with Thin Tunnel Barrier." *Electronics Letters* 51(9) (2015), pp. 718–720.

34. W. Li, and J. C. S. Woo. "Vertical P-TFET with a P-type SiGe Pocket." *IEEE Transactions on Electron Devices* 67(4) (2020), pp. 1480–1484.

35. S. Blaeser et al. "Novel SiGe/Si Line Tunneling TFET with High Ion at Low VDD and Constant SS". In: *2015 IEEE International Electron Devices Meeting (IEDM)*. IEEE, 2015, pp. 22–3.

36. H. Simhadri, S. S. Dan, R. Yadav, and A. Mishra. "Double-Gate Line-Tunneling Field-Effect Transistor Devices for Superior Analog Performance." *International Journal of Circuit Theory and Applications* 49(7) (2021), pp. 2094–2111.

37. S. Hariprasad, and S. S. Dan. "Superior Analog Performance Due to Source-Gate Overlap in Vertical Line-Tunneling FETs and Their Circuits." *Silicon* (2022), pp. 1–10.

5 Phase transition materials for low-power electronics

Sameer Yadav, Pravin Neminath Kondekar, and Bhaskar Awadhiya

CONTENTS

5.1 INTRODUCTION

The rapid growth of AI/ML-based implantable, wearable, and portable electronic devices has kept the spotlight on ultra-low-power electronics advancements using emerging materials, devices, circuits, and architectures. Dimensional scaling and the never-ending progress of modern technology have sparked a plethora of exploratory computing and data storage studies. Several device architectures are being investigated to enable ultra-low-power electronics circuit operation using sub-60 mV/decade (sub-kT/q) switching [1–3]. For a few decades, the most researched steep switching devices have been tunnel FET (TFET) [4], negative capacitance FET (NCFET) [5], and hyper/phase FET [6–8]. To provide steep switching characteristics, TFETs use quantum tunneling modulation through a barrier. However, when

DOI: 10.1201/9781003359234-5

123

designing TFETs, it is difficult to achieve the same ON current in n-type and p-type tunnel FETs. In contrast to a TFET, an NCFET enables sub-60 mV/decade switching by employing the ferroelectric materials as a sandwiched layer between the gate metal and gate oxide in conventional MOSFETs. Because of the ferroelectric's negative capacitance, a voltage amplification action is performed, resulting in a lower subthreshold swing and a greater ON current (I_{ON}). The voltage step-up action, on the other hand, is followed by an increased gate capacitance, which may negate the advantages of higher I_{ON}. One such family of materials that promotes dimensional scaling at lower technology nodes is known as "phase transition/correlated/threshold switching materials" (PTM). The unique property of abruptly switching from insulator to metal or metal to insulator upon triggering from stimuli (electrical, thermal, pressure, strain, optical) in these materials helps in achieving sub-60 mV/decade (sub-KT/q) switching in emerging devices (hyper FET/phase FET). Phase transition in the PTM family can occur due to a variety of physical processes such as filamentary ion diffusion, dimerization, electron-electron correlation, and so on. These materials' unique electrical properties can be used to build innovative logic/memory devices and circuits for next-generation electronics. Hyper/phase FET achieves steep switching by utilizing the property of insulator-to-metal transitions on the electrical triggering of an augmented PTM [9]. Transistor switching is aided by the PTM's abrupt current-driven switching from the insulating to the metallic state and vice versa. As a hysteretic device, however, it creates a complicated design space for low-power electronics applications.

This chapter begins in Section 5.2 by discussing the material perspective of PTMs, their history, theory, physical mechanisms behind the transition, and key features needed for low-power steep switching. Section 5.3 discusses emerging applications of these materials in low-power steep switching, digital circuits, memory designing, non-Boolean computing (coupled oscillatory dynamical systems, neuromorphic computing, and in-memory computing), and other novel circuit applications. Several novel concepts, methods, and device-circuit co-design frameworks for using PTMs in the design of low-power logic and memory applications are also discussed in this section. We also give an insight into the use of phase transition materials in hybrid devices like 2D MoS_2, negative capacitance-based phase FET, hybrid phase change-tunnel FET, etc. for achieving ultra-low steep switching devices. Section 5.4 concludes the chapter, followed by the future scope for PTMs in low-power electronics in Section 5.5.

5.2 PHASE TRANSITION MATERIAL PERSPECTIVE

PTMs are from transition metal oxide families and show an abrupt change in resistivity upon a trigger from stimuli []. Electrical [10–12], thermal [13], mechanical (pressure, strain) [14, 15], and optical stimuli [16] cause PTMs to exhibit sudden changes in resistance due to insulator-metal and metal-insulator transitions. It has been suggested by various researchers that transition in the PTM family occurs due to various physical processes. Some material exhibits transition due to filamentary ion diffusion [6, 19], while some follow electron-electron correlation [17, 18]-based transition or transition using dimerization [20]. Materials with a broad range of hysteresis, thermal stability, and resistivity are found in the PTM family [40–43] (e.g. VO_2, V_2O_3, V_2O_4, TiO, Ti_2O_3, $SmNiO_3$, Cu-doped HfO_2, doped chalcogenide, NbO_2 etc.). Furthermore, new PTMs are being thoroughly investigated, and innovative ways to tailor their properties, such as strain, have been revealed [44]. Such approaches and a large range of PTMs show promise in terms of down-selecting and optimizing PTMs for specific applications. Note that materials with strongly correlated electrons undergo similar transitions and are referred to as "correlated materials" (CM) [45, 46]. Transition threshold values separate the high and low resistance states of all these transitioning materials.

5.2.1 THEORIES BEHIND MIT IN PTM

In literature, metal-insulator transition (MIT) in PTM is explained by various theories, namely, Mott-MIT, Peierls-MIT, and Anderson-MIT. Conventional band theory (e.g. Bloch-Wilson insulator or band insulator) did not predict the carrier enhancement in the PTM materials. Mott's theory [22] suggests that the effect of the electron-electron interactions plays a vital role in a phase transition. It mathematically stated that in a PTM, MIT occurs when the electron carrier density (n) is greater than the critical carrier density (n_c) and provides the relation between Bohr radius a_H and n_c: $n_c^{1/3} a_H \approx 0.2$. PTMs that follow Mott's theory are known as Mott-Hubbard or Mott-MIT insulators. MIT phenomenon occurring through electron-lattice or electron-phonon interaction is known as Peierls-MIT [23]. PTMs falling in this mechanism undergo lattice structural changes accompanied by conductivity changes. In the 1950s, Anderson [24] discovered that crystal lattice defects result in the insulating state in PTMs, and PTMs following this mechanism are called Anderson-MIT.

5.2.2 CONTROLLING PARAMETERS FOR MIT IN PTM

PTM can also be classified in terms of triggering parameters, i.e. control of metal-to-insulator transitions. The most discussed case is temperature controlling, where, by varying the temperature i.e. (heating/cooling), MIT is triggered; see Figure 5.1. Second is bandwidth controlling where triggering with internal and external pressure in material MIT occurs. For example, using substitutional doping with different-sized atom pressure can be exerted in materials like $RNiO_3$ (R = Pr, Nd, and Sm). The third is the band-filling control whereby changing the doping level either with acceptor or donor MIT can be triggered, e.g. with manganites and cuprates. Also, there are some PTMs where any two or all three triggering parameters control MIT. Bandwidth or temperature can be used to control transition in RNiO3-type PTMs.

5.2.3 SPECIAL FOCUS ON VO_2 AS PTM

There are certain materials like vanadium oxides where the primary physical mechanism is still debatable. A thin film of VO_2 showcases a four-fold change in resistance upon triggering from electrical stimuli followed by band structure changes. Its MIT

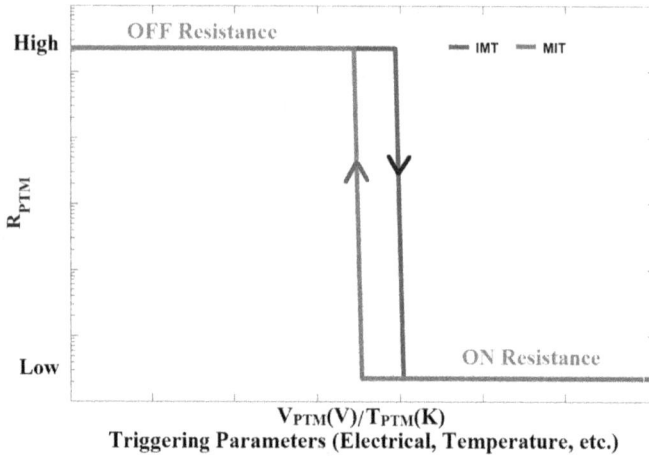

FIGURE 5.1   Generalized resistance transitions from insulator to metal (IMT) or metal-to-insulator (MIT) profile by various triggering, especially electrical or temperature, that are useful for low-power electronics

occurs in a temperature range of 341 K–344 K. It is shown in [25] that the structural property of VO_2 also changes during MIT, i.e. the insulating structure (monoclinic phase) changes to a metallic structure (rutile phase). During insulator to metal transition, the $3d_{||}$ band divided and formed filled lower energy (bonding band $3d_{||}$) and empty higher energy (antibonding band $3d_{||}^{*}$). The antibonding $3d_{\pi}^{*}$ further moves to higher energy. This results in a bandgap of about 0.6–0.7eV in VO_2. Thus, the MIT phenomenon in VO_2 raises the question of whether it is a Peierls or a Mott-Hubbard insulator, i.e. structural-change-induced MIT or carrier-induced MIT. In 1975, Zylbersztejn and Mott [26] suggested that VO_2 cannot be considered a Mott-Hubbard insulator. Then, performing LDA calculations on a VO_2 monoclinic M1 structure, Wentzcovitch et al. [27] in 1994, suggested that VO_2 is a band insulator with a semi-metal nature and has a much fewer number of carriers. Later, Rice et al. [28] claimed that VO_2 is a Mott-Hubbard insulator by performing a calculation on the M2 insulating phase rather than the M1 insulating phase considered by Wentzcovitch. Using ultrafast spectroscopy on VO_2 and studying its structural properties, Cavalleri et al. [29] stated that VO_2 is not a Mott-Hubbard insulator. Kim et al. [30] suggested VO_2 is a Mott-Hubbard insulator using femtosecond pump-probe measurements in a metallic rutile structure. Both theories have been accepted in the literature, and both should be considered while designing PTM-based devices, especially VO_2. There has also been debate about which controlling mechanism transition happens in VO_2: temperature-assisted electrical transition or electric field-assisted electrical transition. While some demonstrate that temperature [31] is the primary factor, others believe that electric-field-driven transitions [32] play a crucial role. Yang et al. [33] draw the conclusion that electric-field-assisted switching is more dominant and that Joule heating may not be adequate for MIT.

The hypothesis put forth by the authors states that a particular threshold voltage, which is said to be temperature-dependent, is needed for phase transitions. The PTM is divided into two categories in the literature [34, 35]: electronic-driven PTM (E-IMT) and thermally driven PTM (T-IMT), and both are analyzed. The development of a compact model or SPICE for using PTM in low-power electronics is still hampered by the lack of understanding of physics. But for low-power electronics at nanoscale dimensions, electric-driven transitions are more dominating than temperature-driven, and many models developed in recent years have considered both thermal- and electric-driven transitions [36–39].

VO$_2$ as PTM has gained the advantage because it can transition near room temperature and gels well with the CMOS VLSI fabrication process flow. In VO$_2$, the phase transition happens at incredibly quick timescales. The phase transition time constants in VO$_2$ have been measured experimentally using 4-D ultrafast electron microscopy, pulsed voltage measurements optical pump probes, time-resolved X-ray diffraction, and terahertz spectroscopy. The timescale of the phase transition is typically at the level of picoseconds or faster, with the exception of pulsed voltage measurements, which may be constrained by the instrumentation's resolution. An intriguing possibility of building an ultrafast switch emerges because the MIT can be activated at sub-picosecond timescales. In addition, due to electrical triggering, PTM has been used in high-performance logic. Hysteresis, which frequently follows a structural transition, produces intriguing candidates for memory devices. To put it another way, the dynamics of the transition could be used to make artificially structured materials with electrically tunable nanoscale metallic and dielectric states. The oscillatory behavior in PTM can be used for nano-oscillators for neuromorphic computing (discussed later) [47, 48].

5.2.4 WORKING OF PTM

The electrical characteristics of these materials, regardless of the underlying physical phenomenon, can be behaviorally generalized as follows. PTM occurs in two phases: insulator and metal, with the metallic phase's resistance (R_{MET}) typically being four times lower than the insulating phase's resistance (R_{INS}). The generalized current-voltage (I-V) response and device geometry for a typical PTM are shown in Figure 5.2. In Figure 5.2(a), the width, length, and thickness of the PTM are W_{PTM}, L_{PTM}, and T_{PTM} respectively. As seen in Figure 5.2(b), when there is no electrical stimulation, PTMs stay in the insulating state. Insulator-to-metal transition (IMT) occurs when a sufficiently high current (I_{C-IMT}) (or voltage V_{C-IMT}) flows through (applied across) the material. Conversely, metal-to-insulator transition (MIT) is triggered when current/voltage is reduced below a certain level (I_{C-MIT} or V_{C-MIT}). Due to the results of different transitions, these materials result in hysteretic characteristics. IMT and MIT occur abruptly (but not instantaneously) [49–51]. The resistivity of metal (ρ_{MET}), insulating states (ρ_{INS}), and the critical current density for IMT (J_{C-IMT}) and MIT transitions (J_{C-MIT}) are device/geometry-independent material properties. As illustrated below, the device-specific parameters (I_{C-IMT}, I_{C-MIT}, R_{MET}, and R_{INS}) can be represented in terms of material-level parameters (and geometric dimensions).

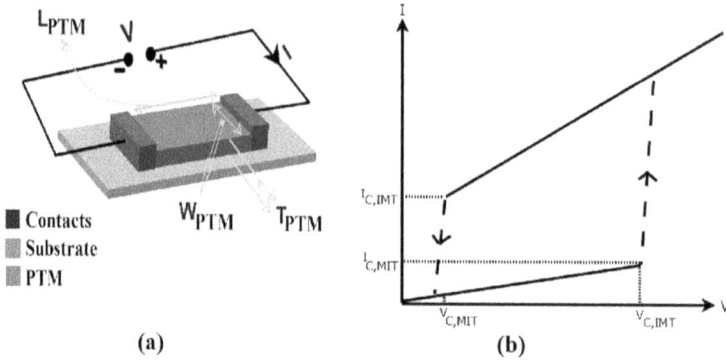

FIGURE 5.2 (a) Two-terminal PTM general structure and (b) its generalized I-V characteristics

$$I_{C\_IMT} = J_{C\_IMT} * W_{PTM} * T_{PTM} \tag{5.1}$$

$$I_{C\_MIT} = J_{C\_MIT} * W_{PTM} * T_{PTM} \tag{5.2}$$

$$R_{MET} = \rho_{MET} \frac{L_{PTM}}{W_{PTM} * T_{PTM}} \tag{5.3}$$

$$R_{INS} = \rho_{INS} \frac{L_{PTM}}{W_{PTM} * T_{PTM}} \tag{5.4}$$

It's worth noting that some materials, particularly those with filamentary conduction [52], may not have a linear relationship between resistance and area ($A_{PTM} = W_{PTM}$ *T_{PTM}). Similarly, the resistance and length (L_{PTM}) relationship may be nonlinear. The effective resistivities, ρ_{MET} and ρ_{INS}, are defined as R_{INS} (A_{PTM}, L_{PTM})*A_{PTM}/ L_{PTM} and R_{MET} (A_{PTM}, L_{PTM})*A_{PTM}/L_{PTM}, respectively, for such materials with a complicated dependency of the resistance on the geometry. $R_{INS/MET}$ (A_{PTM}, L_{PTM}) is the selector's insulating/metallic resistances, which are non-linearly dependent on A_{PTM} and L_{PTM}. Effective resistivity, in general, can be a function of geometry and not only a constant parameter, as can be seen. The voltage across the selector can also influence resistivity. Furthermore, J_{C-IMT} and J_{C-MIT} may be functions of the area due to similar effects. Furthermore, some PTMs can have unipolar electrical properties (responding simply to the positive or negative polarity of voltage). To conclude, the electrical properties of PTMs are characterized by (a) abrupt transitions, (b) a high resistance ratio, and (c) hysteresis.

5.3 APPLICATIONS OF PTM IN LOW-POWER ELECTRONICS

As discussed earlier, PTM behaves as an ultrafast switch and has oscillatory behavior due to its abrupt switching behavior. As shown in Figure 5.3, this ultrafast switching

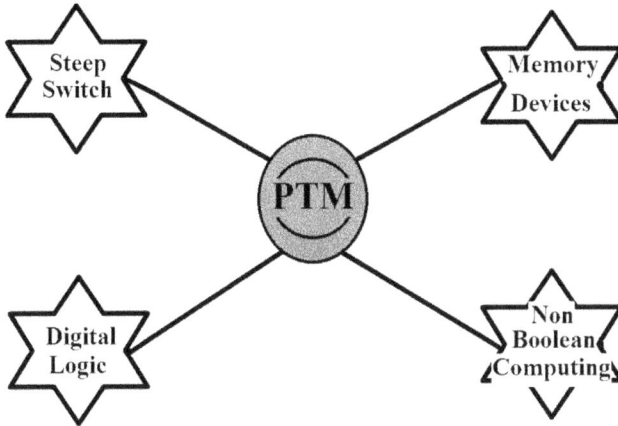

FIGURE 5.3 Some application areas of PTM in low-power electronics

find benefits in various low-power applications including two- or three-terminal devices, steep switching to harness power advantage in lower technology nodes, low-power novel digital logic memory design, and non-Boolean computing architecture. The subsections that follow go over emerging phase transition-based electronic devices (Mott-FET), steep switching (Phase FET/hyper FET), logic switches, memory, non-Boolean computing architectures, etc.

5.3.1 PTM AS TWO- OR THREE-TERMINAL ELECTRONIC DEVICES

Novel and low-power oxide electronics are made possible due to the potential to electrically induce MIT in PTM-based two- or three-terminal device configurations at or close to room temperature. Electric oscillations, abrupt resistance transitions with voltage sweeping, and nonlinear S-shaped I-V curves were all reported at the beginning of the 1970s, and these occurrences laid the groundwork for the phenomenon of abrupt MIT switching behavior in two-terminal PTMs, especially VO_2. A number of simple switch devices based on the nonlinear I-V properties of VO_2 have also been demonstrated. Initially, the MIT in VO_2 was not associated with electrical triggering (E-MIT); rather temperature-assisted triggering was explored. MIT can be caused by an electric field in VO_2, according to Stefanovich et al. [18], although some researchers believe that current-assisted heating can trigger MIT. Using theoretical-based simulations, it has been shown that Joule heating occurring due to current leakage cannot trigger MIT [35]. Several groups have studied E-MIT [32, 53–56]. A three-terminal Mott field effect transistor (Mott-FET) may be produced as a result of a field-driven phase transition [57] and can offer important information about the physical mechanism of the transition. For a Mott-FET, the channel is made up of a Mott insulator, and the channel switches between an insulating and metallic state through gate terminal control [58–62]. This mechanism may also have benefits because a metallic channel can have much higher electron carrier concentrations.

5.3.2   STEEP SWITCHING

As transistor dimensions are shrinking, the difficulty to lower the subthreshold swing [63] below the Boltzmann limit is a significant and long-standing problem. As a result, a worldwide search is underway for an ideal switch that can overcome Boltzmann's limit and offer SS of less than 60mV per decade. The subthreshold swing (SS), which is represented in the expressions in Equations 5.5–5.7, is the gate voltage needed to shift the drain current by one order of magnitude; while a transistor operates in the subthreshold region.

$$SS = \frac{\partial V_{GS}}{\partial \log_{10} I_{DS}} = \overbrace{\frac{\partial V_{GS}}{\partial \varphi_s}}^{m} * \overbrace{\frac{\partial \varphi_s}{\partial \log_{10} I_{DS}}}^{n} \tag{5.5}$$

$$m = 1 + \frac{C_{dep}}{C_{ox}} \tag{5.6}$$

$$n = 2.3 * \frac{KT}{q} \tag{5.7}$$

In these equations, V_{GS} represents the applied gate voltage, I_{DS} is the subthreshold region drain-to-source current, C_{OX}, and C_{DEP} is oxide and depletion region capacitances. The body factor of a transistor is m, while the transport factor is n, which quantifies the change in drain-to-source current with surface potential (φ_s) and represents the channel conduction process. Even if somehow C_{OX} tends to infinity, the subthreshold swing of a MOSFET cannot be scaled below 60mV/decade. There are three ways to solve this problem; i) reducing the m factor, ii) reducing the n factor, or (iii) reducing both the m and n factor.

Phase transition FET makes use of two-terminal resistive switching devices (PTM) (e.g. Mott insulator, correlated materials) coupled with the existing FET to achieve steep switching by reducing transport factor. The negative differential resistance effect occurring due to volatile resistive switching is used within the PTM, which, when coupled in series connection to the source or drain or gate of the baseline device in PTFETs, decreases the leakage off-state current and achieves ultra-low steep switching. In 2015, Shukla et al. [1] were the first to introduce the unique concept of PTFET. VO_2 was used as a PTM in the PTFET. By connecting the VO_2 material-based device [i.e. PTM] to the source contact of the FET, the PTFET is designed and implemented (as shown in Figure 5.4(a)). The channel resistance is in an insulating state (high) when the PTFET is switched off because the external gate voltage is not sufficient to induce MIT in PTM. The effective V_{GS} and V_{DS} of the FET are both reduced by this PTM device. The PTFET channel starts conducting once the gate biasing is increased, and this results in a decrease in the resistance. As a result, the PTM device's externally applied voltage is raised. When the external supply voltage

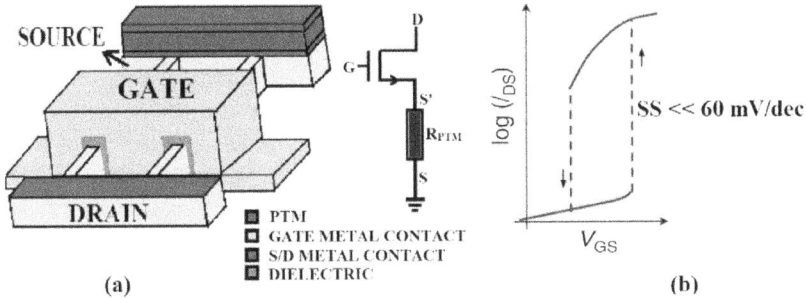

FIGURE 5.4 Source-connected PTFET (a) device and its circuit equivalent with FinFET as baseline transistor and (b) its generalized I_D -V_{GS} characteristics showing steep switching and improved ON-to-OFF current ratio

to the PTM device exceeds the threshold potential (minimum voltage to turn on the PTM), the resistance of the PTM drops quickly, causing the applied drain bias voltage in the PTFET channel to drop. The off-state leakage current is reduced as a result of the aforementioned operations, but the on-state drive current is maintained. Because of the abrupt switching in VO_2, steep switching characteristics can be acquired. Figure 5.4(b) illustrates this concept. Shukla et al. show that the on-to-off current ratio of n-type and p-type PTFET transistors is increased by 20% and 60%, respectively, in their work with comparison to baseline FET. In 2016, Frougier et al. [9] introduced monolithically integrated PTFET using VO_2 after the emergence of PTFET with fabrication steps. VO_2 was formed on the baseline device's source contact, utilizing DC sputtering in this study. At 300 K, the on-to-off current ratio was enhanced by 36%, and the SS was reduced to 8 mV per decade. In 2017, Aziz et al. [64, 65] investigated VO_2-based PTF in FET for low-power devices using the SPICE model. Due to reliable resistive switching at near room temperature ($\approx 340K$) and low ON resistance of VO_2 material, it has been used in PTFET. But this room temperature-resistive switching in VO_2 restricts its uses in higher temperature products. Also, in comparison to the filament-based PTM device, VO_2 has a lower resistance ratio, which results in the search for novel materials to resolve the technological challenge.

Resistive switches based on filament physical processes like CBRAM [66] are also used as PTM by modulating the current flow by varying its compliance current. The weak/strong formation of filament at low/high compliance current (around 10–100 μA) achieves the threshold-switching characteristics to be implemented. In 2016, TiO_2-based PTM was used in designing of a phase FET device by Song et al. [8], which achieved SS (around 10mV/decade) at low V_{DD}= 0.25V and reduced the OFF current (less than 1pA). However, it was only used as a PTM when the compliance current was less than 10 μA and hence limits the ON current. Furthermore, the PTM device requires an inherent delay (1 μsec) during switching. In the same year, Lim et al. [67] studied the developed PTFETs by integrating CuSx-based PTM and Si-H-based PTM in series connection to the baseline transistor's drain contact. At around 10 μA compliance current, the PTM can be turned on at low VDD ($\approx 0.25V$) and display ≈ 6 orders of resistive switching.

Jeong et al. [68] demonstrated atomic threshold-switching FET, which utilized AgTi/HfO$_2$-based PTM in series connection to the drain contact of 2D-MoS$_2$ baseline FET. Song et al. [69] utilized the Ag-SiTe as PTM and discussed the annealing effect. In 2016, Shukla et al. proposed the PTFET with a HfO$_2$-based PTM [6]. At a 100 μA compliance current, the HfO$_2$-based PTM acts as a threshold selector. They achieved a threshold voltage of ≈1.5V and a low off-state leakage current of ≈10pA. Furthermore, TS has a 58 ns turn-on time and a 67 ns turn-off time. The HfO$_2$-based PT-FET exhibits a ≈50% improvement in the on-to-off current ratio and higher thermal stability (≈90°C). Later, in 2017, by attaching the PTM in series connection to the gate contact of the conventional transistor, Park et al. proposed NbO$_2$-based PTFET [70]. Despite the fact that the on-to-off current ratio does not improve noticeably, the PTM device has been able to achieve high off-state leakage current (≈1 μA), with steep switch characteristics. In addition, the PTM requires only 10 ns for recovery (from a low to high resistance) and has no current flow restriction. However, turning on the PTM device necessitates additional resistance, resulting in an area penalty issue in the arrangement. Lee et al. discussed the three different PTM threshold switching in steep switching [71]. Oh et al. [72] demonstrated the role of the AgSe electrode and bipolar pulse forming in PTM to achieve steep switching.

The PTM device in PTFET can be linked to any of the three contacts (i.e. source, drain, or gate) of the conventional transistor in a series connection to use the negative differential resistance effect produced by PTM. As shown in Figure 5.4, the source-connected PTFET (referred to as S-PTFET) lowers both voltages (drain-to-source and gate-to-source). On the other hand, PTM connected to the drain of MOSFET (referred to as D-PTFET) solely lowers the drain-to-source voltage. Similarly, PTM connected in gate contact (referred to as G-PTFET) merely lowers the gate-to-source voltage. S-PTFET achieves a better reduction in the OFF current with the drawback of a slight reduction of the ON current of PTFET. The off-state leakage current and on-state current are reduced the least by the D-PTFET. The G-PTFET combines the best features of both D-PTFET and S PTFETs. It does, however, require an additional resistor. In 2017, Vitale et al. [73] looked at VO$_2$-based S-PTTFET and G-PTTFET with TFET as a conventional transistor. However, due to the high off-state leakage current, VO$_2$ is unable to display all advantages and disadvantages of the suggested device architectures. Also, a ferroelectric tunnel junction (FTJ) [74] can behave as PTM in phase FET. Shin et al. [75] also show Pb(Zr$_{0.52}$Ti$_{0.48}$)O$_3$-based PTFET. The ON current and inherent switching delay of PTFET need to be properly explored for efficient functioning. Various filament-based PTM-based PTFET devices have demonstrated a slow switching time (0.1–1 μs) and a low on-state current (10-100 μA). Recently, some hybrid devices using PTFET have been suggested to provide ultra-low SS. In 2022, Yadav et al. [76, 77] utilized negative capacitance in gate and PTM in the source of the conventional FinFET and reported SS of about 4mV/dec. Similarly, Vitale et al. [73] used PTM with TFET to achieve ultra-steep switching.

5.3.3 Digital logics/circuits

As shown in Figure 5.5(a), a design by Aziz et al. using a 14 nm FinFET transistor as the conventional technology and monolithically connecting PTM in the source

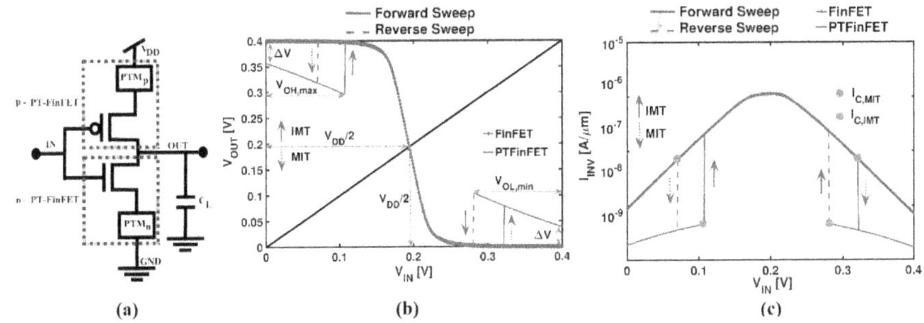

FIGURE 5.5 (a) PTFinFET-based inverter schematic, (b) its voltage-transfer-characteristic, and (c) current characteristic of PTFET inverter at 14 nm FinFET technology node with VO_2 as PTM

CMOS-based logic utilized n-PT-FinFET and p-PT-FinFET. [65]. The authors showed that a well-designed PTFinFET-based logic achieved performance benefits in power dissipation and speed in comparison to baseline FinFET for lower supply voltages. Also, being an exploratory device, a proper device-circuit co-design framework is needed in PTFET-based circuits.

In 2021, Cheng et al. [78] designed a memristive hybrid inverter utilizing a threshold switch and achieves sub-pW-leakage and hysteresis-free CMOS circuits. Yadav et al [79] discussed circuit advantages and drawbacks of hybrid negative capacitance-based phase FET. Despite having several benefits at the device level, PTFET appears to pose a limitation when designing logic circuits, necessitating additional care [80–82]. Firstly, because of the various PTM thresholds, the PTFET has a hysteresis feature in its device characteristics (Figure 5.4(b)).

Hysteresis imposes some restrictions on PTFET when designing PTFET-based logics, as it propagates from device to circuit, resulting in an unusual VTC of the inverter (Figure 5.5(b), (c)). However, Aziz et al. [65] showed with proper selection and tuning of both PTM (PTM_n and PTM_p) functional logic can be achieved. Secondly, due to PTM's finite metallic resistance, it has the tendency to lower the ON current of PTFET [77, 79]. This has an impact on the PTFET logic's speed (delay) performance. Thirdly, PTFET logic, as shown in Figure 5.5(b), demonstrates that the high insulating resistance of PTM typically decreases static output voltage (logic 1 and 0), which should be VDD and GND as in the case of a conventional CMOS inverter. This leads to signal degradation and power penalties during low input switching frequencies, as opposed to high input switching frequencies [80–82]. The output voltage decreases even further as logic gates are cascaded. In addition to power advantages, PTFET has delay instabilities for different input switching frequencies compared to FET. Additionally, the RO energy-delay product varies for various supply voltages in PTFET. These drawbacks appear to be problematic when building PTFET logics and should be given more attention in the future.

5.3.4 Memory Devices

The inherent hysteresis, abrupt switching, and memristive device-type behavior in PTM find their applications in state-of-the-art memory devices like STT-MRAM, SRAMs, cross-point array memory, and DRAMs. Therefore, PTM has been explored by various researchers for low-power memory applications. In order to improve performance, in 2015, Aziz et al. [83] used VO_2 as PTM in parallel connection with the MTJ device in the read path. This increased the cell tunneling magnetoresistance and read stability in multi-port MRAM. In the same year in another study, Aziz et al. [84] improved STT-MRAMs by utilizing parallelly connected optimized MTJ and PTM and achieved better stability and read efficiency. In 2016, Srinivasa et al. [85] propose an SRAM designed with PTM films to achieve lower power dissipation, higher write ability and read stability. Cha et al. [86] used NbO_2 (PTM) as a selector in a mushroom device structure and studied its scaling effects in cross-point array memory applications. They concluded that filamentary conducting paths created during the forming process have a significant impact on IMT behavior by analyzing the scaling trend of the threshold current. By reducing the conducting path inside the NbO_2 layer, the findings hold the promise of improving the performance of the selector device. In 2018, Aziz et al. [87] again designed spin-transfer torque (STT) MRAM non-volatile memory with a threshold switch to enhance the read operation and discussed its design space using a device-circuit co-design framework. In 2019, Shen et al. [88] proposed compact PTM-assisted single-ended 7T-SRAM, 8T-differential-SRAM, and 2T-DRAM with separate read-write ports and achieved performance improvements. In 2021, Nibhanupudi et al. [89], designed a heterogeneous 6T-SRAM bit cell utilizing PTM in series connection with the gate of the cross-coupled pull-down cell and achieved decreased read access time, lower power dissipation, with little increment in write time and improved retention stability compared to standard SRAM cell. In the same year, Raman et al. [90] used a bipolar threshold selector and capacitive-coupled assisted method in FeFET memory to lower the write voltage. In 2022, Ambrosi et al. [91] used SiNGeCTe, an arsenic-free chalcogenide material, for low voltage selector applications in a cross-point memory architecture based on a two-terminal 1T1R memory cell for high-density and 3D compatible embedded memory and studied its reliability.

5.3.5 Non-Boolean Computing Architectures

The Von Neumann architecture has been the foundation for computing and information processing for the last few decades. However, as AI and machine learning applications grow, there are some computationally challenging issues, such as associative processing (for example, computer vision) and combinatorial optimization where the traditional paradigm is fundamentally inadequate because memory and computation are done separately. Neuromorphic computing is a recent and active research areas for low-power computing architecture that makes use of spiking neural networks (SNN) to address the above problem of memory and computing separation. By using dynamic systems, e.g. coupled oscillators in forming analog co-processor systems, synchronized dynamics with inherent parallelism can be incorporated into these systems, which improves upon the traditional CMOS microprocessors. The instability of abrupt switching in PTM like VO_2 during insulating and metallic states is used to form coupled oscillators. Negative feedback can be achieved by taking advantage of the instability through a series connection of MOSFET and PTM, and a low-power relaxation oscillator [92] can be formed. The computational fabric is based on the synchronization dynamics of the oscillators, and a capacitive-coupling scheme is employed to allow the exchange of reactive power among the oscillators while preventing them from interfering with each other's quiescent point. These capacitive-coupled VO_2 oscillators provide an experimental test bed for tackling difficult computational problems because of the dynamics of their phase synchronization. Shukla et al. [93] demonstrated these PTM-based oscillators for solving computational problems in saliency detection.

PTM's inherent stochastic property is useful for creating SNN neurons that are fast and low-power efficient for neuromorphic computing. Transition metal oxides have proven to be potential candidates due to the occurrence of MIT for making low-cost and energy-efficient SNN. In 2017, Parihar et al. [94] showed the VO_2 as a neuron and used its stochastic nature in making a biomimetic computational kernel that can be used for solving optimization and ML problems. In 2021, Zhang et al. [95] varied the oxygen concentration in $La_{1-x}Sr_xCoO_3$ using density functional theory to determine the bias voltage for transition and suggested methods to reduce bias for transition which can be useful for optimizing and designing neurons in neuromorphic computing. Carapezzi et al. [96–98] used TCAD simulation to model a thermally induced transition in VO_2 and used mixed mode TCAD and SPICE simulation to study the VO_2 dynamics in oscillatory behavior for neuromorphic applications. In 2020, Moatti et al. [99] uses VO_2 as Mott memory and studied the volatile and non-volatile behavior of MIT in VO_2 by tuning oxygen vacancies to provide a path for neuromorphic applications.

5.3.6 OTHER APPLICATIONS

Recently, the unique properties of abrupt switching and stochastic nature in PTM have been explored in various analog circuits to solve supply voltage droop problems [100] in making novel multipliers [101], a pseudo-random number generator [102], and a power-efficient design of sense amplifier [103].

5.4 CONCLUSION

The abrupt, volatile, non-volatile, ultrafast electrical switching, and oscillatory behavior, or stochastic nature, in phase transition material due to MIT in metal oxides and some metamaterial make it a useful class of material for low-power electronics. For a few decades, the special focus has been on VO_2 as PTM due to its transition near room temperature and its fabrication easiness with the existing CMOS process. Recently, many kinds of PTM have been explored as two-terminal threshold selectors in steep switching, logic, memory, and neuromorphic applications. Many TCAD and SPICE models have been developed to evaluate their efficacy in a device (phase FET/hyper FET, Mott-FET) and circuits. PhaseFET achieves steep switching and achieves power-efficient switches in low-power applications. Also, hysteresis in PTM finds applications in SRAMs, DRAMs, MRAMs, and crossbar memories.

Stochastic and oscillatory behavior finds its application in neuromorphic computing to form nano-oscillator and help in solving computationally hard problems.

5.5 FUTURE OUTLOOK

Being an exploratory emerging material and emerging device, there is a need for a device-circuit co-design methodology for the designing of devices and circuits. Also, there is a need to find a mechanism to reduce hysteresis in PTM for digital circuits. However, for memory applications, a hysteresis window should be increased for advantages. The fabrication complexity of the existing CMOS process needs to be researched in the future, along with study of the reliability and endurance of PTM and devices associated with PTM. Debates about the physics and mechanisms of transition (E-IMT, T-IMT) in PTM hinder the development of proper TCAD and compact and SPICE models for device and circuit simulation. These new exploratory devices have distinct potential and constraints that must be thoroughly investigated before moving forward with commercialization. For each of these ideas and methodologies, a comprehensive device-circuit co-design needs to be carried out to assess the potential ramifications and feasibility.

REFERENCES

1. Shukla, N., Thathachary, A.V., Agrawal, A., Paik, H., Aziz, A., Schlom, D.G., Gupta, S.K., Engel-Herbert, R., & Datta, S. (2015). A steep-slope transistor based on abrupt electronic phase transition. *Nature Communications, 6,* 7812.
2. Si, M., Jiang, C., Su, C.J., Tang, Y., Yang, L., Chung, W., Alam, M.A., & Ye, P.D. (2017). Sub-60 mV/dec ferroelectric HZO MoS2 negative capacitance field-effect transistor with internal metal gate: The role of parasitic capacitance. *2017 IEEE International Electron Devices Meeting (IEDM),* 23.5.1-23.5.4.
3. Khatami, Y., & Banerjee, K. (2009). Steep subthreshold slope n- and p-type tunnel-FET devices for low-power and energy-efficient digital circuits. *IEEE Transactions on Electron Devices, 56*(11), 2752–2761.
4. Ionescu, A.M., & Riel, H.E. (2011). Tunnel field-effect transistors as energy-efficient electronic switches. *Nature, 479*(7373), 329–337.
5. Salahuddin, S.S., & Datta, S. (2008). Use of negative capacitance to provide voltage amplification for low power nanoscale devices. *Nano Letters, 8*(2), 405–410.
6. Shukla, N., Grisafe, B., Ghosh, R.K., Jao, N., Aziz, A., Frougier, J., Jerry, M., Sonde, S., Rouvimov, S., Orlova, T.A., Gupta, S.K., & Datta, S. (2016). Ag/HfO2 based threshold switch with extreme non-linearity for unipolar cross-point memory and steep-slope phase-FETs. *2016 IEEE International Electron Devices Meeting (IEDM),* 34.6.1-34.6.4.
7. Verma, A., Song, B., Meyer, D.A., Downey, B.P., Wheeler, V.D., Xing, H.G., & Jena, D. (2016). Demonstration of GaN HyperFETs with ALD VO 2. *74th Annual Device Research Conference (DRC),* 1–2.
8. Song, J., Woo, J., Lee, S., Prakash, A., Yoo, J., Moon, K., & Hwang, H. (2016). Steep slope field-effect transistors with Ag/TiO2-based threshold switching device. *IEEE Electron Device Letters, 37*(7), 932–934.
9. Frougier, J., Shukla, N., Deng, D., Jerry, M., Aziz, A., Liu, L.N., Lavallee, G.P., Mayer, T.S., Gupta, S.K., & Datta, S. (2016). Phase-Transition-FET exhibiting steep switching slope of 8mV/decade and 36% enhanced ON current. *2016 IEEE Symposium on VLSI Technology,* 1–2.

10. Scherwitzl, R., Zubko, P., Lezama, I.G., Ono, S., Morpurgo, A.F., Catalán, G., & Triscone, J. (2010). Electric-field control of the metal-insulator transition in ultrathin NdNiO3 films. *Advanced Materials*, *22*(48), 5517–5520.

11. Zhou, Y., Chen, X., Ko, C., Yang, Z., Mouli, C., & Ramanathan, S. (2013). Voltage-triggered ultrafast phase transition in vanadium dioxide switches. *IEEE Electron Device Letters*, *34*(2), 220–222.

12. Adda, C., Lee, M., Kalcheim, Y., Salev, P., Rocco, R., Vargas, N.M., Ghazikhanian, N., Li, C., Albright, G., Rozenberg, M.J., & Schuller, I.K. (2022). Direct observation of the electrically triggered insulator-metal transition in V3O5 far below the transition temperature. *Physical Review X, 12(1),* 011025.

13. Zhang, Y., & Ramanathan, S. (2011). Analysis of "on" and "off" times for thermally driven VO$_2$ metal-insulator transition nanoscale switching devices. *Solid-State Electronics*, *62*(1), 161–164.

14. Devidas, T.R., Chandra Shekar, N.V., Sundar, C.S., Chithaiah, P., Sorb, Y.A., Bhadram, V.S., Chandrabhas, N., Pal, K., Waghmare, U.V., & Rao, C.N. (2014). Pressure-induced structural changes and insulator-metal transition in layered bismuth triiodide, BiI3: A combined experimental and theoretical study. *Journal of Physics – Condensed Matter*, *26*(27), 275502.

15. Wu, J., Gu, Q., Guiton, B.S., de Leon, N.P., Ouyang, L., & Park, H. (2006). Strain-induced self organization of metal-insulator domains in single-crystalline VO$_2$ nanobeams. *Nano Letters*, *6*(10), 2313–2317.

16. Beteille, F., & Livage, J. (1998). Optical switching in VO$_2$ thin films. *Journal of Sol-Gel Science and Technology*, *13*(1/3), 915–921.

17. Sakuma, R., Miyake, T., & Aryasetiawan, F. (2008). First-principles study of correlation effects in VO$_2$. *Physical Review. Part B*, *78*(7), 075106.

18. Stefanovich, G.B., Pergament, A., & Stefanovich, D.G. (2000). Electrical switching and Mott transition in VO$_2$. *Journal of Physics: Condensed Matter*, *12*(41), 8837–8845.

19. Luo, Q., Xu, X., Liu, H., Lv, H., Gong, T., Long, S., Liu, Q., Sun, H., Banerjee, W., Li, L., Lu, N., & Liu, M. (2015). Cu BEOL compatible selector with high selectivity (>107), extremely low off-current (~pA) and high endurance (>1010). *2015 IEEE International Electron Devices Meeting (IEDM)*, 10.4.1-10.4.4.

20. Streltsov, S.V., & Khomskii, D.I. (2014). Orbital-dependent singlet dimers and orbital-selective Peierls transitions in transition-metal compounds. *Physical Review. Part B*, *89*(16), 161112.

21. Yang, Z., Ko, C., & Ramanathan, S. (2011). Oxide electronics utilizing ultrafast metal-insulator transitions. *Annual Review of Materials Research*, *41*(1), 337–367.

22. Mott, N.F. (1949). The basis of the electron theory of metals, with special reference to the transition metals. *Proceedings of the Physical Society. Section A,* 62(7), 416.

23. Peierls, R.E., & Roberts, L.D. (1956). Quantum theory of solids. *Physics Today*, *9*(5), 29–29.

24. Anderson, P.W. (1958). Absence of diffusion in certain random lattices. *Physical Review*, *109*(5), 1492–1505.

25. Westman, S., Lindqvist, I., Sparrman, B., Nielsen, G., Nord, H., & Jart, A. (1961). Note on a phase transition in VO$_2$. *Acta Chemica Scandinavica*, *15*, 217–217.

26. Zylbersztejn, A., & Mott, N.F. (1975). Metal-insulator transition in vanadium dioxide. *Physical Review. Part B*, *11*(11), 4383–4395.

27. Wentzcovitch, R.M., Schulz, W.W., & Allen, P.B. (1994). VO$_2$: Peierls or Mott-Hubbard? A view from band theory. *Physical Review Letters*, *72*(21), 3389–3392.

28. Rice, T.M., Launois, H., & Pouget, J.P. (1994). Comment on "VO2: Peierls or Mott-Hubbard? A view from band theory". *Physical Review Letters*, *73*(22), 3042.

29. Cavalleri, A., Dekorsy, T., Chong, H., Kieffer, J., & Schoenlein, R.W. (2004). Evidence for a structurally-driven insulator-to-metal transition in VO 2: A view from the ultrafast timescale. *Physical Review. Part B, 70*(16), 161102.
30. Kim, H., Lee, Y.W., Kim, B.J., Chae, B.G., Yun, S.J., Kang, K., Han, K., Yee, K.J., & Lim, Y. (2006). Monoclinic and correlated metal phase in VO(2) as evidence of the Mott transition: Coherent phonon analysis. *Physical Review Letters, 97*(26), 266401.
31. Ruzmetov, D.A., Narayanamurti, V., Ramanathan, S., & Senanayake, S.D. (2008). Correlation between metal-insulator transition characteristics and electronic structure changes in vanadium oxide thin films. *Bulletin of the American Physical Society, 77*(19), 195442.
32. Ruzmetov, D.A., Gopalakrishnan, G., Deng, J., Narayanamurti, V., & Ramanathan, S. (2009). Electrical triggering of metal-insulator transition in nanoscale vanadium oxide junctions. *Journal of Applied Physics, 106*(8), 083702.
33. Yang, Z., Hart, S., Ko, C., Yacoby, A., & Ramanathan, S. (2011). Studies on electric triggering of the metal-insulator transition in VO 2 thin films between 77 K and 300 K. *Journal of Applied Physics, 110*(3), 033725.
34. Lin, J., Alam, K., Ocola, L.E., Zhang, Z., Datta, S., Ramanathan, S., & Guha, S. (2017). Physics and technology of electronic insulator-to-metal transition (E-IMT) for record high on/off ratio and low voltage in device applications. *2017 IEEE International Electron Devices Meeting (IEDM)*, 23.4.1-23.4.4.
35. Gopalakrishnan, G., Ruzmetov, D.A., & Ramanathan, S. (2009). On the triggering mechanism for the metal–insulator transition in thin film VO$_2$ devices: Electric field versus thermal effects. *Journal of Materials Science, 44*(19), 5345–5353.
36. Lin, J., Ramanathan, S., & Guha, S. (2018). Electrically driven insulator–metal transition-based devices—Part I: The electrothermal model and experimental analysis for the DC characteristics. *IEEE Transactions on Electron Devices, 65*(9), 3982–3988.
37. Lin, J., Ramanathan, S., & Guha, S. (2018). Electrically driven insulator–metal transition-based devices—Part II: Transient characteristics. *IEEE Transactions on Electron Devices, 65*(9), 3989–3995.
38. Dasgupta, A., Verma, A., & Chauhan, Y.S. (2019). Analysis and compact modeling of insulator–metal transition material-based PhaseFET including hysteresis and multidomain switching. *IEEE Transactions on Electron Devices, 66*(1), 169–176.
39. Amer, S., Hasan, M.S., Adnan, M.M., & Rose, G.S. (2019). SPICE modeling of insulator metal transition: Model of the critical temperature. *IEEE Journal of the Electron Devices Society, 7*, 18–25.
40. O'Hara, A., & Demkov, A.A. (2015). Nature of the metal-insulator transition in NbO2. *Physical Review. Part B, 91*(9), 094305.
41. Hansmann, P., Toschi, A., Sangiovanni, G., Saha-Dasgupta, T., Lupi, S., Marsi, M., & Held, K. (2013). Mott–Hubbard transition in V2O3 revisited. *Physica Status Solidi (B), 250*(7), 1251–1264.
42. Wang, Y., Liu, Q., Long, S., Wang, W., Wang, Q., Zhang, M., Zhang, S., Li, Y., Zuo, Q., Yang, J., & Liu, M. (2010). Investigation of resistive switching in Cu-doped HfO2 thin film for multilevel non-volatile memory applications. *Nanotechnology, 21*(4), 045202.
43. Shukla, N., Joshi, T., Dasgupta, S., Borisov, P., Lederman, D., & Datta, S. (2014). Electrically induced insulator to metal transition in epitaxial SmNiO3 thin films. *Applied Physics Letters, 105*(1), 012108.
44. Cao, J.J., Ertekin, E., Srinivasan, V., Fan, W., Huang, S., Zheng, H., Yim, J.W., Khanal, D.R., Ogletree, D.F., Grossman, J.C., & Wu, J. (2009). Strain engineering and one-dimensional organization of metal-insulator domains in single-crystal vanadium dioxide beams. *Nature Nanotechnology, 4*(11), 732–737.

45. Morosan, E., Natelson, D., Nevidomskyy, A.H., & Si, Q. (2012). Strongly correlated materials. *Advanced Materials*, *24*(36), 4896–4923. https://doi.org/10.1002/adma.201202018.

46. Basov, D., Averitt, R.D., Marel, D.V., Dressel, M., & Haule, K. (2011). Electrodynamics of correlated electron materials. *Reviews of Modern Physics*, *83*(2), 471–541.

47. Taketa, Y., Kato, F., Nitta, M., & Haradome, M. (1975). New oscillation phenomena in VO_2 crystals. *Applied Physics Letters*, *27*(4), 212–214.

48. Fisher, B. (1978). Voltage oscillations in switching VO_2 needles. *Journal of Applied Physics*, *49*(10), 5339–5341.

49. Pashkin, A., Kubler, C., Ehrke, H., Lopez, R., Halabica, A., Haglund, R.F., Jr., Huber, R., & Leitenstorfer, A. (2011). Ultrafast insulator-metal phase transition in VO_2 studied by multiterahertz spectroscopy. *Physical Review. Part B*, *83*(19), 195120.

50. Jerry, M., Shukla, N., Paik, H., Schlom, D.G., & Datta, S. (2016). Dynamics of electrically driven sub-nanosecond switching in vanadium dioxide. *2016 IEEE Silicon Nanoelectronics Workshop (SNW)*, 26–27.

51. Markov, P., Marvel, R.E., Conley, H.J., Miller, K.J., Haglund, R.F., & Weiss, S.M. (2015). Optically monitored electrical switching in VO_2. *ACS Photonics*, *2*(8), 1175–1182.

52. Duchêne, J., Terraillon, M., Pailly, P., & Adam, G.B. (1971). Filamentary conduction in VO_2 coplanar thin-film devices. *Applied Physics Letters*, *19*(4), 115–117.

53. Kim, J., Ko, C., Frenzel, A.J., Ramanathan, S., & Hoffman, J.E. (2010). Nanoscale imaging and control of resistance switching in VO_2 at room temperature. *Applied Physics Letters*, *96*(21), 213106.

54. Ko, C., & Ramanathan, S. (2008). Observation of electric field-assisted phase transition in thin film vanadium oxide in a metal-oxide-semiconductor device geometry. *Applied Physics Letters*, *93*(25), 252101.

55. Chae, B.G., Kim, H., Youn, D., & Kang, K. (2005). Abrupt metal–insulator transition observed in VO_2 thin films induced by a switching voltage pulse. *Physica. Part B*, *369*(1–4), 76–80.

56. Boriskov, P.P., Velichko, A., Pergament, A., Stefanovich, G.B., & Stefanovich, D.G. (2002). The effect of electric field on metal-insulator phase transition in vanadium dioxide. *Technical Physics Letters*, *28*(5), 406–408.

57. Kim, H., Chae, B.G., Youn, D., Maeng, S., Kim, G., Kang, K., & Lim, Y. (2004). Mechanism and observation of Mott transition in VO_2-based two- and three-terminal devices. *New Journal of Physics*, *6*, 52–52.

58. Kim, B.J., Lee, Y.W., Choi, S., Yun, S.J., & Kim, H. (2010). VO_2 thin-film varistor based on metal-insulator transition. *IEEE Electron Device Letters*, *31*, 14–16.

59. Ruzmetov, D.A., Gopalakrishnan, G., Ko, C., Narayanamurti, V., & Ramanathan, S. (2010). Three-terminal field effect devices utilizing thin film vanadium oxide as the channel layer. *Journal of Applied Physics*, *107*(11), 114516.

60. Hormoz, S., & Ramanathan, S. (2010). Limits on vanadium oxide Mott metal–insulator transition field-effect transistors. *Solid-State Electronics*, *54*(6), 654–659.

61. Zhou, C., Newns, D.M., Misewich, J.A., & Pattnaik, P. (1996). A field effect transistor based on the Mott transition in a molecular layer. *Applied Physics Letters*, *70*(5), 598–600.

62. Newns, D.M., Misewich, J.A., Tsuei, C.C., Gupta, A., Scott, B.A., & Schrott, A.G. (1998). Mott transition field effect transistor. *Applied Physics Letters*, *73*(6), 780–782.

63. Theis, T.N., & Solomon, P.M. (2010). It's time to reinvent the transistor! *Science*, *327*(5973), 1600–1601.

64. Aziz, A., Shukla, N., Datta, S., & Gupta, S.K. (2017). Steep switching hybrid phase transition FETs (hyper-FET) for low power applications: A device-circuit co-design perspective–part I. *IEEE Transactions on Electron Devices*, *64*(3), 1350–1357.

65. Aziz, A., Shukla, N., Datta, S., & Gupta, S.K. (2017). Steep switching hybrid phase transition FETs (hyper-FET) for low power applications: A device-circuit co-design perspective—Part II. *IEEE Transactions on Electron Devices, 64*(3), 1358–1365.

66. Sun, H., Liu, Q., Li, C., Long, S., Lv, H., Bi, C., Huo, Z., Li, L., & Liu, M. (2014). Direct observation of conversion between threshold switching and memory switching induced by conductive filament morphology. *Advanced Functional Materials, 24*(36), 5679–5686.

67. Lim, S., Yoo, J., Song, J., Woo, J., Park, J., & Hwang, H. (2016). Excellent threshold switching device (Ioff ~ 1 pA) with atom-scale metal filament for steep slope (< 5 mV/dec), ultra low voltage (Vdd = 0.25 V) FET applications. *2016 IEEE International Electron Devices Meeting (IEDM)*, 34.7.1-37.7.4.

68. Jeong, S., Han, S., Lee, H., Eom, D., Youm, G., Choi, Y., Moon, S., Ahn, K.J., Oh, J., & Shin, C. (2021). Abruptly-switching MoS2-channel atomic-threshold-switching field-effect transistor with AgTi/HfO2-based threshold switching device. *IEEE Access, 9*, 116953–116961.

69. Song, B., Xu, H., Liu, S., Liu, H., & Li, Q. (2018). Threshold switching behavior of Ag-SiTe-based selector device and annealing effect on its characteristics. *IEEE Journal of the Electron Devices Society, 6*, 674–679.

70. Park, J., Lee, D., Yoo, J., & Hwang, H. (2017). NbO2 based threshold switch device with high operating temperature (>85°C) for steep-slope MOSFET (~2mV/dec) with ultra-low voltage operation and improved delay time. *2017 IEEE International Electron Devices Meeting (IEDM)*, 23.7.1-23.7.4.

71. Lee, S., Yoo, J., Park, J., & Hwang, H. (2020). Understanding of the abrupt resistive transition in different types of threshold switching devices from materials perspective. *IEEE Transactions on Electron Devices, 67*(7), 2878–2883.

72. Oh, S., Lee, S., & Hwang, H. (2021). Improved turn-off speed and uniformity of atomic threshold switch device by AgSe electrode and bipolar pulse forming. *IEEE Journal of the Electron Devices Society, 9*, 864–867.

73. Vitale, W.A., Casu, E.A., Biswas, A.K., Rosca, T., Alper, C., Krammer, A., Luong, G.V., Zhao, Q., Mantl, S., Schüler, A., & Ionescu, A.M. (2017). A steep-slope transistor combining phase-change and band-to-band-tunneling to achieve a sub-unity body factor. *Scientific Reports, 7*(1), 355.

74. Yoon, C., Lee, J.H., Lee, S., Jeon, J.H., Jang, J.T., Kim, D.H., Kim, Y.H., & Park, B.H. (2017). Synaptic plasticity selectively activated by polarization-dependent energy-efficient ion migration in an ultrathin ferroelectric tunnel junction. *Nano Letters, 17*(3), 1949–1955.

75. Shin, J.S., Ko, E., & Shin, C. (2018). Analysis on the operation of negative differential resistance FinFET with Pb(Zr$_{0.52}$Ti$_{0.48}$)O$_3$ threshold selector. *IEEE Transactions on Electron Devices, 65*(1), 19–22.

76. Yadav, S., Upadhyay, P., Awadhiya, B., & Kondekar, P.N. (2022). Ferroelectric negative-capacitance-assisted phase-transition field-effect transistor. *IEEE Transactions on Ultrasonics, Ferroelectrics, and Frequency Control, 69*(2), 863–869.

77. Yadav, S., Upadhyay, P., Awadhiya, B., & Kondekar, P.N. (2021). Design and analysis of improved phase-transition FinFET utilizing negative capacitance. *IEEE Transactions on Electron Devices, 68*(2), 853–859.

78. Cheng, B., Emboras, A., Passerini, E., Lewerenz, M., Koch, U., Wu, L., Liao, J., Ducry, F., Aeschlimann, J., Jang, T., Luisier, M., & Leuthold, J. (2021). Threshold switching enabled sub-pW-leakage, hysteresis-free circuits. *IEEE Transactions on Electron Devices, 68*(6), 3112–3118.

79. Yadav, S., Kondekar, P.N., Upadhyay, P., & Awadhiya, B. (2022). Negative capacitance based phase-transition FET for low power applications: Device-circuit co-design. *Microelectronics Journal, 123*, 105411.
80. Avedillo, M.J., & Núñez, J. (2017). Insights into the operation of hyper-FET-based circuits. *IEEE Transactions on Electron Devices, 64*(9), 3912–3918.
81. Jiménèz, M., Núñez, J., & Avedillo, M.J. (2020). Hybrid-phase-transition FET devices for logic computation. *IEEE Journal on Exploratory Solid-State Computational Devices and Circuits, 6*(1), 1–8.
82. Núñez, J., & Avedillo, M.J. (2019). Power and speed evaluation of hyper-FET circuits. *IEEE Access, 7*, 6724–6732.
83. Aziz, A., Shukla, N., Datta, S., & Gupta, S.K. (2015). Read optimized MRAM with separate read-write paths based on concerted operation of magnetic tunnel junction with correlated material. *2015 73rd Annual Device Research Conference (DRC)*, 43–44.
84. Aziz, A., Shukla, N., Datta, S., & Gupta, S.K. (2015). COAST: Correlated material assisted STT MRAMs for optimized read operation. *2015 IEEE/ACM International Symposium on Low Power Electronics and Design (ISLPED)*, 1–6.
85. Srinivasa, S.R., Aziz, A., Shukla, N., Li, X., Sampson, J., Datta, S., Kulkarni, J.P., Narayanan, V., & Gupta, S.K. (2016). Correlated material enhanced SRAMs with robust low power operation. *IEEE Transactions on Electron Devices, 63*(12), 4744–4752.
86. Cha, E., Park, J., Woo, J., Lee, D., Prakash, A., & Hwang, H. (2016). Comprehensive scaling study of NbO2 insulator-metal-transition selector for cross point array application. *Applied Physics Letters, 108*(15), 153502.
87. Aziz, A., & Gupta, S.K. (2018). Threshold switch augmented STT MRAM: Design space analysis and device-circuit co-design. *IEEE Transactions on Electron Devices, 65*(12), 5381–5389.
88. Shen, Z., Srinivasa, S.R., Aziz, A., Datta, S., Narayanan, V., & Gupta, S.K. (2019). SRAMs and DRAMs with separate read–write ports augmented by phase transition materials. *IEEE Transactions on Electron Devices, 66*(2), 929–937.
89. Nibhanupudi, S.S., Raman, S.R., & Kulkarni, J.P. (2021). Phase transition material-assisted low-power SRAM design. *IEEE Transactions on Electron Devices, 68*(5), 2281–2288.
90. Raman, S.R., Nibhanupudi, S.S., Saha, A.K., Gupta, S.K., & Kulkarni, J.P. (2021). Threshold selector and capacitive coupled assist techniques for write voltage reduction in metal-ferroelectric-metal field-effect transistor. *IEEE Transactions on Electron Devices, 68*(12), 6132–6138.
91. Ambrosi, E., Wu, C., Lee, H., Lee, C., Hsu, C., Chang, C., Lee, T., & Bao, X. (2022). Reliable low voltage selector device technology based on robust SiNGeCTe arsenic-free chalcogenide. *IEEE Electron Device Letters, 43*(10), 1673–1676.
92. Shukla, N., Parihar, A.S., Cotter, M., Barth, M., Li, X., Chandramoorthy, N., Paik, H., Schlom, D.G., Narayanan, V., Raychowdhury, A., & Datta, S. (2014). Pairwise coupled hybrid vanadium dioxide-MOSFET (HVFET) oscillators for non-boolean associative computing. *2014 IEEE International Electron Devices Meeting*, 28.7.1-28.7.4.
93. Shukla, N., Tsai, W., Jerry, M., Barth, M., Narayanan, V., & Datta, S. (2016). Ultra low power coupled oscillator arrays for computer vision applications. *2016 IEEE Symposium on VLSI Technology*, 1–2.
94. Parihar, A.S., Jerry, M., Datta, S., & Raychowdhury, A. (2017). Stochastic IMT (insulator-metal-transition) neurons: An interplay of thermal and threshold noise at bifurcation. *Frontiers in Neuroscience, 12*: 210.

95. Zhang, S., Vo, H., & Galli, G. (2021). Predicting the onset of metal–insulator transitions in transition metal oxides—A first step in designing neuromorphic devices. *Chemistry of Materials*, 33(9), 3187–3195.

96. Carapezzi, S., Delacour, C., & Todri-Sanial, A. (2022). Simulation toolchain for neuromorphic oscillatory neural networks based on beyond-CMOS vanadium dioxide devices. *2022 IEEE International Conference on Flexible and Printable Sensors and Systems (FLEPS)*, 1–4.

97. Carapezzi, S., Delacour, C., Boschetto, G., Corti, E., Abernot, M., Nejim, A., Gil, T., Karg, S.F., & Todri, A. (2021). Multi-scale modeling and simulation flow for oscillatory neural networks for edge computing. *2021 19th IEEE International New Circuits and Systems Conference (NEWCAS)*, 1–5.

98. Carapezzi, S., Boschetto, G., Delacour, C., Corti, E., Plews, A., Nejim, A., Karg, S.F., & Todri-Sanial, A. (2021). Advanced design methods from materials and devices to circuits for brain-inspired oscillatory neural networks for edge computing. *IEEE Journal on Emerging and Selected Topics in Circuits and Systems*, 11(4), 586–596.

99. Moatti, A., Sachan, R., & Narayan, J. (2020). Volatile and non-volatile behavior of metal–insulator transition in VO_2 through oxygen vacancies tunability for memory applications. *Journal of Applied Physics*, 128(4), 045302.

100. Teja, S., & Kulkarni, J.P. (2018). *Soft-FET: Phase Transition Material Assisted Soft Switching Field Effect Transistor for Supply Voltage Droop Mitigation*. DAC.

101. Aziz, A., Shukla, N., Seabaugh, A.C., Datta, S., & Gupta, S.K. (2018). Cockcroft-Walton Multiplier based on unipolar Ag/HfO2/Pt threshold Switch. *2018 76th Device Research Conference (DRC)*, 1–2.

102. Jerry, M., Ni, K., Parihar, A.S., Raychowdhury, A., & Datta, S. (2018). Stochastic insulator-to-metal phase transition-based true random number generator. *IEEE Electron Device Letters*, 39(1), 139–142.

103. Aziz, A., Li, X., Shukla, N., Datta, S., Chang, M., Narayanan, V., & Gupta, S.K. (2017). Low power current sense amplifier based on phase transition material. *2017 75th Annual Device Research Conference (DRC)*, 1–2.

6 Impact of total ionizing dose effect on SOI-FinFET with spacer engineering

*Abhishek Ray, Alok Naugarhiya,
and Guru Prasad Mishra*

CONTENTS

6.1 INTRODUCTION

The improvement in transistor performance has fueled continuous growth in the semiconductor industry, but as physical feature sizes have been scaled down, the emergence of short-channel effects (SCEs) causes a threat to the scalability of future devices [1]. Power consumption and heat dissipation in integrated circuits have become a serious challenge with a significant reduction in the channel length of the conventional metal oxide semiconductor field-effect transistor (MOSFETs). To overcome the scaling challenges of planar MOSFET, multigate FETs are proposed [2, 3]. The multigate devices have better controllability over the channel and maintain the electrostatic integrity of devices [4, 5]. In multigate devices, FinFET is the most popular one due to its simple structure and gate wrapped over the channel. Due to the 3-D and tri-gate structure, FinFET has better electrostatic integrity and good subthreshold region performance. For low-power space applications, most of the semiconductor industries have adopted 3-D FinFETs for integrated circuit (IC) manufacturing [6–11].

DOI: 10.1201/9781003359234-6

Types of Radiation Effect

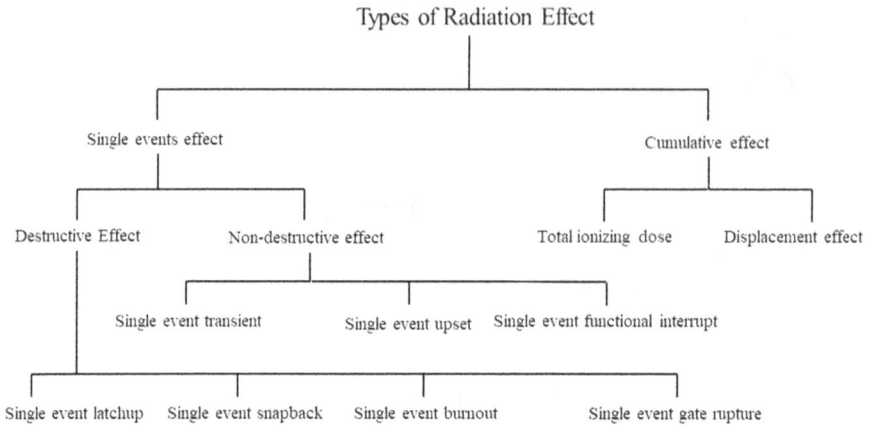

FIGURE 6.1   Classification of radiation effects [29–34]

Reliability and safety are the major concern for the electronic system design of satellites. The system performances are affected and degraded due to the presence of heavy ionized radiation in space [12, 13]. These radiation effects are categorized into two types: 1) single-event effects (SEEs) or soft error and 2) cumulative effects. Classification of the radiation effects is illustrated in Figure 6.1. In the 1960s, the radiation sensitivity of MOS devices was first discovered at a naval laboratory [14]. In the 1980s, single-event effects were analyzed for the first time on digital circuits [15, 16]. There are two types of single-event effects: destructive and non-destructive. Generally, these effects hamper the performance of digital circuits. The non-destructive type effects affect the performance of systems for a short span of time [17]. In low-power digital circuits, heavy ionized particles hamper the performance temporarily [18–21]. Many research groups are working to make a radiation-hardened SRAM cell for satellites [18–25]. The single-event transient (SET) and single-event upset (SEU) alter the storage bits after striking highly energized particles. These effects can be suppressed by adding an extra transistor/circuit element or making the device radiation-hardened by itself [26–28]. The device's radiation hardness is analyzed by the total ionizing dose effect (TID).

After irradiation of the semiconductor device, trap charges are accumulated in the oxides and semiconductor/insulator interfaces. These charges shift the threshold voltage toward negative and degrade the device's OFF-state performance [33, 35]. The radiation impact of FinFET also depends on fin geometry. The maximum TID degradation in bulk FinFET is observed for the narrow fin width (W_{FIN}) and short-channel length (L_g) [36–39]. The TID response of bulk FinFET is similar to planar MOSFETs. Radiation build-up trap charges in shallow trench isolation (STI) trigger lateral parasitic transistors that degrade the OFF-state current. The long channel bulk FinFET shows lower TID degradation, because of the existence of a weak parasitic conduction path STI to substrate [40–44], while SOI-FinFET shows better radiation tolerability for narrow fin width and short L_g [38, 45]. After this, the irradiation threshold voltage shift reduces by increasing the surrounding temperature [46]. The long channel with wider fin width devices are highly susceptible to ionized radiation. In 2020, a FinFET of a compound semiconductor

(InGaAs) fin with a modified gate stack showed a low subthreshold leakage current for 10 keV X-rays [39]. Zhexuan Ren et al. reported TID analysis of pMOS FinFET for different biases and geometry. In the worst-bias or ON-state condition, pMOS FinFET showed the maximum TID degradation [47]. The TID and low-frequency analysis, bulk and SOI n-FinFET for different fin widths have been reported. For the SOI devices, a minimal threshold voltage shift is noticed after a 2-Mrad dose [38]. Stefano Bonaldo et al. reported InGaAs FinFETs with gate stack (HfO_2/Al_2O_3 dielectrics) for low noise analysis of the device, up to a radiation dose of 500 krad [37]. For ultra-high radiation, 1 Grad dose 16-nm bulk n- and p-FinFETs are analyzed for the different channel lengths [36]. In 2022, Ray et al. reported TID analysis of optimized SOI n-FinFET for ultra-high radiation of 2 Mrad dose. The concept of workfunction modulation was incorporated with conventional FinFET for enhancing the pre- and post-radiation performances of devices [45]. Due to high radiation tolerability, narrow W_{FIN}, and short L_g, SOI-FinFET is widely used for space application [48].

This chapter presents a radiation-hardened SOI n-FinFET with spacer engineering. The pre- and post-radiation of different spacers is studied and compared with conventional SOI-FinFET. Under a radiation-prone environment, trap charges are accumulated in the oxide and Si/SiO$_2$ interfaces. These charges make the threshold voltage negative and increase the leakage current of the device. The interface trap charges after irradiation create the parasitic conduction path between the buried oxide (BOX) layer and substrate. This strong parasitic conduction degrades the performance of the subthreshold region of operation [49]. Proposed spacer engineering with high k=25 maintains the low leakage and better subthreshold region of operation. Here, for different values of k – k = 3.7 (SiO$_2$), k = 7.5 (Si$_3$N$_4$), k = 9 (Al$_2$O$_3$), k = 9.14 (AlN), and k = 24 (HfO$_2$) – TID analysis is examined and compared. The TID analysis of spacer SOI-FinFET for different dielectrics is presented for the first time.

6.2 RADIATION-HARDENED DEVICE STRUCTURE AND SIMULATION SETUP

A 30 nm gate length SOI-FinFET with spacer engineering is designed for a radiation environment. Figure 6.2(a) and (b) illustrate the 2-D conventional and proposed

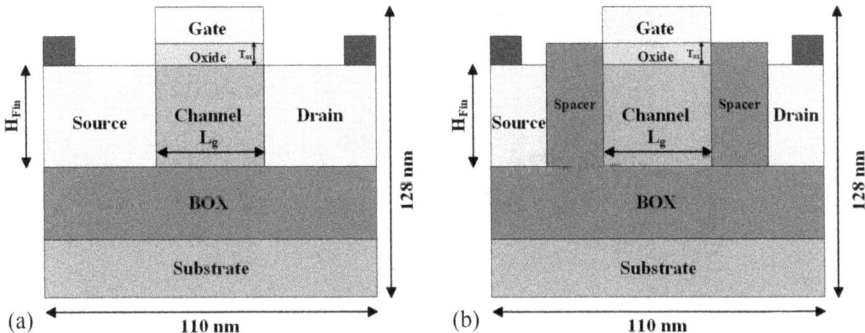

FIGURE 6.2   2-D Schematic of (a) conventional and (b) proposed structure

TABLE 6.1

Parameters used in the simulation

Parameters	Values
Gate oxide thickness (T_{ox}) (nm)	3
Fin width (W_{FIN}) (nm)	10
Channel length (L_g) (nm)	30
Device thickness (nm)	26
Gate workfunction (eV)	4.65
Buried oxide thickness (BOX) (nm)	30
Spacer region length (nm)	20

device structure. Fin height (H_{FIN}) of 70 nm and fin width (W_{FIN}) of 10 nm is considered for the device design. The composition of tungsten and nitride material is used as the gate electrode with workfunction of 4.65 eV. The n-type doping concentration of 2×10^{18} cm^{-3} is used for drain and source doping. In the proposed device, different spacers, such as SiO_2, Si_3N_4, AlN, Al_2O_3, and HfO_2, are used to obtain the best result. Table 6.1 gives an overview of the parameters used in the simulation.

A simulated 3-D structure is shown in Figure 6.3. Lombardi mobility, Auger and Shockley-Read-Hall (SRH) recombination, Fermi, trap, and high mobility physics model with drift-diffusion model level 1 (DDML1) are incorporated in the simulation. For the total ionizing dose analysis, an advanced TID model is used [50]. A SOI-FinFET with a fin width of 15 nm and channel length of 30 nm is simulated for simulator validation. Figure 6.4 illustrates the calibration of the simulation and experimental result [51]. For the total ionizing dose analysis, the device is simulated in a worst-bias condition in order to make the device ready for the TID simulation

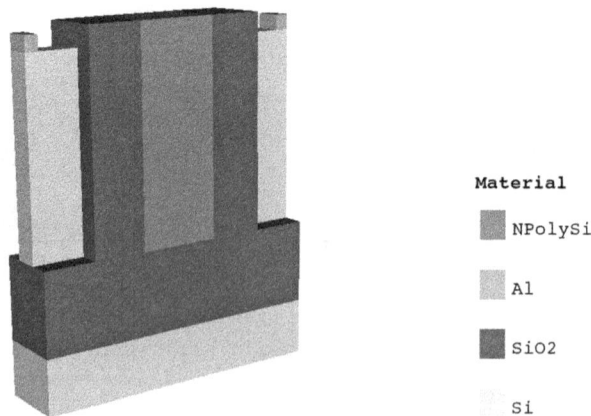

Material

NPolySi

Al

SiO2

Si

FIGURE 6.3   Simulated 3-D structure of the proposed device

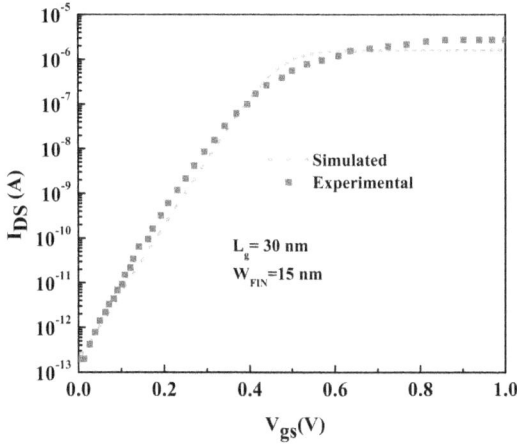

FIGURE 6.4 Calibration of simulation result with the experimental result.

under ON-state bias ($Vg = 1$ V and $V_D = V_S = 0$) condition. The maximum device degradation is observed in the worst-bias condition [47].

6.3 RESULTS AND DISCUSSION

Different spacer materials with dielectric constants of 3.7 (SiO_2), 7.5 (Si_3N_4), 9 (Al_2O_3), 9.14 (AlN), and 24 (HfO_2) are incorporated in the study to examine the impact of radiation. Figure 6.5 depicts the I_{DS} vs V_{GS} comparison of conventional SOI-FinFET with different spacers. Spacer engineering improves the OFF-state performance with a slight increment in threshold voltage. This improvement in the proposed device is noticed because in OFF-state vertical electric field is high. But

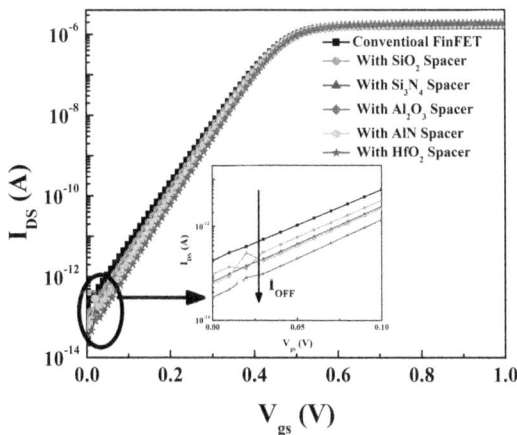

FIGURE 6.5 I_D vs V_{GS} comparisons of the conventional device with different spacers

the ON-current of the device is not affected due to zero electric field [52]. From the graph, a 1-decade lower leakage current (I_{OFF}) is observed for the HfO$_2$ spacer as compared to a conventional device.

The irradiation trap charges are accumulated in the oxide and insulator/semiconductor interface. Because of these charges, the threshold voltage of the device is shifting toward a negative direction with respect to radiation dose. TID responses of all devices are explained below.

6.3.1 TID RESPONSE

The classical TID responses of all devices are investigated for the worst-bias/ ON-state (Vg = 1 V and V_D = V_S = 0) condition. Here, the TID response for doses ranging from 0 to 2000 krad of the device is investigated.

6.3.1.1 Impact of radiation of spacer

Figure 6.6(a) shows the I_D vs V_{GS} comparison with the SiO$_2$ spacer. The typical TID degradation of the device is observed with respect to radiation dose. The OFF-state current comparison of the conventional and proposed device with SiO$_2$ is depicted in Figure 6.6(b). The SiO$_2$ spacer-based device is more radiation-tolerant as compared to conventional SOI-FinFET. SiO$_2$ spacer shows a 1-decade less I_{OFF} than the conventional n-type SOI-FinFET for higher radiation dose (2000 krad). From Figure 6.6(b), it is observed that the device with SiO$_2$ spacer-based device shows 72%, 70%, 81%, 75%, and 64% less I_{OFF} as compared to conventional FinFET with respect to different radiation doses of 100 krad, 500 krad, 1000 krad, 1500 krad, and 2000 krad, respectively. But a nearly 90% shift is observed in the threshold voltage for pre- to post-radiation (2000 krad) of the SiO$_2$ spacer. As the value of the dielectric constant varies from lower to higher, a lower shift in V_{th} is obtained. The higher value of the dielectric constant increases the depletion region because of the presence of a fringing electric field inside the spacers [53, 54]. Parameters such as built-in potential

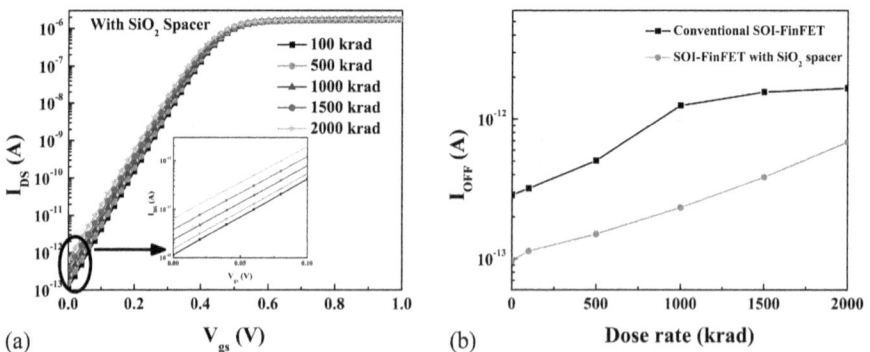

(a)

(b)

FIGURE 6.6 (a) Post-radiation transfer characteristics of SiO$_2$ spacer with respect to radiation dose and (b) I_{OFF} current comparison of conventional and SiO$_2$ spacer device

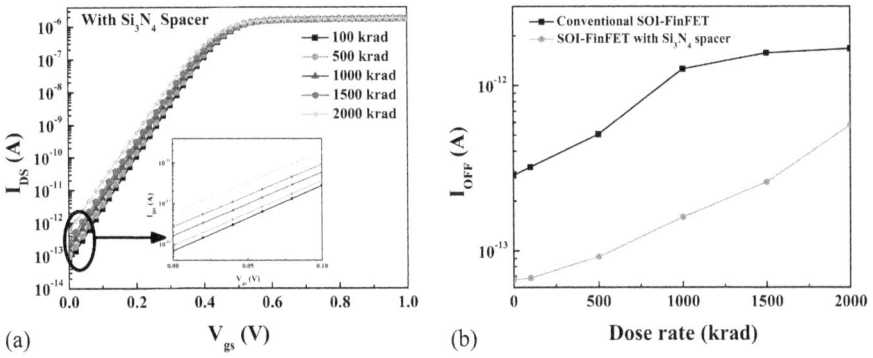

FIGURE 6.7 (a) Post-radiation transfer characteristics of Si_3N_4 spacer with respect to radiation dose and (b) I_{OFF} current comparison of conventional and Si_3N_4 spacer device

are directly dependent on threshold voltage, and built-in potential is proportional to depletion width.

Post-radiation transfer characteristics of the Si_3N_4 spacer are illustrated in Figure 6.7(a). Figure 6.7(b) shows OFF-state current comparisons of conventional SOI-FinFET and Si_3N_4 spacer-based SOI-FinFET. The Si_3N_4 spacer-based device maintains low leakage of 10^{-14} up to radiation of 500 krad. For the pre-rad condition, Si_3N_4 spacer-based FinFET shows a 64% improvement in leakage current as compared to conventional SOI-FinFET devices. As per the radiation doses of 100 krad, 500 krad, 1000 krad, 1500 krad, and 2000 krad, degradation in I_{OFF} of 2%, 35%, 42%, 38%, and 54%, respectively, is observed from the graph in Figure 6.7(b).

Figure 6.8(a) and (b) shows the I_D vs V_{GS} post-radiation curve of the Al_2O_3 spacer and OFF-state current comparison of conventional and the Al_2O_3 spacer-based device with respect to dose rate. As the value of the dielectric constant increases, the I_{OFF} of the device goes lower. This happens because a higher dielectric constant

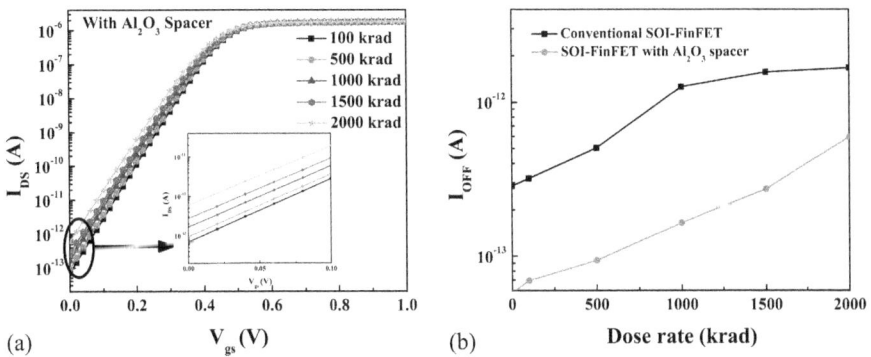

FIGURE 6.8 (a) Post-radiation transfer characteristics of Al_2O_3 spacer with respect to radiation dose and (b) I_{OFF} current comparison of conventional and Al_2O_3 spacer device

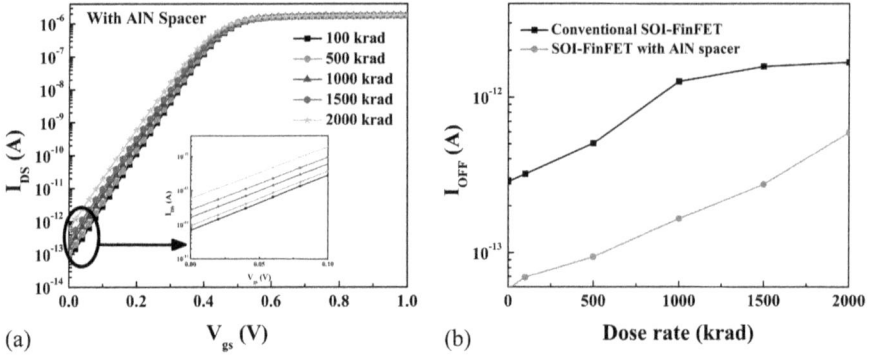

(a)   (b)

FIGURE 6.9   (a) Post-radiation transfer characteristics of AlN spacer with respect to radiation dose and (b) I_{OFF} current comparison of conventional and AlN spacer device

spacer has a strong fringing electric field [55, 56]. The classical TID degradation in I_{OFF} of an Al_2O_3 spacer-based device of 26%, 41%, 40%, and 54% is observed for radiation doses of 500 krad, 1000 krad, 1500 krad, and 2000 krad, respectively.

As the value of dielectric constant (k) is increasing, subthreshold performance and leakage current are improving due to the high vertical electric field at OFF-state bias condition. This improvement in both parameters is also observed for the post-radiation condition. Figure 6.9(a) and (b) illustrates the post-radiation transfer characteristics of the AlN spacer-based device and the I_{OFF} comparison of the spacer-based device and conventional device. The AlN spacer-based device maintains almost the same current switching ratio (10^8) for pre and post (500 krad) radiation. From Figure 6.9(b), it is observed that the AlN spacer-based device shows an 83%, 81%, 86%, 82%, and 69% improvement in I_{OFF} as compared to the conventional device. Post-radiation transfer characteristics with an HfO_2 spacer are depicted in Figure 6.10(a). Figure 6.10(b) shows the I_{OFF} comparison of the HfO_2 spacer with conventional

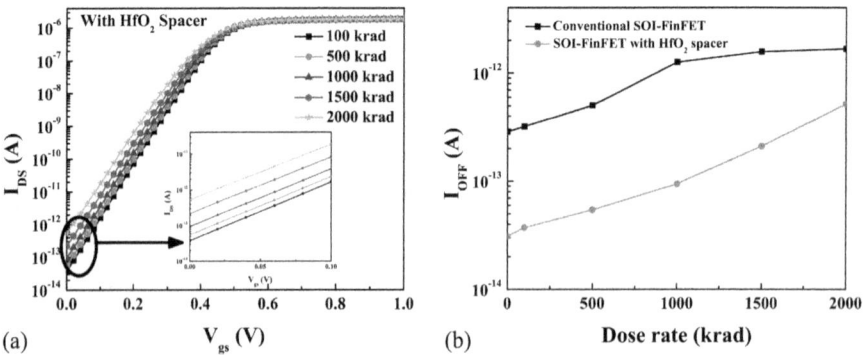

(a)   (b)

FIGURE 6.10   (a) Post-radiation transfer characteristics of HfO_2 spacer with respect to radiation dose and (b) I_{OFF} current comparison of conventional and HfO_2 spacer device

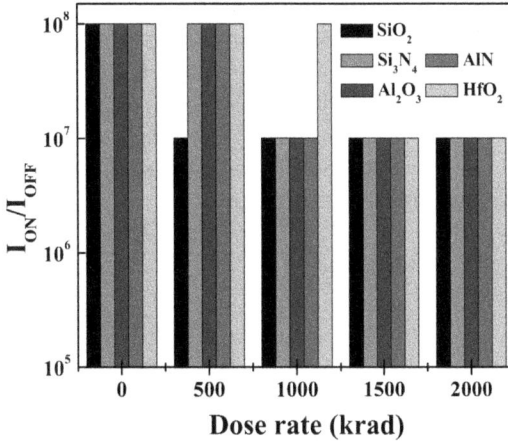

FIGURE 6.11 I_{ON}/I_{OFF} comparisons of all device with respect to radiation doses

device with respect to radiation doses. As per the above results improvement in I_{OFF} and better current switching ratio is obtained with HfO_2 spacer-based device. The high dielectric constant gives a strong fringing field near the channel which yields lower I_{OFF} [11, 57, 58]. The HfO_2 spacer-based device shows an 83%, 91%, 89%, 92%, 86%, and 73% improvement in I_{OFF} as compared to the conventional device for different radiation doses of 100 krad, 500 krad, 1000 krad, 1500 krad, and 2000 krad. The larger spacer length of 20 nm is used in this work for better subthreshold region performance. The proposed device with an HfO_2 spacer maintains an 11% higher V_{th} than the conventional SOI-FinFET. The current switching ratio (I_{ON}/I_{OFF}) comparison of all spacer-based devices is illustrated in Figure 6.11. From the graph, it is observed that in pre-rad conditions, all spacer-based devices show almost the same I_{OFF}. As the radiation dose is increasing, the OFF-state performance and sub-threshold device performances are degrading with respect to dose rate. But the HfO_2 spacer maintains better I_{OFF} and higher V_{th}. Among all the spacer-based devices, HfO_2 spacer-based devices maintain the same decade of I_{OFF} after a high radiation dose of 1000 krad. This shows that the HfO_2 spacer-based device shows superior performance for pre- and post-radiation conditions.

6.3.1.2 Radiation-induced interface trap charges

After the irradiation, trap charges are accumulated in semiconductor/insulator interfaces. These charges shift the threshold voltage toward negative and degrade the leakage current. As the radiation dose increases, interface trap charges are also increased. The irradiation gate terminal also loses its controllability over the channel. These observations confirm that the device in a radiation-prone environment loses its electrostatic integrity. The proposed spacer engineering reduces the parasitic components and improves the electrostatic device performance. This improvement is noticed because spacers have a strong fringing field. Also, a higher dielectric constant shows better gate controllability for both pre- and post-radiation conditions.

This enhances the circuit's digital performance [56, 58]. The interface charge density spectrum after a 2000 krad dose of all spacer-based devices (SiO$_2$, Si$_3$N$_4$, Al$_2$O$_3$, AlN, and HfO$_2$) is depicted in Figures 6.12 (a), (b), (c), (d) and (e), respectively. In Figures 6.12 (a), (b), (c), (d), and (e), it is noticed that interface trap charges are reducing as the value of the dielectric constant is increasing.

The HfO$_2$ spacer-based device SOI n-FinFET shows a minimum interface trap charge density (/cm^2) as compared to others.

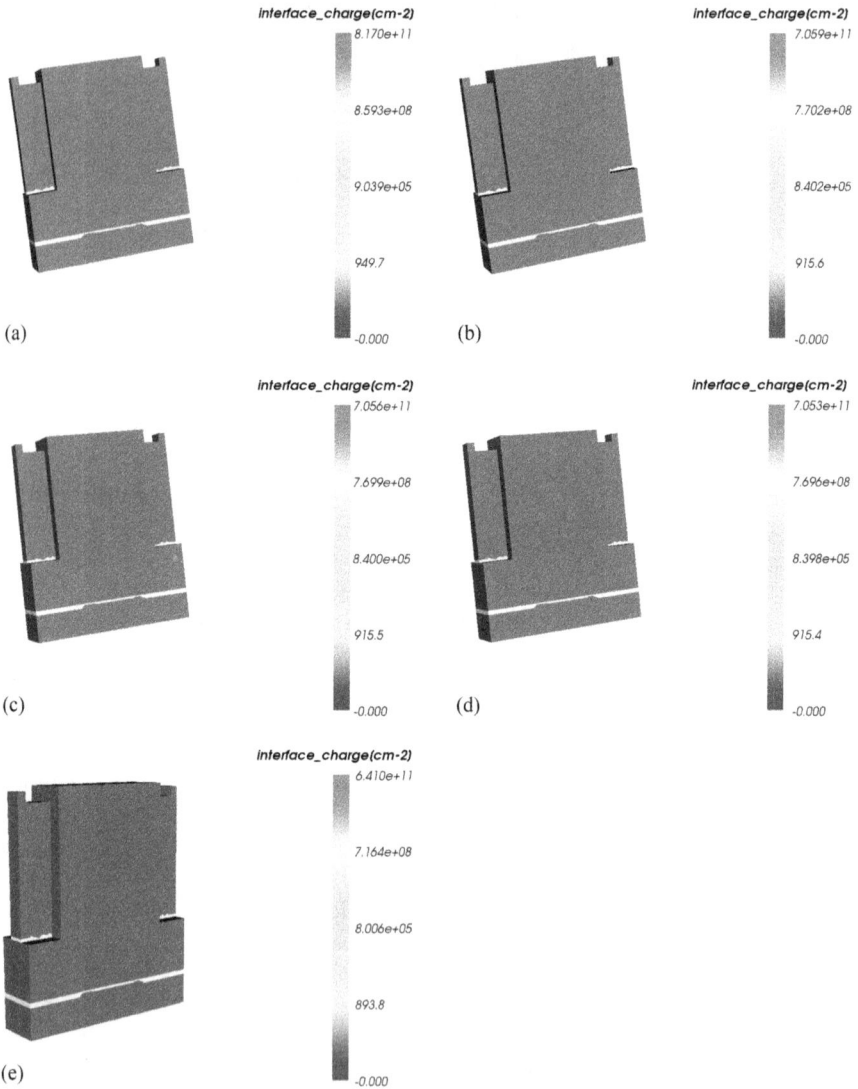

FIGURE 6.12   Interface trap charges spectrum of (a) SiO$_2$ spacer (b) Si$_3$N$_4$ spacer, (c) Al$_2$O$_3$ spacer, (d) AlN spacer, and (e) HfO$_2$ spacer SOI-FinFET, after 2000 krad

6.3.1.3   Shift in threshold voltage after irradiation

At the time of radiation-induced on the device, trap charges are accumulated in the oxide and semiconductor/insulator interfaces. These trap charges make the device threshold voltage negative or shift toward the negative axis. The interface trap charges are a minimal contributor to the shifting of V_{th} [55]. Here, a 20 nm symmetric spacer with a different dielectric constant with the worst-bias condition ($V_g = 1$ V and $V_S = V_D = 0$) is investigated to analyze the V_{th} shift. Figure 6.13 shows the V_{th} shift for different values of k. The maximum shift of 0.231 V for pre- to post-radiation (2000 krad) conditions is noticed for the SiO_2 spacer. The proposed dimension and engineering help to maintain a positive threshold voltage. The symmetric length and higher dielectric constant of the spacer increase the source/channel depletion region. Due to this enlargement in the depletion region, a higher device V_{th} is maintained [52, 54–56]. For the pre-rad condition, HfO_2 spacer-based device shows a high V_{th} of 0.27 V i.e., 20 mV more as compared to the SiO_2 spacer device. After a 100 krad radiation dose, the maximum V_{th} shift of 128 mV is obtained for the SiO_2 spacer. Moreover, Si_3N_4, Al_2O_3, and AlN spacer-based devices show almost the same V_{th} shift of 120 mV. Also, a minimum V_{th} shift of 93 mV is obtained for HfO_2 spacer-based device. After the high radiation 2000 krad dose, a 42%, 30%, 27%, and 24% lower shift in V_{th} is observed for the HfO_2 spacer as compared to SiO_2, Si_3N_4, Al_2O_3, and AlN spacer-based devices. Even after a very high radiation (2000 krad) dose, the proposed HfO_2 spacer-based SOI n-FinFET with symmetric spacer length shows less TID degradation as compared to conventional and other proposed devices.

6.3.1.4   Radiation affected subthreshold swing

In the radiation-prone environment, the subthreshold swing (SS) of the device is degrading as the radiation dose increases. The typical degradation in SS is observed due to radiation-induced trap charges [31]. The subthreshold swing of all spacer-based

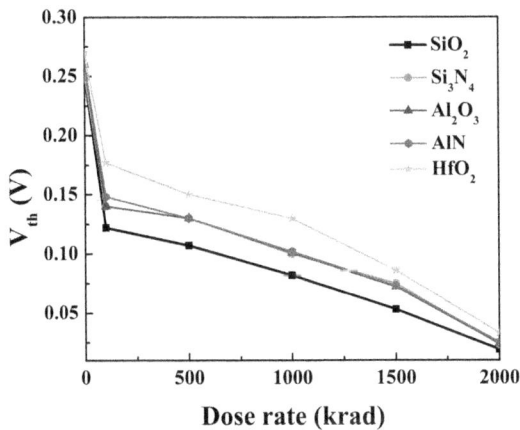

FIGURE 6.13 V_{th} shift for different spacer (SiO_2, Si_3N_4, Al_2O_3, AlN, and HfO_2) devices

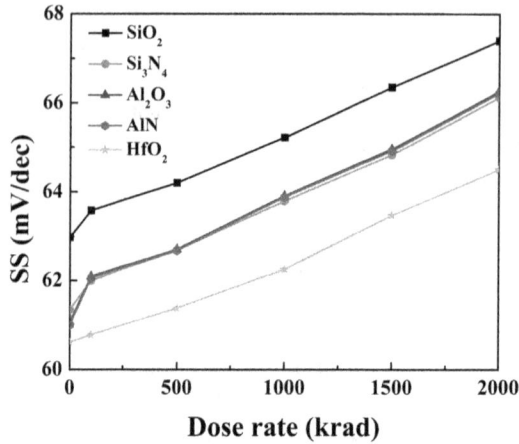

FIGURE 6.14 Subthreshold swing for different radiation dose

devices is calculated and compared for different radiation doses (0 to 2000 krad) and depicted in Figure 6.14.

For pre- and post-radiation conditions, Si_3N_4, Al_2O_3, and AlN spacer-based devices show almost the same SS, while the HfO_2 spacer-based device shows a 4.2% improvement in SS as compared to the SiO_2 spacer-based SOI-FinFET. Figure 6.14 shows, in the HfO_2 spacer-based device, a 2.6%, 2.5%, and 2.4% improvement in SS after a radiation dose of 2000 krad as compared to other proposed spacer-based devices (Si_3N_4, Al_2O_3, and AlN), respectively. The proposed device with different spacers (Si_3N_4, Al_2O_3, AlN, and HfO_2) shows a nearly 2% (Si_3N_4, Al_2O_3, and AlN) and 4.5% (HfO_2) improvement in SS as compared to conventional SOI-FinFET. This improvement is noticed because of the strong fringing electric field through the spacers [58].

Table 6.2 shows a parameter comparison of proposed and conventional devices after a radiation dose of 1000 krad. The results observed for the HfO_2 spacer-based device show good radiation tolerability and reliability.

TABLE 6.2
Parameter comparison of proposed devices and conventional device after 1000 krad dose

Parameter	I_{OFF}	V_{th}	SS	I_{ON}/I_{OFF}
Conventional	1.26×10^{-12}	0.064	66.46	10^6
SiO_2	2.34×10^{-13}	0.082	65.22	10^7
Si_3N_4	1.65×10^{-13}	0.10	63.91	10^7
Al_2O_3	1.64×10^{-13}	0.101	63.88	10^7
AlN	1.63×10^{-13}	0.102	63.79	10^7
HfO_2	9.44×10^{-14}	0.132	62.26	10^8

6.4 CONCLUSION

The impact of TID on SOI n-FInFET with spacer engineering is presented and investigated. For the pre-radiation condition, a 1-decade improvement is observed for all the spacer-based devices (3.7 (SiO_2), 7.5 (Si_3N_4), 9 (Al_2O_3), 9.14 (AlN), and 24 (HfO_2)), and a 9.6% higher threshold voltage is observed for the HfO_2 spacer device as compared to a conventional device. For the pre- and post-radiation conditions, Si_3N_4, Al_2O_3, and AlN spacer-based devices show nearly the same values for both pre- and post-radiation conditions. At a higher radiation dose of 1000 krad, the proposed device with an HfO_2 spacer-based device maintains the almost same leakage current (10^{-14} A) as the pre-radiation condition. After the 2000 krad dose, the HfO_2 spacer-based device shows a 42%, 30%, 27%, and 24% lower shift in V_{th} as compared to SiO_2, Si_3N_4, Al_2O_3, and AlN spacer-based devices, respectively. Moreover, HfO_2 spacer-based FinFET shows a 23%, 9.2%, 13%, and 12% improvement in I_{OFF} as compared to SiO_2, Si_3N_4, Al_2O_3, and AlN spacer-based devices, respectively. These improvements in the results replicate that for radiation doses of 500 krad, the proposed SiO_2, Si_3N_4, Al_2O_3, and AlN spacer-based devices appear to be good enough to sustain a radiation-prone environment. For the higher radiation dose (2000 krad), the proposed HfO_2 spacer-based SOI n-FinFET shows the best radiation-tolerant capability. This suggests that the proposed HfO_2 spacer-based device is reliable for the design of memories and processors of satellites.

REFERENCES

1. K. J. Kuhn, "Considerations for ultimate CMOS scaling," *IEEE Trans. Electron Dev.*, 59(7), pp. 1813–1828, 2012.
2. M. Daga and G. P. Mishra, "Subthreshold performance improvement of underlapped FinFET using workfunction modulated dual-metal gate technique," *Silicon*, 13, pp.1541–1548 2020.
3. M. Daga and G. P. Mishra, "Improvement in electrostatic efficiency using workfunction modulated dual metal gate FinFET," *Mater. Today Proc.*, 43, pp. 3443–3446, 2020.
4. T. Ludwig, I. Aller, V. Gemhoefer, J. Keinelt, E. Nowak, R. V. Josh, A. Mueller, and S. Tomaschko, "FinFET technology for future microprocessors," *IEEE Int. SOI Conf.*, 20 pp. 33–34, 2003.
5. J. P. Colinge, "Multi-gate SOI MOSFETs," *Microelectron. Eng.*, 84(9–10), pp. 2071–2076, 2007.
6. F. Balestra, S. Cristoloveanu, M. Benachir, J. Brini, and T. Elewa, "Double-gate silicon-on-insulator transistor with volume inversion: A new device with greatly enhanced performance," *IEEE Electron Dev. Lett.*, 8(9), pp. 410–412, 1987.
7. D. Hisamoto, T. Kaga, and E. Takeda, "Impact of the vertical SOI 'DELTA' structure on planar device technology," *IEEE Trans. Electron Devices*, 38(6), pp. 1419–1424, 1991.
8. C.-H. Jan, U. Bhattacharya, R. Brain, S.-J. Choi, G. Curello, G. Gupta, W. Hafez, M. Jang, M. Kang, K. Komeyli, T. Leo, N. Nidhi, L. Pan, J. Park, K. Phoa, A. Rahman, C. Staus, H. Tashiro, C. Tsai, P. Vandervoorn, L. Yang, J.-Y. Yeh, and P. Ba, "A 22nm SoC platform technology featuring 3-D tri-gate and high-k/metal gate, optimized for ultra low power, high performance and high density SoC applications," *Tech. Dig. Int. Electron Dev. Meet. IEDM*, pp. 44–47, 2012.

9. H.-J. Cho, H. S. Oh, K. J. Nam, Y. H. Kim, K. H. Yeo, W. D. Kim, Y. S. Chung, Y. S. Nam, S. M. Kim, W. H. Kwon, M. J. Kang, I. R. Kim, H. Fukutome, C. W. Jeong, H. J. Shin, Y. S. Kim, D. W. Kim, S. H. Park, H. S. Oh, J. H. Jeong, S. B. Kim, D. W. Ha, J. H. Park, H. S. Rhee, S. J. Hyun, D. S. Shin, D. H. Kim, H. Y. Kim, S. Maeda, K. H. Lee, Y. H. Kim, M. C. Kim, Y. S. Koh, B. Yoon, K. Shin, N. I. Lee, S. B. Kangh, K. H. Hwang, J. H. Lee, J.-H. Ku, S. W. Nam, S. M. Jung, H. K. Kang, J. S. Yoon, and E. S. Jung, "Si FinFET based 10nm technology with multi Vt gate stack for low power and high performance applications," *Dig. Tech. Pap. Symp. VLSI Technol.*, 2016, pp. 12–13, 2016.

10. M. L. Huang, S. W. Chang, M. K. Chen, Y. Oniki, H. C. Chen, C. H. Lin, W. C. Lee, C. H. Lin, M. A. Khaderbad, K. Y. Lee, Z. C. Chen, P. Y. Tsai, L. T. Lin, M. H. Tsai, C. L. Hung, T. C. Huang, Y. C. Lin, Y.-C. Yeo, S. M. Jang, H. Y. Hwang, H. C.-H. Wang, and C. H. Diaz, "High Perform In0.53Ga0.47As FinFETs fabricated on 300 mm Si substrate," *Dig. Tech. Pap. - Symp. VLSI Technol.*, 2016, pp. 52–53, 2016.

11. A. B. Sachid, M. C. Chen, and C. Hu, "Bulk FinFET with Low-κ spacers for continued scaling," *IEEE Trans. Electron Dev.*, 64(4), pp. 1861–1864, 2017.

12. S. Kuboyama and T. Tamura, "Development of MOS transistors for radiation-hardened large scale integrated circuits and analysis of radiation-induced degradation," *J. Nucl. Sci. Technol.*, 31(1), pp. 24–33, 1994.

13. L. W. Massengill, B. K. Choi, D. M. Fleetwood, R. D. Schrimpf, K. F. Galloway, M. R. Shaneyfelt, T. L. Meisenheimer, P. E. Dodd, J. R. Schwank, Y. M. Lee, R. S. Johnson, and G. Lucovsky, "Heavy-ion-induced breakdown in ultra-thin gate oxides and high-k dielectrics," *Trans. Nucl. Sci.*, 48(6), pp. 1904–1912, 2001.

14. J. Raymond, E. Steele, and W. Chang, "Radiation effects in metal-oxide-semiconductor transistors," *IEEE Trans. Nucl. Sci.*, 12(1), pp. 457–463, 1965.

15. W. A. Hanna, W. A. Kolasinski, M. T. Marra, W. A. Hanna, "Techniques of microprocessor testing and SEU-rate prediction," *IEEE Trans. Nucl. Sci.*, 32(6), pp. 4219–4224, 1985.

16. L. Adams, E. J. Daly, and C. Sansoe, "The SEU risk assessment of Z80A, 8086 and 80C86 microprocessors intended for use in a low altitude polar orbit," *IEEE Trans. Nucl. Sci.* 33(6), pp. 1626–1631, 1986.

17. Y. Lv, Q. Wang, H. Ge, T. Xie, and J. Chen, "A highly reliable radiation hardened 8T SRAM cell design," *Microelectron. Reliab.*, 125(March), p. 114376, 2021.

18. C. I. Kumar and B. Anand, "A highly reliable and energy efficient radiation hardened 12T SRAM cell design," *IEEE Trans. Dev. Mater. Reliab.*, 20(1), pp. 58–66, 2020.

19. L. Wen, Y. Zhang, and P. Wang, "Radiation-hardened, read-disturbance-free new-Quatro-10T memory cell for aerospace applications," *Very Large Scale Integr. Syst. IEEE Trans.*, 28(8), pp. 1935–1939, 2020.

20. L. Artola, G. Hubert, and T. Rousselin, "Single-event latchup modeling based on coupled physical and electrical transient simulations in CMOS technology," *IEEE Trans. Nucl. Sci.*, 61(6), pp. 3543–3549, 2014.

21. V. Ferlet-Cavrois, L. W. Massengill, and P. Gouker, "Single event transients in digital CMOS - A review," *IEEE Trans. Nucl. Sci.*, 60(3), pp. 1767–1790, 2013.

22. C. Yaqing, H. Pengcheng, S. Qian, L. Bin, and Z. Zhenyu, "Characterization of single-event upsets induced by high LET heavy ions in 16 nm bulk FinFET SRAMs," *IEEE Trans. Nucl. Sci.*, 9499(5), pp. 1176–1181, 2021.

23. L. H. Brendler, A. L. Zimpeck, F. L. Kastensmidt, C. Meinhardt, and R. Reis, "Voltage scaling influence on the soft error susceptibility of a FinFET-based circuit," *2021 IEEE 12th Lat. Am. Symp. Circuits Syst. LASCAS*, pp. 4–7, 2021.

24. A. Ray, A. Naugarhiya, and G. P. Mishra, "Influence of SEU effects in low-doped double gate MOSFETs," *Materials Today: Proceedings,* 43 pp. 1–4, 2020.

25. C. Peng, J. Huang, C. Liu, Q. Zhao, S. Xiao, X. Wu, Z. Lin, J. Chen, and X. Zeng, "Radiation-hardened 14T SRAM bitcell with speed and power optimized for space application," *IEEE Trans. Very Large Scale Integr. Syst*, 27(2), pp. 407–415, 2019.

26. Y. Han, T. Li, X. Cheng, L. Wang, J. Han, Y. Zhao, and X. Zeng, "Radiation Hardened 12T SRAM with crossbar-based peripheral circuit in 28nm CMOS technology," *IEEE Trans. Circuits Syst. I Regul. Pap.*, 68(7), pp. 2962–2975, 2021.

27. A. Haran, E. Keren, D. David, N. Refaeli, R. Giterman, M. Assaf, L. Atias, A. Teman, and A. Fish, "Single-event upset tolerance study of a low-voltage 13T radiation-hardened SRAM bitcell," *IEEE Trans. Nucl. Sci.*, 67(8), pp. 1803–1812, 2020.

28. Q. Zhao, C. Peng, J. Chen, Z. Lin, and X. Wu, "Novel write-enhanced and highly reliable RHPD-12T SRAM cells for space applications," *IEEE Trans. Very Large Scale Integr. Syst*, 28(3), pp. 848–852, 2020.

29. H. L. Hughes and J. M. Benedetto, "Radiation effects and hardening of MOS technology: Devices and circuits," *IEEE Trans. Nucl. Sci.*, 50(3), pp. 500–521, 2003.

30. J. R. Schwank and, M. R. Shaneyfelt, "Radiation effects in SOI technologies," *IEEE Trans. Nucl. Sci.*, 50(3), pp. 522–538, 2003.

31. J. R. Schwank, M. R. Shaneyfelt, D. M. Fleetwood, J. A. Felix, P. E. Dodd, P. Paillet, and V. Ferlet-Cavrois, "Radiation effects in MOS oxides," *IEEE Trans. Nucl. Sci.*, 55(4), pp. 1833–1853, 2008.

32. E. Simoen, M. Gaillardin, P. Paillet, R. A. Reed, R. D. Schrimpf, M. L. Alles, F. El-Mamouni, D. M. Fleetwood, A. Griffoni, and C. Claeys, "Radiation effects in advanced multiple gate and silicon-on-insulator transistors," *IEEE Trans. Nucl. Sci.*, 60(3), pp. 1970–1991, 2013.

33. M. Gaillardin, C. Marcandella, M. Martinez, M. Raine, P. Paillet, O. Duhamel, and N. Richard, "Total ionizing dose effects in multiple-gate field-effect transistor," *Semicond. Sci. Technol.*, 32(8), 2017.

34. Q. Huang and J. Jiang, "An overview of radiation effects on electronic devices under severe accident conditions in NPPs, rad-hardened design techniques and simulation tools," *Prog. Nucl. Energy*, 114(March), pp. 105–120, 2019.

35. M. Gaillardin, C. Marcandella, M. Martinez, O. Duhamel, T. Lagutere, P. Paillet, M. Raine, N. Richard, F. Andrieu, S. Barraud, and M. Vinet, "Total ionizing dose response of multiple-gate nanowire field effect transistors," *IEEE Trans. Nucl. Sci.*, 64(8), pp. 2061–2068, 2017.

36. T. Ma, S. Bonaldo, S. Mattiazzo, A. Baschirotto, C. Enz, A. Paccagnella, and S. Gerardin, "TID degradation mechanisms in 16-nm bulk FinFETs irradiated to ultra-high doses," *IEEE Trans. Nucl. Sci.*, 68(8), pp. 1571–1578, 2021.

37. S. Bonaldo, S. E. Zhao, A. O'Hara, M. Gorchichko, X. Zhang, S. Gerardin, A. Paccagnella, N. Waldron, N. Collaert, V. Putcha, D. Linten, S. T. Pantelides, R. A. Reed, R. D. Schrimpf, and D. M. Fleetwood, "Total-ionizing-dose effects and low-frequency noise in 16-nm InGaAs FinFETs with HfO2/Al2O3 Dielectrics," *IEEE Trans. Nucl. Sci.*, 67(1), pp. 210–220, 2020.

38. M. Gorchichko, Y. Cao, X. Zhang, D. Yan, H. Gong, S. E, Zhao, P. Wang, R. Jiang, C. Liang, D. M. Fleetwood, R. D. Schrimpf, R. A. Reed, and D. Linten, "Total-ionizing-dose effects and low-frequency noise in 30-nm gate-length bulk and SOI FinFETs with SiO2/HfO2 gate dielectrics," *IEEE Trans. Nucl. Sci.*, 67(1), pp. 245–252, 2020.

39. S. E. Zhao, S. Bonaldo, P. Wang, X. Zhang, N. Waldron, N. Collaert, V. Putcha, D. Linten, S. Gerardin, A. Paccagnella, R. D. Schrimpf, R. A. Reed, and D. M. Fleetwood, "Total-ionizing-dose effects on InGaAs FinFETs with modified gate-stack," *IEEE Trans. Nucl. Sci.*, 67(1), pp. 253–259, 2020.

40. K. Castellani-Coulié, D. Munteanu, J. L. Autran, V. Ferlet-Cavrois, P. Paillet, and J. Baggio, "Simulation analysis of the bipolar amplification induced by heavy-ion irradiation in double-gate MOSFETs," *IEEE Trans. Nucl. Sci.*, 52(6), pp. 2137–2143, 2005.

41. F. El Mamouni, E. X. Zhang, R. D. Schrimpf, D. M. Fleetwood, R. A. Reed, S. Cristoloveanu, and W. Xiong, "Fin-width dependence of ionizing radiatION-induced subthreshold-swing degradation in 100-nm-gate-length FinFETs," *IEEE Trans. Nucl. Sci.*, 56(6), pp. 3250–3255, 2009.

42. M. L. Alles, R. D. Schrimpf, R. A. Reed, L. W. Massengill, R. A. Weller, M. H. Mendenhall, D. R. Ball, K. M. Warren, T. D. Loveless, J. S. Kauppila, and B. D. Sierawski, "Radiation hardness of FDSOI and FinFET technologies," *Proc. IEEE Int. SOI Conf.*, 615, pp. 9–10, 2011.

43. E. X. Z. Chatterjee, B. L. Bhuva, R. A. Reed, M. L. Alles, N. N. Mahatme, D. R. Ball, R. D. Schrimpf, D. M. Fleetwood, D. Linten, E. Simôen, J. Mitard, C. Claeys, C. Claeys, "Geometry dependence of total-dose effects in bulk FinFETs," *IEEE Trans. Nucl. Sci.*, 61(6), pp. 2951–2958, 2014.

44. B. D. Gaynor and S. Hassoun, "Fin shape impact on FinFET leakage with application to multithreshold and ultralow-leakage FinFET design," *IEEE Trans. Electron Devices*, 61(8), pp. 2738–2744, 2014.

45. A. Ray, A. Naugarhiya, and G. P. Mishra, "Analysis of total ionizing dose response of optimized fin geometry work function modulated SOI-FinFET," *Microelectron. Reliab.*, 134(April), p. 114549, 2022.

46. T. D. Haeffner, R. F. Keller, R. Jiang, B. D. Sierawski, M. W. McCurdy, E. X. Zhang, R. W. Mohammed, D. R. Ball, M. L. Alles, R. A. Reed, R. D. Schrimpf, and D. M. Fleetwood, "Comparison of total-ionizing-dose effects in bulk and SOI FinFETs at 90 and 295 K," *IEEE Trans. Nucl. Sci.*, 66(6), pp. 911–917, 2019.

47. Z. Ren, X. An, G. Li, G. Chen, M. Li, Y. Gang, Q. Guo, X. Zhang, and R. Huang, "TID response of bulk si PMOS FinFETs: Bias, fin width, and orientation dependence," *IEEE Trans. Nucl. Sci.*, 67(7), pp. 1320–1325, 2020.

48. P. E. Dodd, M. R. Shaneyfelt, J. R. Schwank, and J. A. Felix, "Current and future challenges in radiation effects on CMOS electronics," *IEEE Trans. Nucl. Sci.*, 57(4), pp. 1747–1763, 2010.

49. M. L. Alles, R. D. Schrimpf, R. A. Reed, L. W. Massengill, R. A. Weller, M. H. Mendenhall, D. R. Ball, K. M. Warren, T. D. Loveless, J. S. Kauppila, and B. D. Sierawski, "Radiation hardness of FDSOI and FinFET technologies," *Proc. IEEE Int. SOI Conf.*, no. 615, pp. 1–2, 2011.

50. G. User, "Cogenda Genius," ver. 2.2, November 2022.

51. J. Riffaud, M. Gaillardin, C. Marcandella, M. Martinez, P. Paillet, O. Duhamel, T. Lagutere, M. Raine, N. Richard, F. Andrieu, S. Barraud, M. Vinet, and O. Faynot, "Investigations on the geometry effects and bias configuration on the TID response of nMOS SOI tri-gate nanowire field-effect transistors," *IEEE Trans. Nucl. Sci.*, 65(1), pp. 39–45, 2018.

52. V. B. Sreenivasulu and V. Narendar, "Performance improvement of spacer engineered n-type SOI FinFET at 3-nm gate length," *AEU Int. J. Electron. Commun.*, 137(May), p. 153803, 2021.

53. V. B. Sreenivasulu and V. Narendar, "Characterization and optimization of junctionless gate-all-around vertically stacked nanowire FETs for sub-5 nm technology nodes," *Microelectron. J.*, 116(August), p. 105214, 2021.

54. P. Xu, F. Ji, P. T. Lai, and J. G. Guan, "Influence of sidewall spacer on threshold voltage of MOSFET with high-k gate dielectric," *Microelectron. Reliab.*, 48(2), pp. 181–186, 2008.

55. J. H. Scofield and D. M. Fleetwood, "Evidence that similar point defects cause 1/f noise and radiation-induced-hole trapping in MOS transisotrs," *Phys. Rev. Lett.*, 64(5), pp. 579–582, 1990.

56. Y. Zhao and Y. Qu, "Impact of self-heating effect on transistor characterization and reliability issues in sub-10 nm technology nodes," *IEEE J. Electron Devices Soc.*, 7(March), pp. 829–836, 2019.

57. Mallikarjunarao, R. Ranjan, K. P. Pradhan, L. Artola, and P. K. Sahu, "Spacer engineered Trigate SOI TFET: An investigation towards harsh temperature environment applications," *Superlattices Microstruct.*, 97, pp. 70–77, 2016.

58. D. Nehra, P. K. Pal, B. K. Kaushik, and S. Dasgupta, "High permittivity spacer effects on junctionless FinFET based circuit/SRAM applications," *18th Int. Symp. VLSI Des. Test, VDAT 2014*, 2014.

7 Scope and challenges with nanosheet FET-based circuit design

Atefeh Rahimifar and Zeinab Ramezani

CONTENTS

7.1 INTRODUCTION

Various solutions have been proposed in recent years to address the problem of short-channel effects (SCEs) in the field effect transistors (FETs) during downscaling [1–4]. The use of Fin-FET devices was one of the most significant solutions used by Intel for the first time in the 22 nm node in 2011 [5]. The structure of the Fin-FET compared to conventional metal oxide semiconductor field effect transistor (MOS-FET) improved the gate's ability to control the channel. Therefore, the Fin-FET devices significantly improved the short-channel effects, particularly when miniaturized in nanometers [6–8]. However, for downscaling to sub-7-nm, Fin-FETs have recently faced numerous challenges, including reduced reducing device performance, increased cost, patterning, and layout [9–11]. In 2017, a new device named nanosheet FET (NS-FET) was introduced. The NS-FET is one of the gate-all-around (GAA) devices that, by increasing the controllability of the gate on the channel, has played an important role in moving toward devices with sub-7-nm dimensions. [12–14]. The NS-FET, by surrounding the transistor channel all around the gate, significantly increases the controllability of the gate. Thus, SCEs reduce, and the efficiency of the transistor improves in the nano regime. NS-FETs are also promising candidates for the technology of 3-nm nodes and even beyond. Figure 7.1 shows the replacement of the NS-FET technology with Fin-FET technology as downscaling continues [12].

DOI: 10.1201/9781003359234-7

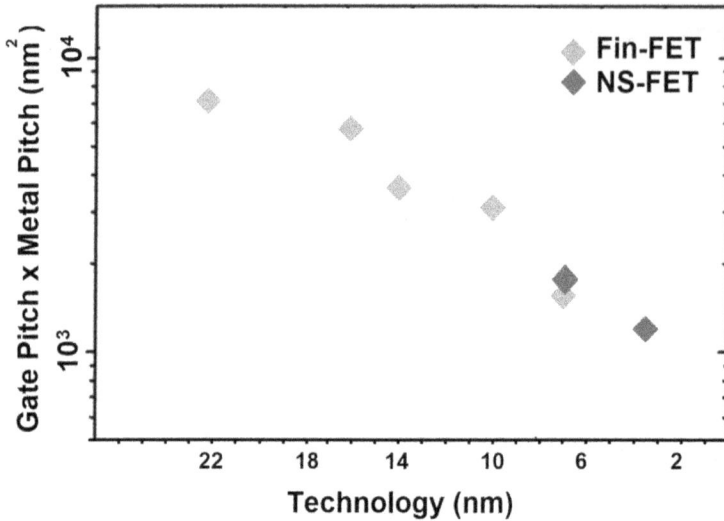

FIGURE 7.1   Replacing of the NS-FET technology with Fin-FET in continuing downscaling [12]

According to research, the use of NS-FET technology resulted in a more than 25% improvement in device performance and a more than 50% reduction in energy consumption [15]. On the other hand, NS-FETs have fewer parasitic capacitors than their counterparts resulting in higher switching capability and better power performance [16–18]. Due to the excellent performance of GAA technology, especially the NS-FETs, the semiconductor industry's behemoths including IBM, Intel, Samsung, and TSMC are moving toward 3-nm and 2-nm NS-FET nodes [5, 19]. Generally, future 3-nm and 2-nm NS-FET nodes are expected to perform 45% better and consume 75% less power consumption than current 7-nm nodes. Therefore, we can anticipate significant advancements in computing platforms that work with quantum computers through cloud environments [15]. This means that NS-FETs are on the verge of taking over the semiconductor world.

With the introduction of the NS-FET as one of the most important candidates for nano regime nodes, the output and electrostatic characteristics of this device must be carefully considered during design. It is also necessary to examine the applications of these devices from the circuit standpoint. In this chapter, while comparing NS-FET with its competitors such as Fin-FET and nanowire FET (NW-FET), the most important design parameters of this device are examined and analyzed from the perspective of temperature variation, changes in the dimensions of nanosheets, the use of high-K gate oxide, etc. in analog/RF and digital applications.

7.2   COMPARISON OF NS-FET WITH OTHER STRUCTURES

Figures 7.2 (a) and (b) shows the structure of NS-FET and stacked NS-FET. In the NS-FET technology, the gate completely surrounds the channel, which increases the

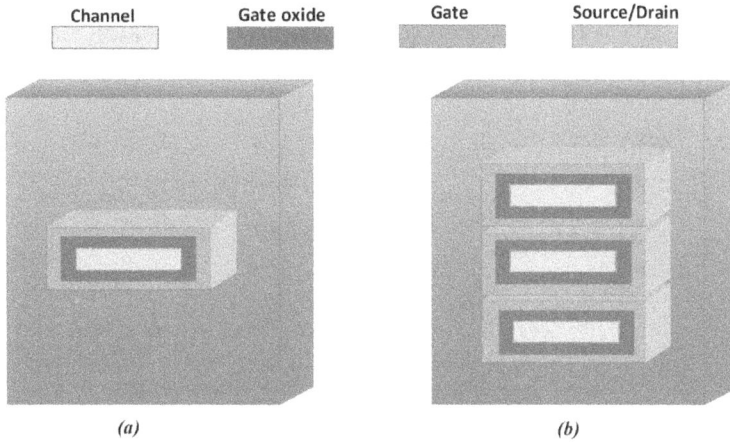

FIGURE 7.2   3D view of the structure of NS-FET (a) single sheet and (b) and stacked sheet

ability of the gate to effectively control the channel. The result shows excellent gate control over short-channel effects. On the other hand, by using several nanosheets as a stack, such as in Figure 7.2 (b), it is possible to improve the output parameters of the device and increase the flexibility in further downscaling [13, 20]. Recent research shows that nanosheets can also be created in multiple stacks [12].

According to recent studies, NS-FET provides more drive current compared to other similar devices such as Fin-FET and NW-FET [21, 22]. The reason is that in NS-FET, the effective width of the channel area (W_{eff}) is larger than other structures and is not limited. Because W_{eff} in NS-FETs is not limited by Fin-pitch or quantization operations, unlike Fin-FETs, designers have more freedom of action in adjusting the width and height of nanoplates for power management and better circuit performance [1, 15, 23]. Figure 7.3 compares the drain current (I_D) in terms of drain-source voltage (V_{DS}) in NS-FET with Fin-FET and NW-FET under similar conditions [24]. As shown in the figure, a higher drain current flows for all gate voltage (V_G) values of the NS-FET, due to the higher W_{eff}.

In order to evaluate the capability of devices in dealing with short-channel effects, it is possible to examine the most critical parameters drain-induced barrier lowering (*DIBL*) and sub-threshold swing (*SS*). *DIBL* is calculated by Equation (7.1), where V_{th} is the threshold voltage. Sub-threshold swing is the rate of change in drain current by one decade in terms of change in gate-source voltage (V_{GS}) and is obtained through the Equation (7.2) [25].

$$DIBL = \left[\frac{V_{th1} - V_{th2}}{V_{DS1} - V_{DS2}} \right] \qquad (7.1)$$

$$SS = \left[\frac{\partial log I_D}{\partial V_{GS}} \right]^{-1} \qquad (7.2)$$

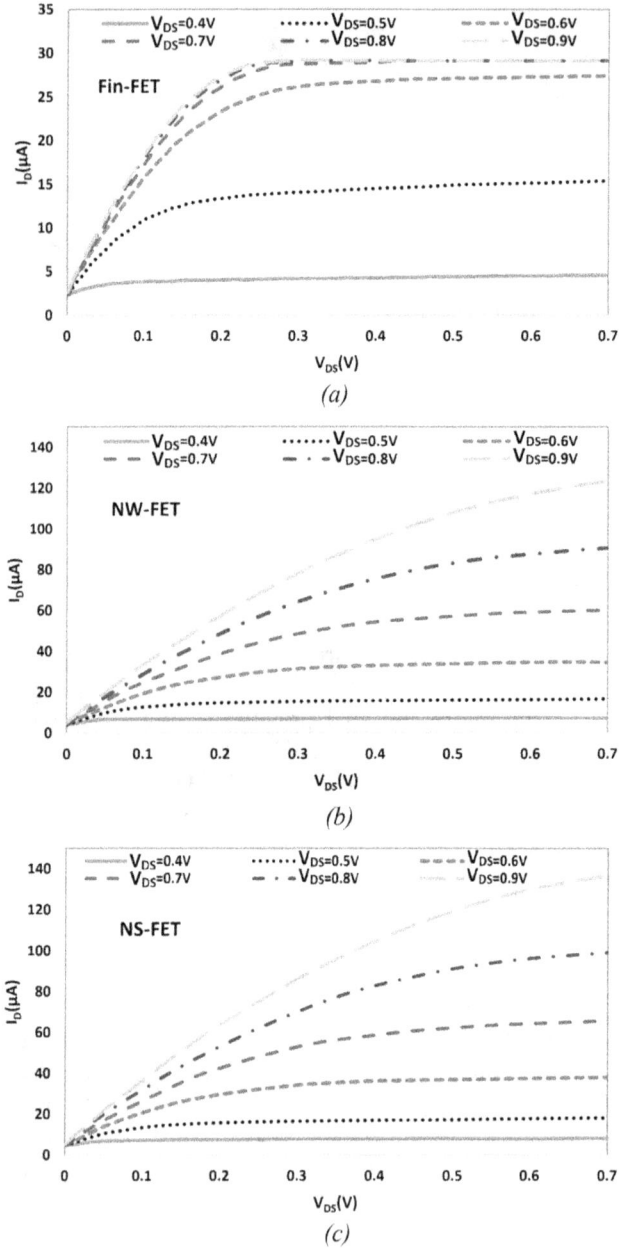

FIGURE 7.3 Output characteristics (I_D–V_{DS}) of (a) Fin-FET, (b) NW-FET, and (c) NS-FET respectively [24]

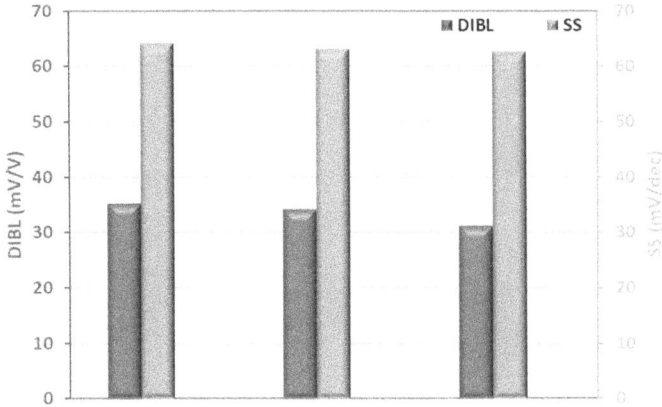

FIGURE 7.4   Comparison of *SS* and *DIBL* in Fin-FET, NW-FET, and NS-FET [24]

Figure 7.4 compares *SS* and *DIBL* for NS-FET, Fin-FET, and NW-FET devices. As shown in the figure, the value of *DIBL* in NS-FET, WN-FET, and Fin-FET is 30.69, 32.06, and 35.14, respectively. It shows *DIBL* in NS-FET is reduced by 12.66% compared to Fin-FET and 4.27% compared to NW-FET. On the other hand, the value of SS in NS-FET has decreased by 1.04% compared to NW-FET and 3.12% compared to Fin-FET [24]. The reason for this superiority of NS-FET compared to both of the other devices is the surrounding of the gate to the channel and as a result, increasing its controlling power over the channel.

Investigating the on current (I_{ON}) and off current (I_{OFF}) of drain is another important parameter of the device output to evaluate its performance. A larger ratio of I_{ON} to I_{OFF} indicates better DC performance of the device and reduces power loss [26]. Figure 7.5 compares I_{ON} and I_{OFF} in NS-FET, NW-FET, and Fin-FET [24]. According to the figure, the I_{OFF} in the NS-FET is significantly reduced compared to the other two structures. On the other hand, I_{ON} in NW-FET has increased by 97.56% compared to Fin-FET, which is the highest value in NS-FET and has increased by 9.31% compared to NW-FET. This significant improvement in I_{ON} and I_{OFF} in NS-FET compared to the other two devices confirms the superiority of NS-FET in downscaling below 7 nm.

Other reasons for the superiority of NS-FET technology include high flexibility in design, faster frequency response, the possibility of supporting multiple threshold voltages, and the possibility of reducing the gate length [12, 27, 28]. On the other hand, the self-heating effects (SHEs) in NS-FET compared to Fin-FET have improved appreciably [29]. Therefore, NS-FETs can be a suitable alternative to Fin-FETs for continuing the downscaling of the gate length [30–32].

In order to design and use the NS-FET in analog and digital integrated circuits, various aspects such as changing the dimensions of the nanosheets, changing the amount of dropped carriers, device dependence on temperature, the engineering of the gate electrode and the channel of the device, the materials used in the construction of the channel and the substrate, the effect of increasing the number of

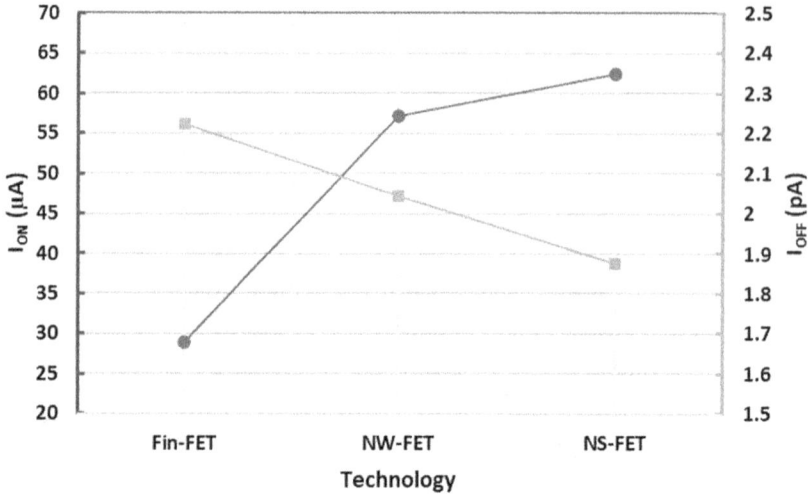

FIGURE 7.5 Comparison of I_{ON} and I_{OFF} in Fin-FET, NW-FET, and NS-FET [24]

nanosheet stacks, and so on, should be studied so that the most optimal device can be designed for each application. In the rest of this chapter, the most important aspects of NS-FET design for use in analog and digital circuits are discussed.

7.3   TEMPERATURE ASSESSMENT OF NS-FET

Temperature evaluation of semiconductor devices is very important, especially in the nano regime, because an increasing temperature can change many output and behavioral characteristics of the device, including the threshold voltage, sub-threshold swing, device speed, and so on. Furthermore, if the temperature of the device network increases, the phonon scattering increases, resulting in a decrease in the device current. On the other hand, the reliability decreases as the temperature dependence increases [33–35]. Since these devices have a wide range of applications in industries that rely on temperature, such as detectors, military, spacecraft, automobile, nuclear, and medicine, temperature evaluation of devices during the design and fabrication process should be considered. The threshold voltage (V_{th}) shows the ability of the device to switch from the *OFF* state to the *ON* [36]. V_{th} is a critical parameter in device scaling to maintain power efficiency.

Figure 7.6 (a) shows a variation of V_{th} and *SS* based on the change in temperature [36]. An increase in temperature causes a decrease in V_{th} and thus an increase in *SS* in the NS-FET. As it is clear from the figure, for lower temperatures, with the reduction of *SS*, the device has less leakage current, which is very important in the design of low-power switches. Another important sub-threshold feature in semiconductor devices is *DIBL*. During the design process, the dependence of *DIBL* on temperature needs to be discussed. Figure 7.6 (b) shows that *DIBL* has a direct and rather linear dependence on temperature and increases

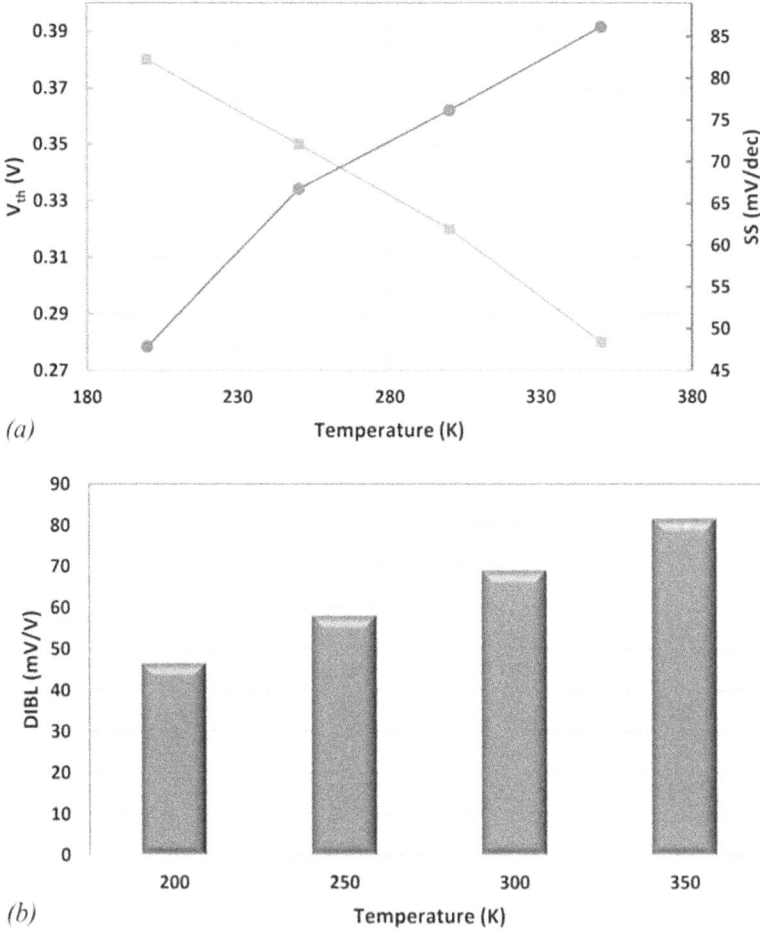

FIGURE 7.6   Variation of (a) V_{th} and SS and (b) $DIBL$ in terms of temperature [36]

with increasing temperature while gate length (L_G) and V_{DS} are constant [36]. During the device design, it is necessary to evaluate the $DIBL$ characteristic and provide solutions to reduce its temperature dependence. Analog/RF communication circuits, digital logic circuits, and memories are integrated and connected together in integrated chips (ICs) and integrated circuits. Therefore, it is important to analyze devices from the point of view of analog/RF circuits. One of the most important criteria for evaluating the analog performance of semiconductor devices is the study of transconductance (g_m) changes. An increase in g_m increases the speed of the transistor, the gate transfer efficiency, and the amplification factor, and improves the performance of the device in logic circuits. g_m has an impact on the amplifier's bandwidth and noise performance. g_m is the variation of the drain current in terms of V_{GS} and it is calculated using Equation (7.3)

[37]. No change or small changes in g_m per temperature change can be reliable evidence to prove the good performance of NS-FET.

$$g_m = \left[\frac{\partial I_D}{\partial V_{GS}} \right] \tag{7.3}$$

Figure 7.7 (a) shows the change of g_m in terms of V_{GS} for different temperatures. The value of g_m increases slightly with decreasing temperature due to a larger I_D. However, the amount of g_m changes compared to the temperature change is relatively small and negligible. This can be one of the reasons for the good performance of NS-FET-based analog circuits in the nano regime. Another important parameter

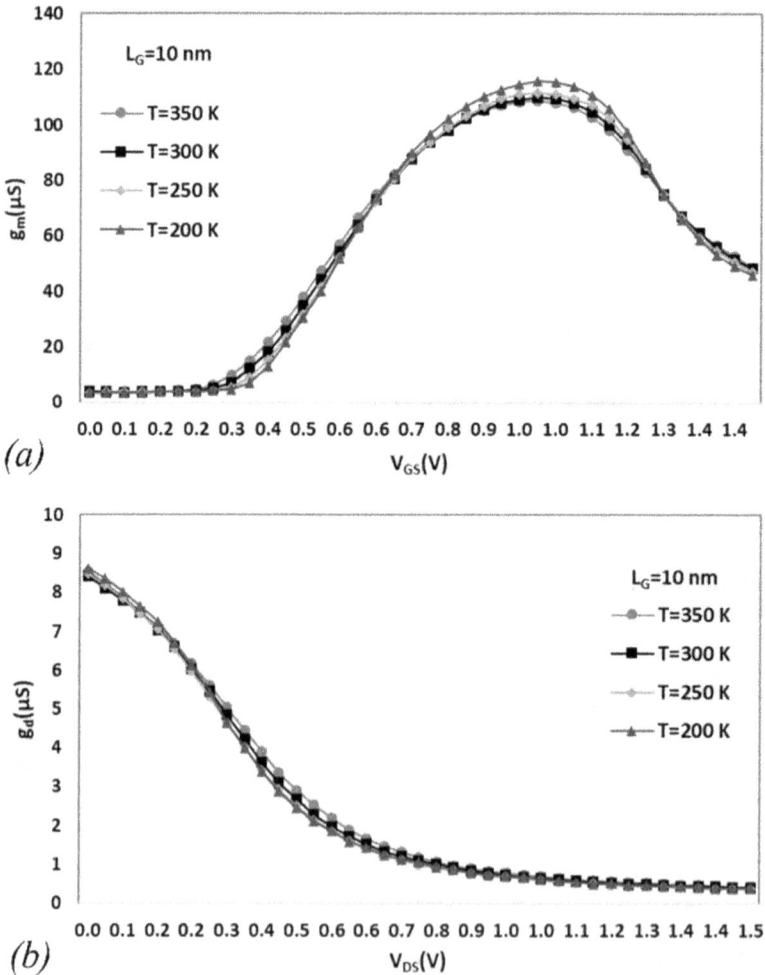

FIGURE 7.7   Variation of (a) transconductance (g_m) and (b) output conductance (g_d) for different temperatures [36]

in the design of semiconductor devices is their output conductivity (g_d), which is required to calculate the intrinsic gain of the device. Equation (7.4) calculates g_d.

$$g_d = \left[\frac{\partial I_D}{\partial V_{DS}} \right]$$

(7.4)

As shown in Figure 7.7 (b), g_d is almost constant and has little changes with increasing temperature. Therefore, it can be expressed that temperature change has no significant effect on g_d.

The total capacity of the gate capacitor (C_{gg}) is equivalent to the gate-drain (C_{gd}) and gate-source (C_{gs}) capacitors [38]. C_{gg} plays a fundamental role in cut-off frequency (f_T) and delays (τ) [36]. The variation of C_{gg} in terms of temperature is described in Figure 7.8

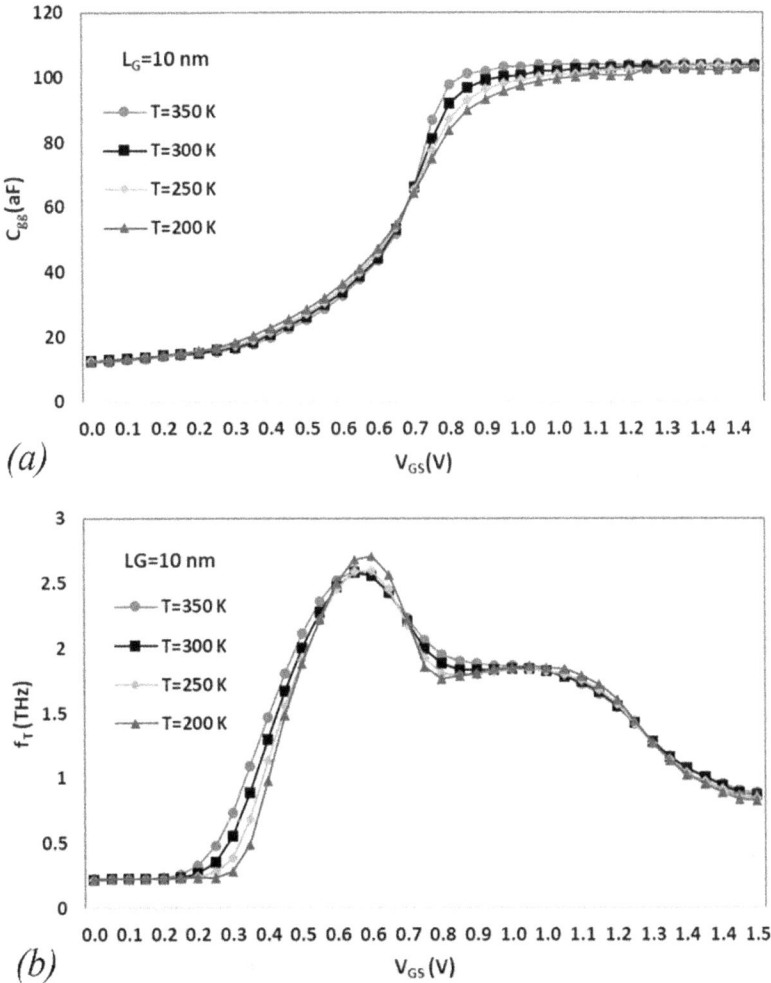

(a)

(b)

FIGURE 7.8   Variation of (a) gate capacitance (C_{gg}) and (b) cut-off frequency (f_T) for different temperatures [36]

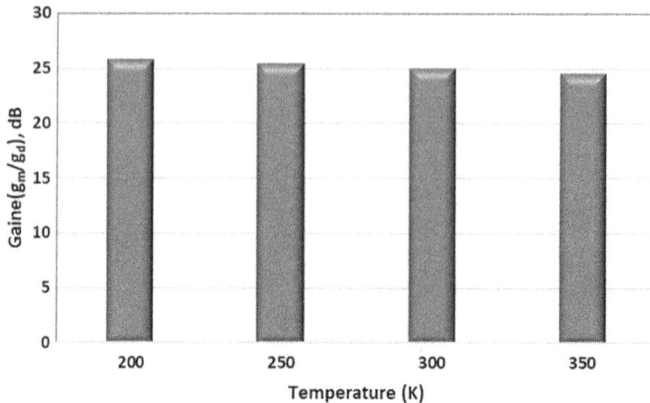

FIGURE 7.9   Variation of *Gain* (g_m / g_d) in terms of temperature [36]

(a). An increase in temperature causes a decrease in the energy band and thus a decrease in the potential barrier of the device. This increases the density of charge carriers in the channel and under the gate, and as a result, the gate capacitor C_{gg} becomes larger [39].

f_T, which is obtained by Equation (7.5) is the frequency at which the current gain is equal to one and is critical to the device's performance in high-frequency operations [40].

$$f_T = \frac{g_m}{2\pi C_{gg}} \tag{7.5}$$

Since the changes of g_m and C_{gg} are very small in terms of temperature changes, it is expected that the cut-off frequency of NS-FET-based RF circuits has a low sensitivity to temperature variations. Figure 7.8 (b) shows that the cut-off frequency changes for different temperatures are relatively small. In this figure, it can be seen that at the temperature of 200 °k, the highest cut-off frequency occurs at V_{GS}=0.6.

The voltage gain of a transistor-based amplifier shows the overall performance of the amplifier circuit. Given that *Gain*= g_m/g_d, any change in g_m or g_d causes a change in *Gain* [36]. Due to the insignificant dependence of g_m and g_d on temperature, as expected and shown in Figure 7.9, the gain changes have decreased relatively little with increasing temperature.

7.4   DOPING ASSESSMENT OF NS-FET

The doping concentration engineering in the channel is one of the primary solutions for improving SCEs and increasing the I_D of MOSFETs [41, 42]. Therefore, it is expected that the drain current and SCEs in the NS-FET be controlled by optimizing the nanosheets' concentration. The doping concentration of nanosheets has a direct effect on the drain current both in the *ON* and *OFF* states. Figure 7.10 describes the effect of increasing doping on the drain current. At lower concentrations, a lower I_{OFF} is observed, and with increasing doping, the value of I_{OFF} increases. However, despite

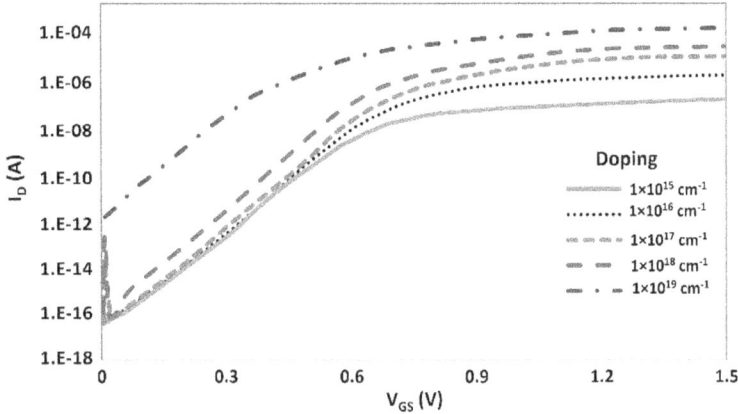

FIGURE 7.10 Variation of I_D for different doping [36]

the increase, relatively little I_{OFF} is observed in NS-FET at high concentrations. On the other hand, it also increases with increasing I_{ON} doping, which is accompanied by a decrease in V_{th}. Therefore, this device with high doping is suitable for use in low-power digital and analog circuits. It should be noted that higher doping increases the Coulomb scattering rate and may decrease the mobility of current carriers [43]. As a result, when designing and manufacturing the device, the performance of the device should be optimized according to the application. So far, some research has been presented considering the effect of varying the concentration of current carriers on the NS-FET drive current which can be considered [44, 45].

7.5 DIMENSION ASSESSMENT OF NANOSHEETS

One of the most important challenges of NS-FET node technology is choosing the appropriate scale of nanosheets. Recently, significant research has been conducted on the effect of nanosheet dimensions on the electrical characteristics of nanosheets. The results showed that by choosing the appropriate dimensions and number of nanosheets, the electrical characteristics of the device can be greatly improved to deal with the SHEs [45–48]. Therefore, it is necessary to evaluate the performance of the device depending on the variation in the width and height of the nanosheets. g_m is one of the most critical characteristics for evaluating the behavior of NS-FETs in analog circuits considering changing the dimensions of the nanosheets. Increasing g_m improves the device's performance in digital logic by increasing transistor speed, gate transfer efficiency, and amplification factor [49]. In order to evaluate g_m in terms of variation in the width (NS_W) and height (NS_H) of the NS-FET nanosheets, one can refer to Figure 7.11 (a). In this figure, the variation of the maximum transconductance value ($g_{m,Max}$) with respect to the different widths and heights of the NS-FET is shown. $g_{m,Max}$ increases almost linearly with increasing width and height of nanosheets. This is obvious, because with the increase in the dimensions of the nanosheets, the number of carriers increases, and as a result, $g_{m,Max}$ increases. An increase in the number of nanosheets also leads to an increase in $g_{m,Max}$.

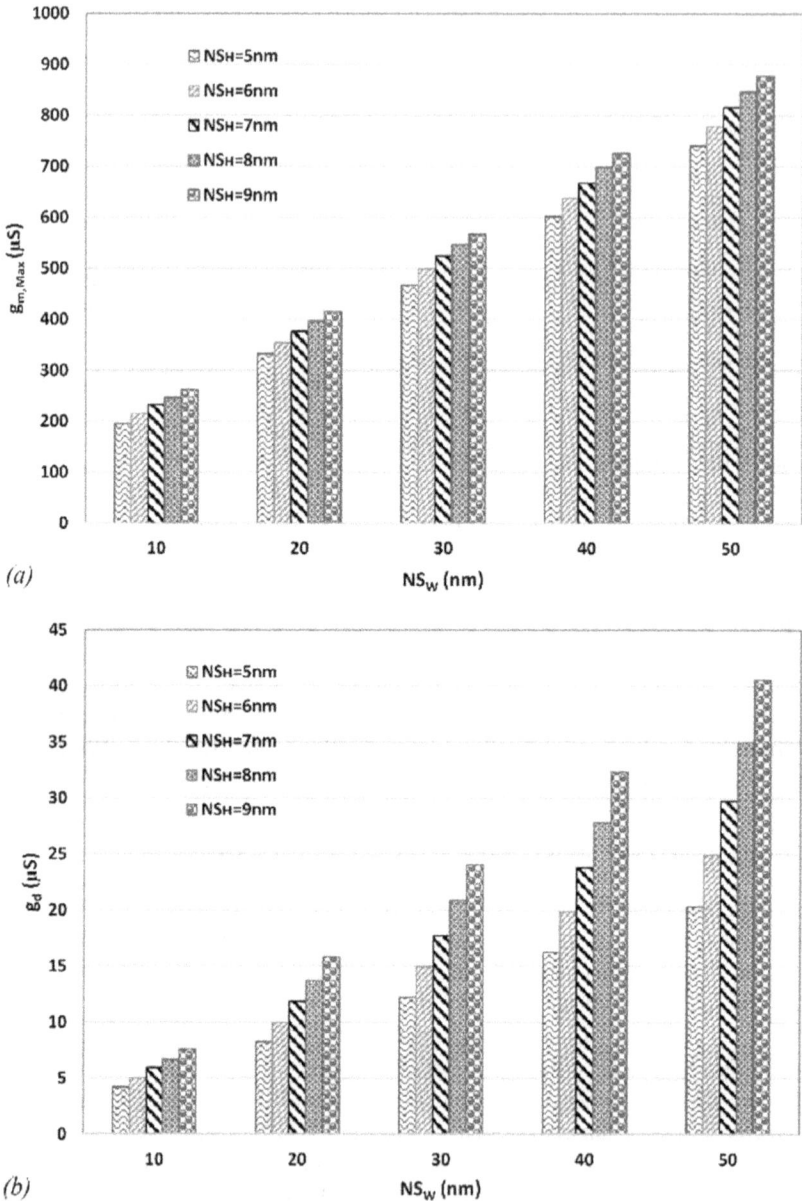

FIGURE 7.11 Variation of (a) transconductance ($g_{m,Max}$) and (b) output conductance (g_d) in terms of dimension variations of nanosheets [24]

Meanwhile, $g_{m,Max}$ for Fin-FET and NW-FET in the same conditions (5-nm technology) is around 100 μs and 50 μs, respectively [50].

Typically, the devices used in the design of analog circuits work in the saturation region, in which I_D is independent of V_{DS}. However, as the device shrinks and the channel length shortens, the effect of V_{DS} on the electrostatics of the channel has an impact on the I_D and, as a result, an increase in g_d [25]. Therefore, it is important to study g_d changes in nano regime devices. Figure 7.11 (b) shows the changes in g_d according to the change in the height of nanosheets. According to this figure, with the increase in the dimensions of the nanosheets, g_d has also increased. Considering that increasing the dimensions of nanosheets improves g_m while decreasing g_d, it is preferable to evaluate and optimize these two important parameters when designing analog devices. Gate capacitors are divided into two types: intrinsic capacitors and parasitic capacitors. As the dimensions of the device are reduced, the parasitic capacitors increase, causing the device's and integrated circuits' speeds to decrease [circuit-3nm]. Two important parameters in the analysis of analog devices are the measurement of the total gate capacitor and gate-drain capacitor. Figures 7.12 (a)

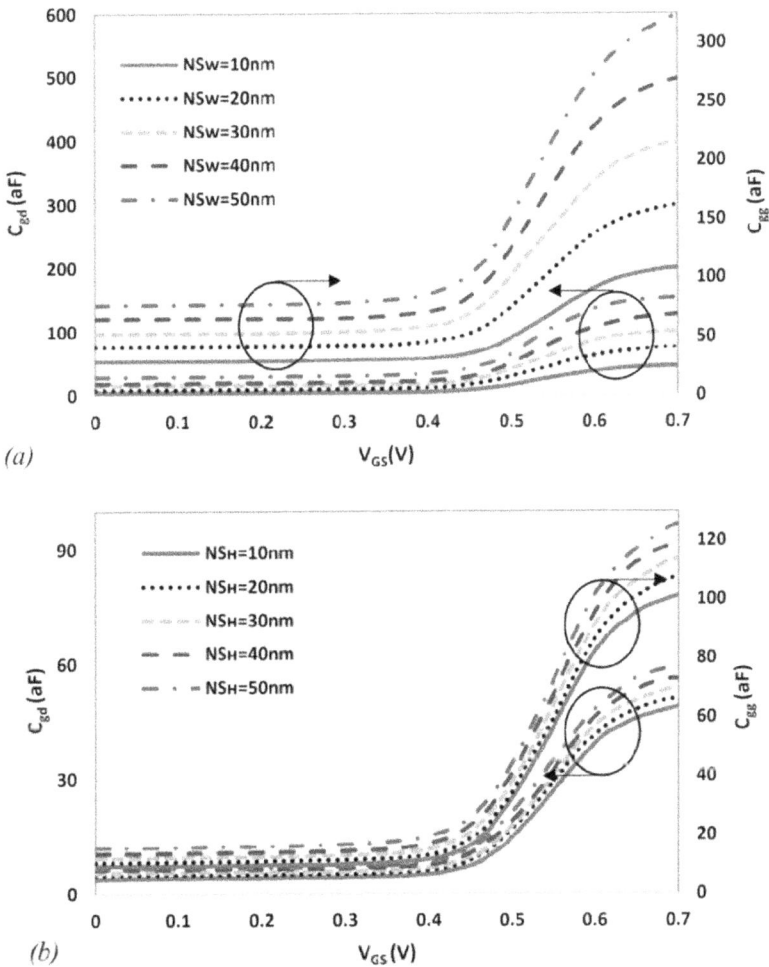

(a)

(b)

FIGURE 7.12 Variation of C_{gg} and C_{gd} in terms of (a) width variation and (b) height variation of nanosheets [24]

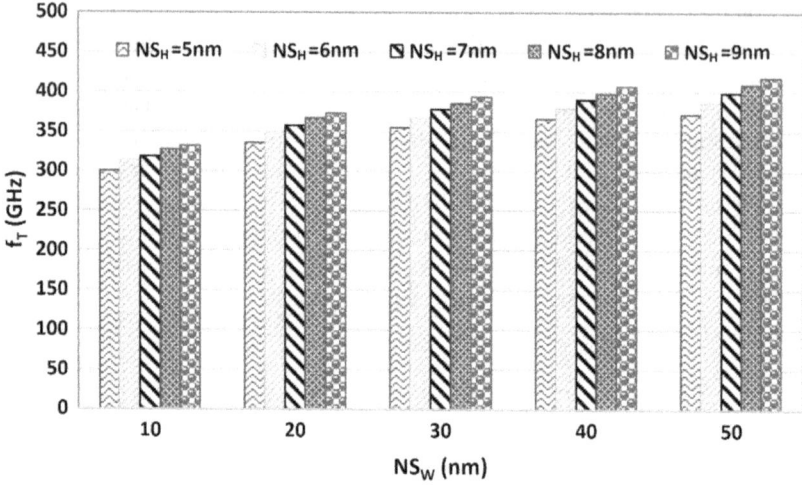

FIGURE 7.13 Variation of cut-off frequency (f_T) in terms of height variation nanosheets [24]

and (b) show the changes of C_{gg} and C_{gd}, according to the changes in the width of nanosheets (NS_W) and the height of nanosheets (NS_H), respectively. Changing the dimensions of nanosheets by varying the active distance causes changes in the parasitic capacitors of the transistor [51]. As the dimensions of the nanosheets decrease, both C_{gg} and C_{gd} capacitors decrease due to smaller out margins and less overlap. One of the challenges in the design of nodes of sub-7-nm is the small capacitance of the devices. This makes accurate measurement of capacitors difficult. Furthermore, changing the g_m and capacitors of semiconductor devices causes a change in the cut-off frequency (f_T). As shown in Figure 7.13, increasing the width and height of nanosheets increases the cut-off frequency. The studies conducted on the dimensions of nanosheets cause significant changes in the analog parameters of the device. Therefore, during the designing and manufacturing of NS-FET devices, optimal dimensions should be obtained by taking into account all aspects.

7.6 ASSESSMENT OF USING HIGH-K DIELECTRIC AS GATE OXIDE

A primary solution for dealing with SCE during downscaling is to reduce the gate oxide thickness (t_{ox}). In the new devices, t_{ox} has reached its critical limit, and further thinning leads to tunneling and increased gate current, resulting in power loss. A cost-effective solution is to replace SiO_2 with a high-K dielectric as the gate oxide. This reduces the SCE while also lowering the gate tunneling current [12, 52–54].

In this section, the effect of using TiO_2 with $K=40$ instead of SiO_2 with $K=3.9$ as a high-K dielectric on the electrical and electrostatic characteristics of NS-FET is investigated [55]. K denotes the dielectric of the dielectric constant. Table 7.1 compares the NS-FET device equipped with TiO_2 with other recent NS-FETs. It can be seen that the use of high-K dielectric improves I_{ON}/I_{OFF} and SS.

TABLE 7.1

Comparison of electrical characteristics in recent works on NS-FET [55]

Channel length (L_C) (nm)	I_{ON} ($\mu A/\mu m$)	I_{ON}/I_{OFF}	SS (mV/dec)	Reference
14	290.5	-	68.10	[56]
12	699	1.2×10^5	71	[57]
12	1410	1.41×10^5	73.9	[44]
12	646	1.24×10^7	68.8	[55] when SiO_2 was used as the gate oxide
12	779	2.5×10^{107}	70	[55] when SiO_2 was used as the gate oxide

Figure 7.14 displays the effect of using high-K dielectric on transconductance and output conductance in the NS-FET. As shown in the figure, the use of high-K dielectric improves g_m and degrades g_d, although the rate of degradation of g_d is relatively small. In addition, for the thinner nanosheets with a larger surface, the device will have higher g_m and lower g_d [55]. Figure 7.15 shows the changes of C_{gg} and C_{gd} in terms of normalized drain current when the gate oxide is SiO_2 and it is TiO_2. According to the figure, using high-K dielectric increases C_{gg} while keeping C_{gd} almost constant. According to Equation (7.5), increasing C_{gg} causes a decrease in f_T. On the other hand, an increase in g_m causes an increase in f_T. Therefore, according to what Figure 7.16 depicts, using a high-K dielectric, the f_T increases compared to the conventional structure with SiO_2 as a gate

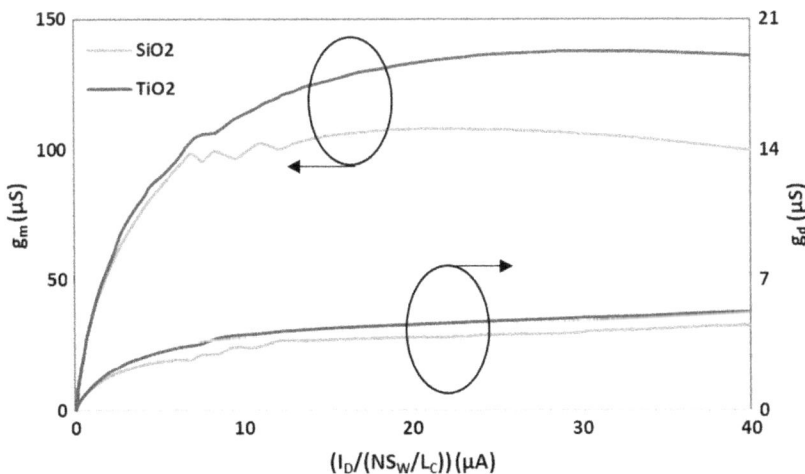

FIGURE 7.14 Comparison of the effect of using high-K dielectric (TiO_2) instead of SiO_2 on variation of g_m and g_d in the NS-FET [55]

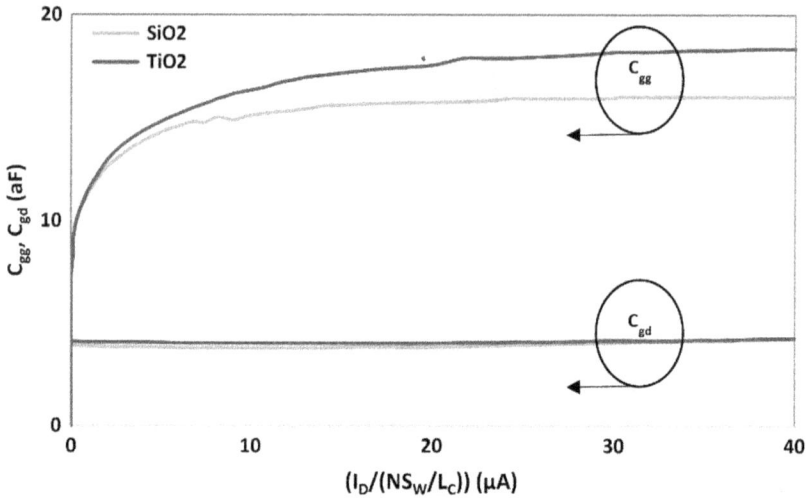

FIGURE 7.15 Comparison of the effect of using high-K dielectric (TiO$_2$) instead of SiO$_2$ on variation of the changes of C_{gg} and C_{gd} in the NS-FET [55]

oxide. Therefore, it can be said that the use of high-K dielectric in NS-FET can relatively improve the analog parameters of the device. This is especially important in low-power applications [55]. According to the results obtained in recent research, the use of high-K dielectric in logic gates and digital circuits has almost no effect on their performance [55, 58].

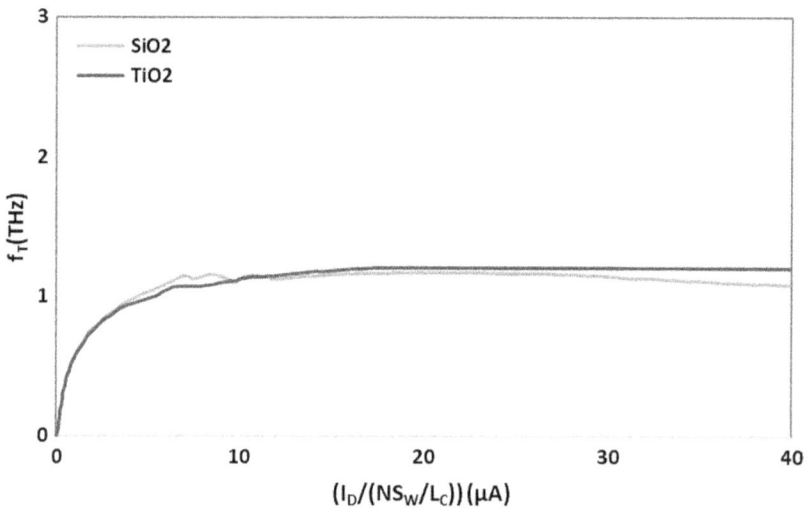

FIGURE 7.16 Comparison of the effect of using high-K dielectric (TiO$_2$) instead of SiO$_2$ on variation of f_T in the NS-FET [55]

7.7   DIGITAL APPLICATIONS

Logic gates are the basis of digital circuits [58]. Therefore, it is important to study the performance of the device from the perspective of logic gates. One of the most important parameters that must be considered to evaluate the performance of logic gates is the gain of the gate. On the other hand, in today's high-speed world of technology, the study of delay is of particular importance. Therefore, the time behavior of digital circuits based on NS-FET should be evaluated and optimized in the different number of circuit stages. In addition, since most digital circuits are designed for low-power applications, the power consumption of digital gates is a hot topic. When designing and manufacturing NS-FET for use in digital applications, it is necessary to evaluate several key parameters such as gain, delay, and power consumption of logic gates. Recent research results show that reducing the channel length reduces the gain [58] and speed of logic gates based on NS-FET [24]. However, due to its novelty, research is still very sparse. Therefore, analyzing the behavior of NS-FETs in the implementation of digital circuits would be beneficial in the future.

7.8   CONCLUSION

By analyzing and evaluating the most recent scientific findings, this study attempted to compare the structure and performance of NS-FET with other competitors in the nano regime. By exploring the conducted research, the reasons for the superiority of the NS-FET over Fin-FET and NW-FET in downscaling to sub-7-nm were described. By comparing the behavior of the electrostatic characteristics of NS-FET with Fin-FET and NW-FET in sub-7-nm nodes, we found that NS-FET can be more effective in dealing with short-channel effects because of controlling the gate over the channel range. In addition, the structure and electrostatic characteristics of NS-FETs were investigated for analog/RF and digital applications. The most important challenges of designing semiconductor devices, including the effect of temperature variation, modifying the dimensions of nanosheets, using high-K dielectric instead of gate oxide, and the amount of doping concentration on the internal and output characteristics of the circuit, were evaluated and discussed. Since nowadays semiconductor devices form the foundation of all electronic circuits in various applications such as analog, digital, dynamic, and static memory types, the structure of the device must be optimized based on the type and requirements of the desired application.

REFERENCES

1. Y. Lee et al., "Design study of the gate-all-around silicon nanosheet MOSFETs," *Semiconductor Science and Technology*, 35(3), p. 03LT01, 2020.
2. L. Cao et al., "Novel channel-first fishbone FETs with symmetrical threshold voltages and balanced driving currents using single work function metal process," *IEEE Transactions on Electron Devices*, 69(11), pp. 5971–5977, 2022.

3. A. A. Orouji, A. Rahimifar, and M. Jozi, "A novel double-gate SOI MOSFET to improve the floating body effect by dual SiGe trench," *Journal of Computational Electronics*, 15(2), pp. 537–544, 2016.

4. M. K. Anvarifard, Z. Ramezani, and I. S. Amiri, "Proposal of an embedded nanogap biosensor by a graphene nanoribbon field-effect transistor for biological samples detection," *Physica Status Solidi (A)*, 217(2), p. 1900879, 2020.

5. P. Ye, T. Ernest, and M. V. Khare, "The nanosheet transistor is the next (and maybe last) step in Moore's law," *IEEE Spectrum*, 30, 2019.

6. T. A. Bhat, M. Mustafa, and M. R. Beigh, "Study of short channel effects in n-FinFET structure for Si, GaAs, GaSb and GaN channel materials," *Nano- Electron. Phys.*, 7(3), p. 3010, 2015.

7. T. Yamashita et al., "Sub-25nm FinFET with advanced fin formation and short channel effect engineering," in *2011 Symposium on VLSI Technology-Digest of Technical Papers*, pp. 14–15, 2011.

8. B. Yu et al., "FinFET scaling to 10 nm gate length," in *Digest International Electron Devices Meeting*, pp. 251–254, 2002.

9. R. Divakaruni and V. Narayanan, "(Keynote) challenges of 10 nm and 7 nm CMOS for server and mobile applications," *ECS Translator*, 72(4), p. 3, 2016.

10. M. G. Bardon et al., "Dimensioning for power and performance under 10nm: The limits of FinFETs scaling," in *2015 International Conference on IC Design & Technology (ICICDT)*, pp. 1–4, 2015.

11. V. Narula and M. Agarwal, "Enhanced performance of double gate junctionless field effect transistor by employing rectangular core–shell architecture," *Semiconductor Science and Technology*, 34(10), p. 105014, 2019.

12. N. Loubet et al., "Stacked nanosheet gate-all-around transistor to enable scaling beyond FinFET," in *2017 Symposium on VLSI Technology*, pp. T230–T231, 2017.

13. S.-D. Kim, M. Guillorn, I. Lauer, P. Oldiges, T. Hook, and M.-H. Na, "Performance trade-offs in FinFET and gate-all-around device architectures for 7nm-node and beyond," in *2015 IEEE SOI-3D-Subthreshold Microelectronics Technology Unified Conference (S3S)*, pp. 1–3, 2015.

14. D. Jang et al., "Device exploration of nanosheet transistors for sub-7-nm technology node," *IEEE Transactions on Electron Devices*, 64(6), pp. 2707–2713, 2017.

15. S. Valasa, S. Tayal, L. R. Thoutam, J. Ajayan, and S. Bhattacharya, "A critical review on performance, reliability, and fabrication challenges in nanosheet FET for future analog/digital IC applications," *Micro and Nanostructures*, 170, p. 207374, 2022.

16. J. Kim, J.-S. Lee, J.-W. Han, and M. Meyyappan, "Single-event transient in FinFETs and nanosheet FETs," *IEEE Electron Device Letters*, 39(12), pp. 1840–1843, 2018.

17. J.-S. Yoon, J. Jeong, S. Lee, and R.-H. Baek, "Systematic DC/AC performance benchmarking of sub-7-nm node FinFETs and nanosheet FETs," *IEEE Journal of the Electron Devices Society*, 6, pp. 942–947, 2018.

18. S. Barraud et al., "Performance and design considerations for gate-all-around stacked-nanowires FETs," in *2017 IEEE International Electron Devices Meeting (IEDM)*, pp. 22–29, 2017.

19. H. Jagannathan et al., "Vertical-transport nanosheet technology for CMOS scaling beyond lateral-transport devices," in *2021 IEEE International Electron Devices Meeting (IEDM)*, pp. 21–26, 2021.

20. S. Bangsaruntip et al., "High performance and highly uniform gate-all-around silicon nanowire MOSFETs with wire size dependent scaling," in *2009 IEEE International Electron Devices Meeting (IEDM)*, pp. 1–4, 2009.

21. N. Abas, S. Dilshad, A. Khalid, M. S. Saleem, and N. Khan, "Power quality improvement using dynamic voltage restorer," *IEEE Access*, 8, pp. 164325–164339, 2020.

22. M.-C. Chen et al., "TMD FinFET with 4 nm thin body and back gate control for future low power technology," in *2015 IEEE International Electron Devices Meeting (IEDM)*, p. 32, 2015.

23. A. Veloso et al., "Nanosheet FETs and their potential for enabling continued Moore's law scaling," in *2021 5th IEEE Electron Devices Technology & Manufacturing Conference (EDTM)*, pp. 1–3, 2021.

24. V. B. Sreenivasulu and V. Narendar, "Design insights of nanosheet FET and CMOS circuit applications at 5-nm technology node," *IEEE Transactions on Electron Devices*, 69(8), pp. 4115–4122, 2022.

25. V. B. Sreenivasulu and V. Narendar, "Characterization and optimization of junctionless gate-all-around vertically stacked nanowire FETs for sub-5 nm technology nodes," *Microelectronics Journal*, 116, p. 105214, 2021.

26. R. M. Imenabadi, M. Saremi, and W. G. Vandenberghe, "A novel PNPN-like Z-shaped tunnel field-effect transistor with improved ambipolar behavior and RF performance," *IEEE Transactions on Electron Devices*, 64(11), pp. 4752–4758, 2017.

27. N. Singh et al., "Ultra-narrow silicon nanowire gate-all-around CMOS devices: Impact of diameter, channel-orientation and low temperature on device performance," in *2006 International Electron Devices Meeting*, pp. 1–4, 2006.

28. C. W. Yeung et al., "Channel geometry impact and narrow sheet effect of stacked nanosheet," in *2018 IEEE International Electron Devices Meeting (IEDM)*, pp. 26–28, 2018.

29. G. Chalia and R. S. Hegde, "Study of self-heating effects in silicon nano-sheet transistors," in *2018 IEEE International Conference on Electron Devices and Solid State Circuits (EDSSC)*, pp. 1–2, 2018.

30. S. Reboh et al., "Imaging, modeling and engineering of strain in gate-all-around nanosheet transitors," in *2019 IEEE International Electron Devices Meeting (IEDM)*, pp. 11–15, 2019.

31. H. Mertens et al., "Vertically stacked gate-all-around Si nanowire transistors: Key process optimizations and ring oscillator demonstration," in *2017 IEEE International Electron Devices Meeting (IEDM)*, pp. 34–37, 2017.

32. S. Barraud et al., "Performance and design considerations for gate-all-around stacked-nanowires FETs," in *2017 IEEE International Electron Devices Meeting (IEDM)*, pp. 22–29, 2017.

33. P. Srinivas, A. Kumar, S. Jit, and P. K. Tiwari, "Self-heating effects and hot carrier degradation in In0. 53Ga0. 47As gate-all-around MOSFETs," *Semiconductor Science and Technology*, 35(6), p. 065008, 2020.

34. I. S. Myeong, M. J. Kang, and H. Shin, "Self-heating and electrothermal properties of advanced sub-5-nm node nanoplate FET," *IEEE Electron Device Letters*, 41(7), pp. 977–980, 2020.

35. C.-C. Chung, H.-Y. Ye, H. H. Lin, W. K. Wan, M.-T. Yang, and C. W. Liu, "Self-heating induced interchannel V_t difference of vertically stacked Si nanosheet gate-all-around MOSFETs," *IEEE Electron Device Letters*, 40(12), pp. 1913–1916, 2019.

36. V. B. Sreenivasulu and V. Narendar, "Design and temperature assessment of junctionless nanosheet FET for nanoscale applications," *Silicon*, 14(8), pp. 3823–3834, 2022.

37. C. K. Pandey, D. Dash, and S. Chaudhury, "Improvement in analog/RF performances of SOI TFET using dielectric pocket," *International Journal of Electronics*, 107(11), pp. 1844–1860, 2020.

38. V. Narendar, "Performance enhancement of FinFET devices with gate-stack (GS) high-K dielectrics for nanoscale applications," *Silicon*, 10(6), pp. 2419–2429, 2018.

39. R. Saha, B. Bhowmick, and S. Baishya, "Temperature effect on RF/analog and linearity parameters in DMG FinFET," *Applied Physics. Part A*, 124(9), pp. 1–10, 2018.

40. N. Vadthiya, P. Narware, V. Bheemudu, and B. Sunitha, "A novel bottom-spacer ground-plane (BSGP) FinFET for improved logic and analog/RF performance," *AEU – International Journal of Electronics and Communications*, 127, p. 153459, 2020.

41. Z. Ramezani, A. A. Orouji, S. A. Ghoreishi, and I. S. Amiri, "A Nano junctionless double-gate MOSFET by using the charge plasma concept to improve short-channel effects and frequency characteristics," *Journal of Electronic Materials*, 48(11), pp. 7487–7494, 2019.

42. Z. Ramezani and A. A. Orouji, "Investigation of veritcal graded channel doping in nanoscale fully-depleted SOI-MOSFET," *Superlattices and Microstructures*, 98, pp. 359–370, 2016.

43. H. Park and B. Choi, "A study on the performance of metal-oxide-semiconductor -field-effect-transistors with asymmetric junction doping structure," *Current Applied Physics*, 12(6), pp. 1503–1509, 2012.

44. D. Nagy, G. Espineira, G. Indalecio, A. J. García-Loureiro, K. Kalna, and N. Seoane, "Benchmarking of FinFET, nanosheet, and nanowire FET architectures for future technology nodes," *IEEE Access*, 8, pp. 53196–53202, 2020.

45. Y. Choi et al., "Simulation of the effect of parasitic channel height on characteristics of stacked gate-all-around nanosheet FET," *Solid-State Electronics*, 164, p. 107686, 2020.

46. A. Goel, A. Rawat, and B. Rawat, "Benchmarking of analog/RF performance of finFET, NW-FET, and NS-FET in the ultimate scaling limit," *IEEE Transactions on Electron Devices*, 69(3), pp. 1298–1305, 2022.

47. V. Jegadheesan, K. Sivasankaran, and A. Konar, "Impact of geometrical parameters and substrate on analog/RF performance of stacked nanosheet field effect transistor," *Materials Science in Semiconductor Processing*, 93, pp. 188–195, 2019.

48. S. Kim et al., "Investigation of electrical characteristic behavior induced by channel-release process in stacked nanosheet gate-all-around MOSFETs," *IEEE Transactions on Electron Devices*, 67(6), pp. 2648–2652, 2020.

49. C.-W. Lee et al., "High-temperature performance of silicon junctionless MOSFETs," *IEEE Transactions on Electron Devices*, 57(3), pp. 620–625, 2010.

50. U. K. Das and T. K. Bhattacharyya, "Opportunities in device scaling for 3-nm node and beyond: FinFET versus GAA-FET versus UFET," *IEEE Transactions on Electron Devices*, 67(6), pp. 2633–2638, 2020.

51. P. Kushwaha, A. Dasgupta, M.-Y. Kao, H. Agarwal, S. Salahuddin, and C. Hu, "Design optimization techniques in nanosheet transistor for RF applications," *IEEE Transactions on Electron Devices*, 67(10), pp. 4515–4520, 2020.

52. H. S. Momose et al., "1.5 nm direct-tunneling gate oxide Si MOSFET's," *IEEE Transactions on Electron Devices*, 43(8), pp. 1233–1242, 1996.

53. J. Robertson and R. M. Wallace, "High-K materials and metal gates for CMOS applications," *Materials Science and Engineering: R: Reports*, 88, pp. 1–41, 2015.

54. H. Wong and H. Iwai, "On the scaling of subnanometer EOT gate dielectrics for ultimate Nano CMOS technology," *Microelectronic Engineering*, 138, pp. 57–76, 2015.

55. S. Tayal et al., "Gate-stack optimization of a vertically stacked nanosheet FET for digital/analog/RF applications," *Journal of Computational Electronics*, 21, pp. 1–10, 2022.

56. S. Gupta and A. Nandi, "Effect of air spacer in underlap GAA nanowire: An analogue/RF perspective," *IET Circuits, Devices and Systems*, 13(8), pp. 1196–1202, 2019.

57. M.-J. Tsai et al., "Fabrication and characterization of stacked poly-Si nanosheet with gate-all-around and multi-gate junctionless field effect transistors," *IEEE Journal of the Electron Devices Society*, 7, pp. 1133–1139, 2019.

58. S. Tayal et al., "Investigation of nanosheet-FET based logic gates at sub-7 nm technology node for digital IC applications," *Silicon*, 14, pp. 1–7, 2022.

8 Scope with TFET-based circuit and system design

P. Suveetha Dhanaselvam, B. Karthikeyan,
P. Vanitha, and P. Anand

CONTENTS

8.1 INTRODUCTION

"System on a chip" (SoC) is an enriching research field that involves miniaturization, designing, and testing of transistors. Former trends in SoC design were concentrated on perfecting the performance of the system without giving substantial consideration to power consumption. Complementary metal oxide semiconductor (CMOS) has been an extensively accepted technology for designing SoCs for decades [1]. CMOS systems has become more pronounced in recent years through technology scaling, especially multigate structures. Even though speed is saturated, the rising number of transistors per chip, a reduction in size that follows Moore's law, led to multi-core processor chips. Technology scaling significantly outstands the system performance, but it also allowed an increase in the complexity of systems in cost-effective ways. One such challenge is the life of the battery and its power consumption. The growth of energy-effective systems is becoming imperative with the extensively increasing use of battery-operated systems. Longer battery life while maintaining efficient performance mandates minimum power consumption. Still, power consumption has become a major constraint in design specifications because of increased leakage with every new technology development. In the advent of designing energy-efficient and low-cost devices overcoming the said constraints, TFET (tunnel field-effect transistor)-based devices have gained a lot of attention among various conventional CMOS devices. Over the past years, there has been a mounting appeal for the TFET, and there is extensive research being done on this transistor.

Applications such as the Internet of Things (IoT), wireless sensor networks (WSN), biosensors, etc., demanded low-power cost-effective reliable devices, which

DOI: 10.1201/9781003359234-8

led to a boom in studies on new circuits and system designs based on steep-slope TFETs [2] to overcome MOSFET's limitation of unavoidable increasing leakage power while maintaining acceptable performance in low voltage operations. Amid those emergent transistor technologies, TFET has become a reliable one due to its compatibility with the CMOS process and negligible leakage current on the order of fA/μm [3].

8.2 TUNNELING FIELD-EFFECT TRANSISTOR

As one of the promising alternatives for the conventional MOSFET, T. Baba et al. developed the tunneling field-effect transistor in 1992. The merits of TFET include a subthreshold swing of less than 60mV/dec, reduced short-channel effects, ultra-low-power operation, and reduced leakage currents. The band-to-band tunneling mechanism is an awesome feature of TFET, which is responsible for the reduction of leakage current and thereby enhances the ON-OFF ratio. TFET also offers high speed and energy efficiency in the domain of ultra-low-power integrated circuits. Tunnel FET can be seen as a proficient substitute for the MOSFET for ultra-low-power and high-speed applications [4]. The construction of TFET is similar to MOSFET with immense variation in the switching mechanism. The basic structure of TFET is a gated PIN diode with a quantum band-to-band tunneling (BTBT) phenomenon. Tunneling greatly increases the operating speed, with an increased I_{ON}/I_{OFF} ratio and low threshold voltage.

A typical TEFT device structure consists of a PIN junction with a p-type source, intrinsic channel, and n-type drain, in which the electrostatic potential of the intrinsic area is controlled by the gate terminal. This is depicted in Figure 8.1. The potential applied at the gate accumulates electrons in the intrinsic channel section. With the gate bias reaching the threshold voltage, BTBT materializes when the conduction band of the intrinsic region aligns with the valence band of the P-region. BTBT

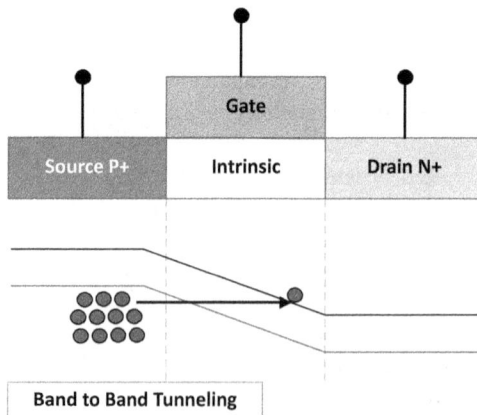

FIGURE 8.1   TFET structure and BTBT

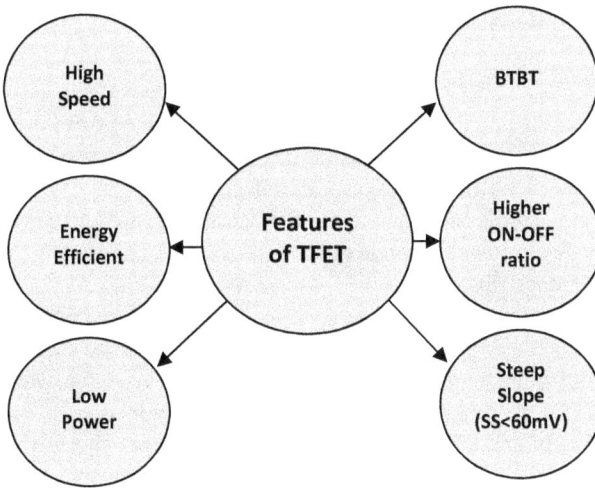

FIGURE 8.2   Characteristics of TFET

involves the variation of the position of the band gap of the intrinsic channel region of the device relative to the source and drain energy levels. When a positive voltage is applied to the gate of TFET, there is a sufficient narrowing of the band gap leading to tunneling, which in turn switches ON the device. In the OFF state, the gate bias is low, close to 0 V, resulting in the misaligned broader band gap of the channel blocking the tunneling. Unlike MOSFET, TFET conducts for both positive and negative values of gate voltage, with the band-to-band tunneling happening at the source channel junction or at the channel drain junction. Thus, TFET is ambipolar in nature [5].

TFE-based processors and digital ICs have a huge potential in today's world of mobile devices functioning on lower power budgets. In addition to the inherent characteristics of the TFET devices, as shown in Figure 8.2, their compatibility and integration with CMOS devices on the same chip make them the most viable devices. The unique characteristics of the tunnel FET direct the technology toward ultra-low-power and compact new circuit topologies [1]. Various research has demonstrated a rational variation of TFET technology including heterojunction and silicon-based TFETs [6]. In hybrid applications, heterojunction TFETs (HTFET) are preferred since the objective is to surpass the performance of conventional MOS devices while reducing leakage currents. Silicon TFETs, alternatively, offer a lower drive current but also provide reduced leakage and are less constrained in reverse-bias operation. In contrast to HTFETs, the fabrication of silicon TFET is well-matched with current CMOS processes offering sensible production yield and seamless on-chip co-integration with standard CMOS devices. Nowadays there is tremendous growth in the market of battery-operated devices, which includes event-trigged novel devices and sensors with long standby cycles demanding scaled-down supply voltages and low leakages, providing a larger scope for TFET-based circuit and system design.

8.3   SCOPE AND APPLICATIONS

8.3.1   TFET-BASED BIOSENSORS

The biosensor is a device that can generate electrical signals from the physiochemical reaction of biomolecules. The sensing process of the targeted biomolecule primarily comprises two different stages: biomolecule detection and transduction [7]. The targeted biomolecules are analyzed at the detection stage and a measurable electrical signal for further processing is generated from the physiochemical reaction in the transduction stage [8].

Life-threatening lethal bio attacks such as the coronavirus position humans on high alert. Invisible and rapidly spreading viruses make people's lives so miserable. Other than this, improvements in weapon technology pave the way for bio wars with advanced bio warheads, which comprise pathogenic viruses or bacteria that spread very silently and can take the lives of innocent people. In the modern-day world, mitigation against biohazards is a huge challenge, and biosensors provide enhancements and refinements for this issue with methodical tactics for biomolecule detection. Biosensor technology has improved significantly since Clark et al. [9] discovered the first enzyme-based biosensor in 1962. With fast, reliable, and accurate detection, biosensors have had widespread applications ranging from the medical field for early-stage disease detection and diagnosis, drug delivery, food processing, environment monitoring, security, and surveillance.

In recent times, FET-based biosensors [10–12] have received a lot of attention from researchers worldwide due to their superior properties like label-free detection, small size, rapid response, reliability, the possibility of on-chip integration for amplification circuitry and sensors, and the possibility for mass production with low cost, high selectivity, and reusability. To detect targeted biomolecules, the oxide layer of the FET is employed with the bio receptors/bio-recognition element. Once these receptors capture the targeted biomolecules, they undergo a conjugation process that generates electrochemical reactions, and these electrochemical reactions lead to the gating effect of the semiconductor device [13]. This gating effect changes the electrical properties of the device and is characterized as the sensitivity parameters for the detection of biomolecules before and after capturing the targeted biomolecules by the receptors. There are many parameters with which we can measure sensitivity, such as current ratios (I_{ON}/I_{OFF}), the shift in threshold voltage (V_T), and the variation of ON current (Ion). Although FET-based biosensors are having a lot of advantages among others, they are facing major issues, such as (a) scaling difficulties and short-channel effects (SECs) experienced by the FET in the process of miniaturization [14], and (b) theoretical limitations on the minimum achievable subthreshold swing (SS > 60 mv/dec). These issues lead to narrowing the device performance and sensitivity, and the thermionic emission of electrons in FET results in high power dissipation. To avoid these problems, researchers have focused the new technology of FET-based biosensors, i.e., TFET-based biosensors, which have low power and superior characteristics due to band-to-band tunneling of carrier and steep subthreshold swing. Another crucial measurable parameter of biosensors is the response time. To have a quick response, the subthreshold swing should be as low as possible. Since

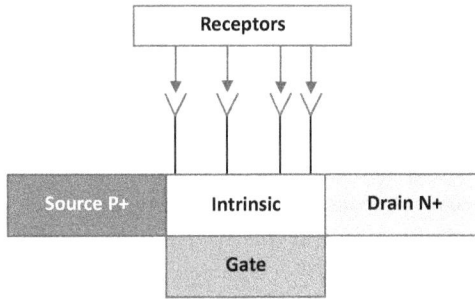

FIGURE 8.3 TFET as biosensors

the TFET can achieve an SS (< 60mv/dec) less than CFET, recently a lot of research focusing on designing TFET-based biosensors. Full details about FET-based biosensors are available in many literature surveys and research articles. Currently, there is a lot of progress in the development of TFET-based biosensors.

TFET-based biosensors consist of an electrode 1-source and an electrode 2-drain, and the region between the two electrodes serves as the bio-recognition element. This element is responsible for receiving the targeted biomolecules and producing electrical activity based on its reception. The working principle of TFET-based biosensors is based on the fluctuations in the charge concentration provided at the surface of the channel. These fluctuations are converted into gate voltage. Following this, at the final stage, there is an increase in the drain current due to the tunneling effect (Figure 8.3).

In 2012, Deblina et al. described a silicon nanowire-based TFET [SiNWTFET] biosensor for ultrasensitive and label-free detection [13]. The biosensor uses a single nanowire to form the PIN structure with a gate, source, and drain. Above the intrinsic channel region, a thin silicon dioxide layer is laid which acts as a receptor to identify the target molecules. The process of detection of biomolecules is done in two different steps. The first step is to increase the surface potential for detecting biomolecules since charge ions are present. The other step is to increase surface potential due to the presence of the tunneling current.

R. Narang et al. [10] proposed the idea of a dielectric-modulated TFET biosensor [15] using the dielectric-modulated FET design idea. In this type of biosensor, a cavity region is created in the dielectric oxide layer of the device to capture the biomolecules. When there is a change in the dielectric constant value, there is a coupling effect between the gate and oxide layer which causes channel bending. This leads to the tunneling effect and causes changes in the drain current. In order to improve the application, a single gate structure can also be replaced by a double gate [16, 17].

8.3.2 TFET-BASED STATIC RANDOM-ACCESS MEMORIES

The TFET also finds extensive applications in memory devices, specifically in static random-access memory (SRAM). The efficiency of the SRAM constructed using

TFET [18] is often compared to the conventional SRAMs. The increased popularity of portable compact devices elevated the requirement for SRAM, and it is popularly used in SoCs and high-performance VLSI circuits. SRAM optimization is of great importance since these memories take up a significant amount of the chip's space. Fine-tuning of performance parameters may yield optimized total chip performance. Complementary MOSFET nanoscale devices face a lot of significant difficulties in terms of performance and power consumption [19]. As for the ever-increasing intra-die parameter variability and power supply scaling, SRAM cell read and write stability is a major challenge in CMOS technology in nanoscale regimes. The TFET has come to light as one of the capable replacements for CMOS with the design of ultra-low-power memories due to negligible leakage current [20]. Optimizing TFET circuits with a focus on SRAM designs to reduce leakage current has gained a lot of attention recently. The noteworthy challenges in designing TFET-based SRAMs include characteristics such as unidirectionality and lower ON current than CMOS, resulting in a high degree of difficulty in sustaining a balance between stable read and write operations, with reduced access times. Thus, TFET SRAM designs have to be explored comprehensively in order to optimize area, stability, and performance.

8T SRAM is conventionally regarded as a more dependable memory unit. Usually, 8T static RAM cells are read from a single side, while 6T SRAM cells are read simultaneously from both sides. This exposes both of the internal nodes of the 6T static RAM cell to the pre-charged bit lines, which makes the 6T SRAM switch its state in an undesirable manner [21]. Hence, the 6T SRAM cell is more vulnerable than the 8T SRAM cell. The 6T SRAM can be made to execute read operations more reliably by incorporating the read technique of 8T SRAM to 6T SRAM. In addition, 6T SRAM has around a 30% smaller area with better power efficiency and is preferred for power-efficient compact TEFT based SRAM designs. The 6T SRAM cell is made up of six transistors, four of which are connected as inverters, where data bits are stored as 1 or 0, while the other two operate as pass transistors controlling the SRAM cell through the bit line. When the word line (WL) is at logic high, the SRAM cell can be accessed. The amount of time taken to read and write determines the speed of SRAMs, in other words, a propagation delay. Noise greatly interferes with the operation of the SRAM, and it may affect the stability of the memory by making it deviate from the intended functions. The static noise margin (SNM) is used to measure the reliability of the memory cells. SNM also reflects the fluctuations with the changes in supply voltage.

Figure 8.4 indicates the 6T TFET SRAM operation with outward transfer n-type transistors [22]. In the TFET symbol, an arrow is used to depict the current flow with the arrowhead at the position of the source. In read operations, both bit lines are pre-charged at GND, and the current flowing from the cell pulls up the bit line on the side storing logic 1, thus creating a potential difference between the bit lines. In write operations, the bit line BLR is pulled up to V_{DD}. In this case, with node V_1 pulled down by bit line BLL to GND, the reverse biased transistor T2 leakage current $I_{leakage}$, which is high with $V_{DS} = -V_{DD}$, aids in carrying out the write operation by flipping the cell through positive feedback. The current flows through the transfer and load transistor pairs in both read and write operation modes. The same transistor

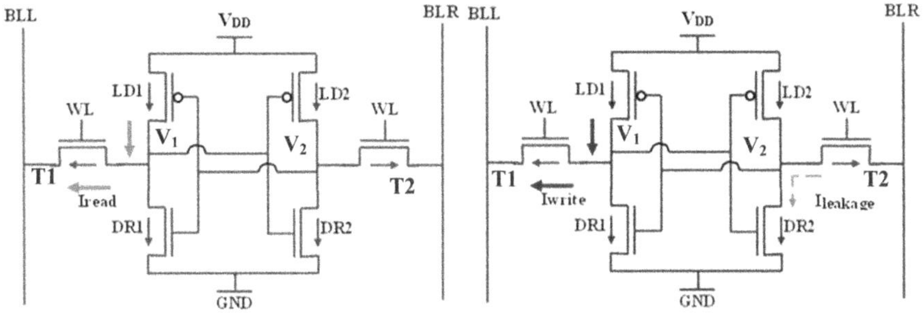

FIGURE 8.4   TFET-based 6T SRAM read and write operation

pair is allowed at the same time to pull down the internal storage node to GND during write operations and forbids this node to discharge during read. Therefore, in the 6T-TFET SRAM design, in contrast to the CMOS, it is impossible to optimize read and write stabilities separately. Consequently, a middle ground between the two has to be found. TFETs are viable in the standard a 6T-SRAM-bit cell-based design.

8.4   CONCLUSION

The structure of TFET, its characteristics, and its scope with specific applications are discussed in this chapter. This will be useful for researchers who have just started their research on TFET. There are numerous other TFET applications to investigate in addition to those covered in this chapter.

REFERENCES

1. N. Gupta, A. Makosiej, A. Amara, A. Vladimirescu, and C. Anghel. *TFET Integrated Circuits: From Perspective Towards Reality*. Cham: Springer International Publishing, 2021, doi: 10.1007/978-3-030-55119-3.
2. M. Alioto, "Ultra-low power design approaches for IoT," in *2014 IEEE Hot Chips 26 Symposium HCS*, pp. 1–57, Aug. 2014, doi: 10.1109/HOTCHIPS.2014.7478801.
3. C. Anghel, Hraziia, A. Gupta, A. Amara, and A. Vladimirescu, "30-nm tunnel FET with improved performance and reduced ambipolar current," *IEEE Trans. Electron Devices*, 58(6), pp. 1649–1654, Jun. 2011, doi: 10.1109/TED.2011.2128320.
4. J. Appenzeller, Y.-M. Lin, J. Knoch, and Ph. Avouris, "Band-to-band tunneling in carbon nanotube field-effect transistors," *Phys. Rev. Lett.*, 93(19), p. 196805, Nov. 2004, doi: 10.1103/PhysRevLett.93.196805.
5. W. Y. Choi and W. Lee, "Hetero-gate-dielectric tunncling field-effect transistors," *IEEE Trans. Electron Devices*, 57(9), pp. 2317–2319, Sep. 2010, doi: 10.1109/TED.2010.2052167.
6. A. M. Ionescu and H. Riel, "Tunnel field-effect transistors as energy-efficient electronic switches," *Nature*, 479(7373), pp. 329–337, Nov. 2011, doi: 10.1038/nature10679.
7. P. Mehrotra, "Biosensors and their applications – A review," *J. Oral Biol. Craniofac. Res.*, 6(2), pp. 153–159, 2016, doi: 10.1016/j.jobcr.2015.12.002.
8. S. P. Mohanty and E. Kougianos, "Biosensors: A tutorial review," *IEEE Potentials*, 25(2), pp. 35–40, Mar. 2006, doi: 10.1109/MP.2006.1649009.

9. L. C. Clark and C. Lyons, "Electrode systems for continuous monitoring in cardiovascular surgery," *Ann. N. Y. Acad. Sci.*, 102, pp. 29–45, Oct. 1962, doi: 10.1111/j.1749-6632.1962.tb13623.x.

10. R. Narang, K. V. S. Reddy, M. Saxena, R. S. Gupta, and M. Gupta, "A dielectric-modulated tunnel-FET-based biosensor for label-free detection: Analytical modeling study and sensitivity analysis," *IEEE Trans. Electron Devices*, 59(10), pp. 2809–2817, Oct. 2012, doi: 10.1109/TED.2012.2208115.

11. D. Sarkar, "Fundamental limitations of conventional-FET biosensors: Quantum-mechanical-tunneling to the rescue," *MIT Media Lab.* https://www.media.mit.edu/publications/fundamental-limitations-of-conventional-fet-biosensors-quantum-mechanical-tunneling-to-the-rescue/ (accessed Dec. 19, 2022).

12. R. Narang, M. Saxena, and M. Gupta, "Comparative analysis of dielectric-modulated FET and TFET-based biosensor," *IEEE Trans. Nanotechnol.*, 14(3), pp. 427–435, May 2015, doi: 10.1109/TNANO.2015.2396899.

13. D. Sarkar and K. Banerjee, "Proposal for tunnel-field-effect-transistor as ultra-sensitive and label-free biosensors," *Appl. Phys. Lett.*, 100(14), p. 143108, Apr. 2012, doi: 10.1063/1.3698093.

14. P. R. Nair and M. A. Alam, "Screening-limited response of nanobiosensors," *Nano Lett.*, 8(5), pp. 1281–1285, May 2008, doi: 10.1021/nl072593i.

15. H. Im, X.-J. Huang, B. Gu, and Y.-K. Choi, "A dielectric-modulated field-effect transistor for biosensing," *Nat. Nanotechnol.*, 2(7), pp. 430–434, Jul. 2007, doi: 10.1038/nnano.2007.180.

16. J.-H. Ahn, S.-J. Choi, J.-W. Han, T. J. Park, S. Y. Lee, and Y.-K. Choi, "Double-gate nanowire field effect transistor for a biosensor," *Nano Lett.*, 10(8), pp. 2934–2938, Aug. 2010, doi: 10.1021/nl1010965.

17. M. Im, J.-H. Ahn, J.-W. Han, T. J. Park, S. Y. Lee, and Y.-K. Choi, "Development of a point-of-care testing platform with a nanogap-embedded separated double-gate field effect transistor array and its readout system for detection of avian influenza," *IEEE Sens. J.*, 11(2), pp. 351–360, Feb. 2011, doi: 10.1109/JSEN.2010.2062502.

18. V. Saripalli, A. Mishra, S. Datta, and V. Narayanan, "An energy-efficient heterogeneous CMP based on hybrid TFET-CMOS cores," in *Proceedings of the 48th Design Automation Conference - DAC 11*, p. 729, 2011, doi: 10.1145/2024724.2024889.

19. A. C. Seabaugh and Q. Zhang, "Low-voltage tunnel transistors for beyond CMOS logic," *Proc. IEEE*, 98(12), pp. 2095–2110, Dec. 2010, doi: 10.1109/JPROC.2010.2070470.

20. X. Yang and K. Mohanram, "Robust 6T Si tunneling transistor SRAM design," *Des. Autom. Test Eur.*, pp. 1–6, Mar. 2011, doi: 10.1109/DATE.2011.5763126.

21. M. Kutila, A. Paasio, and T. Lehtonen, "Comparison of 130 nm technology 6T and 8T SRAM cell designs for Near-Threshold operation," in *2014 IEEE 57th International Midwest Symposium on Circuits and Systems (MWSCAS)*, pp. 925–928, Aug. 2014, doi: 10.1109/MWSCAS.2014.6908567.

22. D. Kim et al., "Low power circuit design based on heterojunction tunneling transistors (HETTs)," in *Proceedings of the 14th ACM/IEEE International Symposium on Low Power Electronics and Design - ISLPED '09*, San Francisco, CA, p. 219, 2009, doi: 10.1145/1594233.1594287.

9 An overview of FinFET-based capacitorless 1T-DRAM

Mitali Rathi and Guru Prasad Mishra

CONTENTS

9.1 INTRODUCTION

The connectivity and interaction of "smart objects" to provide automatic services constitute the fast-developing field known as the "Internet of Things" (IoT). New markets require new memory specifications, which for the Internet of Things should support extreme downsizing and drastically reduced energy usage [1–3]. There is a huge requirement for low-power devices and devices with minimum areas. Developers must therefore re-evaluate their design objectives by utilizing memory in novel and creative ways. The vast memories (i.e., DRAM, Flash memory, SRAM) used are charge-based, hence storing charge is the most important topic of concern [4, 5]. A comparison of features of capacitorless 1T-DRAM, 1T1C DRAM, and SRAM memory technology is given in Table 9.1 [6].

The DRAM, consisting of a capacitor and a transistor, has shown very good reliability results and has been integral for decades. But it faces technological and physical challenges due to the shrinkage of device feature size [6, 7]. For sufficient charge storage, it needs a deep trench capacitor, or there is a need for stacking the capacitor. So, it needs to be scaled a million times to meet today's market requirements. However, downscaling of transistors is very easy and essential for all applications and is still going on. But downscaling of capacitors is very difficult. In addition, the

TABLE 9.1

Comparison of capacitorless 1T-DRAM, with other memories based on some features

Features	DRAM	SRAM	1T-DRAM
Feature size	8F2	100F2	4F2
Storage node	Capacitor	Flip flop	Floating region
Cell complexity	One transistor, one capacitor	Six transistors	One transistor
Speed	Fast	Very fast	Fast
Read	destructive	Non-destructive	Non-destructive

fabrication of capacitors is complex. Capacitors are the heart of the DRAM, i.e., it is the storage region of the DRAM. Hence there is a need to replace the capacitor (storage body) of the DRAM and think of some other alternative storage units for the charges [5, 8–10].

9.1.1 CAPACITORLESS 1T-DRAM

Suppressing the capacitor is all that is required to affect an irreversible paradigm change and to meet technical requirements. The concept of capacitorless DRAM, i.e., single transistor DRAM, was introduced more than 20 years ago in 1993 by Hsing-jen Wann and Chenming Hu in an IEDM meeting [11]. They proposed a capacitorless DRAM (CDRAM) cell on an SOI substrate with a small cell area, large read current, and simplicity in fabrication. Then many researchers worked on this concept. In the year 2002, a simpler 1T-DRAM was introduced [6]. It takes advantage of the body charging effect of PDSOI MOSFETs and only utilizes three signal lines and a single channel. The main functioning of 1T-DRAM uses the concept of the floating body effect. The floating body of the FET was used to store the majority of carriers. That's why 1T-DRAMs are also called floating body DRAMs (FBRAMS). Various device structures based on capacitorless 1T-DRAM were proposed with different working principles, such as the surrounding gate MOSFET with vertical channel-based capacitorless DRAM, DG 1T quantum well DRAM, Si/SiGe double heterojunction bipolar transistor-based 1-T DRAM, GaP-silicon transistor for 1T-DRAM, 1T-DRAM with an electron-bridge channel, etc. [3, 12–23].

Recently polycrystalline silicon-based capacitorless DRAM has also been proposed in MOSFET as well as FinFET. Grain boundaries are formed in polycrystalline silicon which contains trap charges [24, 25]. Holes are stored under the polysilicon region using the band-to-band tunneling mechanism. Various material-based DRAM has also been introduced using band engineering [8, 26–30]. Carrier lifetime engineering has also been done for the DRAM operation [31].

9.1.2 OPERATION OF CAPACITORLESS 1T-DRAM

In capacitorless 1T-DRAM, holes are accumulated in the floating body during program operation by high impact ionization, or GIDL [31]. In impact ionization, the hot electron injection process is involved. This affects the trapped charges adversely. But faster programming speed and a large sensing window need high impact ionization, which can be achieved by applying high programming voltage [6, 21, 32]. Another method is GIDL programming. In the GIDL method, hot electrons are not generated for the programming operation. Holes are generated using by the band-to-band tunneling (BTBT) mechanism. But GIDL current is not enough to charge the body, hence there is a need to supply an additional voltage for band banding [33–35].

Sense margin: This is the difference between the read1 current (current after program), and read0 current (current after erase). It is also known as the programming window of the capacitorless 1T-DRAM. It is denoted by SM.

Retention time: Retention time is the critical time when the DRAM loses half of its initial charge. In other words, it can be quoted as the time when the sense margin becomes half of its maximum value. It is denoted by t_{ret}. And as the device thickness is shrinking, the retention time is decreasing. So, it is a major topic of concern for capacitorless1T-DRAM with scaling.

9.1.3 SCALING CHALLENGES

The transistor widely used for DRAM memory was MOSFET. But downscaling of MOSFET is very difficult beyond a certain limit, since short-channel effects (SCEs) i.e., drain-induced barrier lowering (DIBL), subthreshold swing (SS), hot carrier effects, etc. arise. These SCEs alter the device performance and increase the leakage current. To overcome this problem, multi-gate devices were introduced. In multi-gate devices, more than one gate controls the channel, hence the electrostatic control over the channel is more, which enhances the device's on-state current (I_{on}) and minimizes the off-state (I_{off}) current, thereby improving the switching ratio of the device and minimizing the power consumption [27, 36, 37].

There are many multi-gate devices such as FinFET, gate-all-around GAA FET, pi-gate, DG-MOSFET, etc. Due to ease of fabrication and simple structure, FinFET has been widely adopted [38–41]. Many companies such as Intel, Global Foundries, TSMC, and Samsung are using FinFET for mobile phone and laptop applications. In FinFET, the channel is wrapped by the gate from two or three sides known as DG-FinFET or TG-FinFET respectively [42–46]. There are many applications where FinFET can be used, such as in mechatronics, as a biosensor, hydrogen gas sensors, and in memories [21, 47–52].

9.1.4 FINFET-BASED CAPACITORLESS 1T-DRAM

In this chapter, we will study FinFET-based capacitorless DRAM. There have been many research breakthroughs in FinFET-based capacitorless 1T-DRAM [32, 48,

53–59]. The scalability of DRAM critically affects the retention time. In order to prevent short-channel effects, as the gate length decreases, the channel impurity concentration increase. Storage charge decreases in the capacitor and the retention period also shorten due to the degraded junction leakage characteristics caused by the rise in channel impurity concentration. Similar effects are caused in the case of 1T-DRAM. The silicon thickness reduces the impurity concentration and also reduces short-channel effects. Hence, retention time decreases because of increasing junction leakage. Additionally, due to the smaller volume of the floating body, the quantity of storage charges also reduces with a smaller gate length. As a result, the sense margin is reduced. Hence it becomes very difficult to scale a single-gate partially depleted silicon-on-insulator MOSFET. To overcome these scaling challenges, E. Yoshida et al. proposed a double gate FinFET DRAM (i.e., DG-FinDRAM) [60, 61]. The front MOS structure works as a typical switching transistor in the DG-FinDRAM, and the back MOS structure acts as the floating body storage node. Memory operations are enabled even with substantially scaled totally depleted FinFETs because of proper reverse-biasing of the MOS structure that stores excess holes in the floating body. Hence the sense margin and retention time improve even at a gate length of less than 100 nm.

Reported, there were two gates controlling the operation of fully depleted DG-Fin-DRAM. This requires a separate gate separation step and an additional bias to store the hole in the floating body. Then a concept of partially depleted SOI FinFET (PDSOI FinFET) based 1T-DRAM was proposed. This allows accumulation of holes in a very thin fin and also no second gate is needed. The fully depleted region of the PDSOI FinFET was used for scalability, and the PDSOI region was used as a floating body for the storage of excess holes. The channel of the FDSOI region was surrounded by a gate, and the extended channel region (i.e., the PDSOI region) was surrounded by an isolation dielectric layer. The holes are generated by impact ionization by high drain potential, and the erase operation is done for forward junction current by supplying low voltage to the drain terminal [32, 53–55].

In this chapter, a novel solution is proposed in which the channel of the fully depleted FinFET is extended and surrounded by isolation dielectric. (In other words, we can say that the extended source and drain region of the PDSOI FinFET is suppressed.) This new structure has less fabrication complexity. In this structure, the extended channel region will be the extra storage region for holes. The operation is carried out by a high impact ionization mechanism. The operation of 1T DRAM has been described in Section 9.1.2.

9.2   DEVICE DESCRIPTION

The simulated structure of the proposed structure is shown in Figure 9.1. Figure 9.1 (a) shows the three-dimensional structure of the device. Figures 9.1(b) and 1(c) depict the front view and top view of the device respectively. The front view is taken as a view along the y- and z-axis, and the top view is taken along x- and z-axis. The extended channel H_{ext} is clearly visible in Figure 9.1(c), which shows the side view of the device along the x- and y-axis. The upper channel is overlapped by a gate from

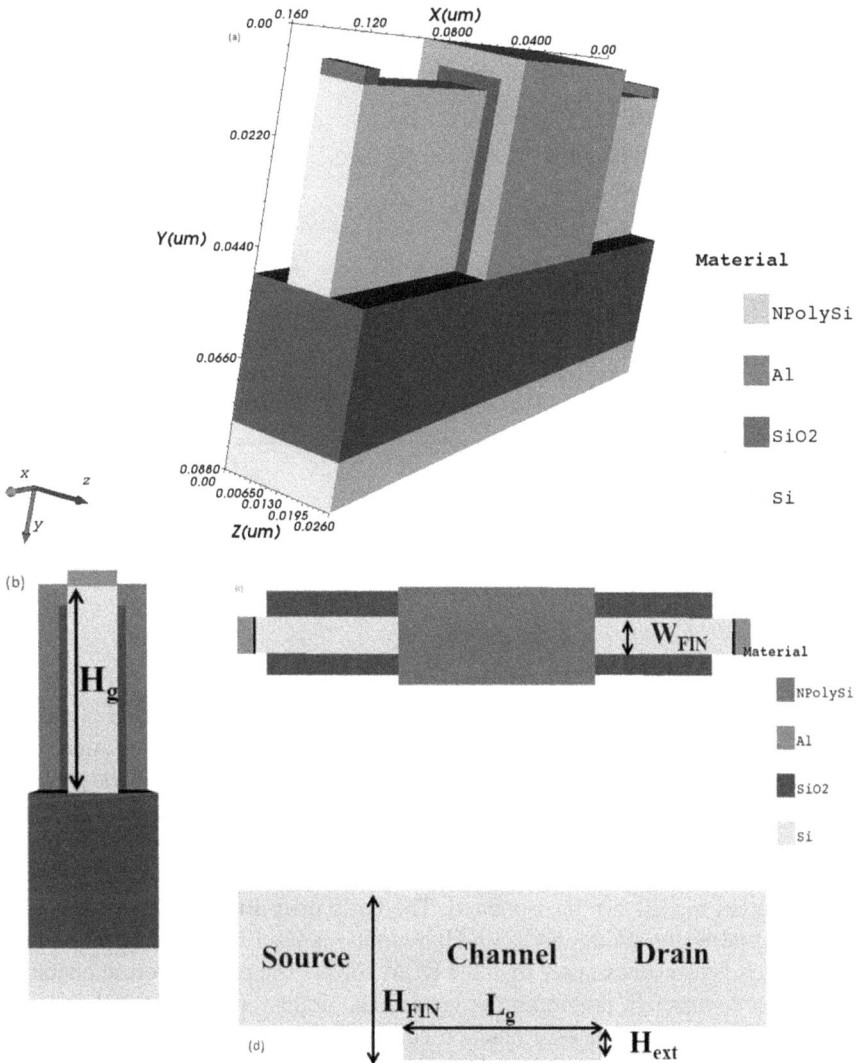

FIGURE 9.1 (a) Simulated 3-D structure of proposed device. (b) Front view, (c) top view, and (d) side view of the proposed device

three sides, and the extended channel region is surrounded by isolation dielectric. The holes are stored in the extended region. This is the floating body of the proposed capacitorless 1T-DRAM. The total fin height is 50 nm, the height of the extended region is 10 nm, and the gate height is 40 nm. Hence, the height of the FDSOI FinFET is 40 nm. The thickness of the fin is taken as 10 nm, and the channel length is 60 nm. The source and drain are highly doped with a doping concentration of 1e20 cm^{-3}. The channel is lightly doped with a concentration of 1e16 cm^{-3}. The isolation

TABLE 9.2

Parameters used in the simulation

Parameters	Values
Fin height H_{FIN} (nm)	50
Fin width W_{FIN} (nm)	10
Channel length L_g (nm)	60
Gate height H_g (nm)	40
Extended fin height H_{ext} (nm)	10
Gate workfunction (eV)	4.75
Isolation oxide	SiO_2 and HfO_2
Channel doping concentration (cm^{-3})	1e16
Source/drain doping concentration (cm^{-3})	1e20

dielectrics used for the study are SiO_2 and HfO_2 with permittivity of 3.9 and 22 respectively. Table 9.2 depicts the geometrical design aspects with their dimensions considered for the simulation of the proposed work.

9.3   SIMULATION SETUP AND MODEL DESCRIPTION

The simulation is performed using the visual TCAD tool Cogenda [62]. High impact ionization models are used for the programming operation. Impact ionization is the process in which electron-hole pairs are generated due to carrier drift in the presence of a high electric field. The Selberherr II model is used for the generation of electron-hole pairs through impact ionization. Lombardi is used as a carrier mobility model to describe carrier mobility in the transistor. Fermi, hole mobility, SRH, and AUGER recombination models are incorporated. The basic drift-diffusion level-1 equation solver is used for program operation. Halfimplicit is used for all transient simulations, which is five times faster than the DDM solver. The program/erase operation is performed with high programming voltage i.e., drain voltage $V_{DS}=2$ V and gate voltage $V_{GS}=1.5$ V and forward junction current with low drain voltage i.e., drain voltage $V_{DS}=-0.5$ V and gate voltage $V_{GS}=1$ V. The P/E pulse duration is taken to be 10 ns as given in international roadmap for devices (International Technology Roadmap for Devices and Systems (IRDS)). The electrons are attracted toward the drain due to this high voltage, and the hole gets accumulated in the floating region which is the storage node for the holes. Then to hold the holes in the storage region, a minimum hold voltage of $V_{GS}= -0.1$ V is supplied to the gate terminal, and the drain voltage is kept at 0 V. The hole concentration and recombination rate are observed after the hold operation. A non-destructive read is performed with a low drain voltage of $V_{DS}=0.3$ V and a gate voltage of $V_{GS}=1.5$ V. The biasing voltages for different DRAM operations are summarized in Table 9.3. Figure 9.2 depicts the transient plot of different operating voltages, i.e., drain to source voltage V_{DS} and gate to source voltage V_{GS} for the operation of the proposed DRAM with a sequence of operation

TABLE 9.3

Operating voltages applied to the terminals

Operation	Drain voltage (V_{DS})	Gate voltage (V_{GS})	Source voltage (V_{SS})
Program (Write 1)	2 V	1.5 V	0 V
Erase (Write 0)	–0.5 V	1 V	0 V
Hold	0 V	–0.1 V	0 V
Read	0.3	1.5 V	0 V

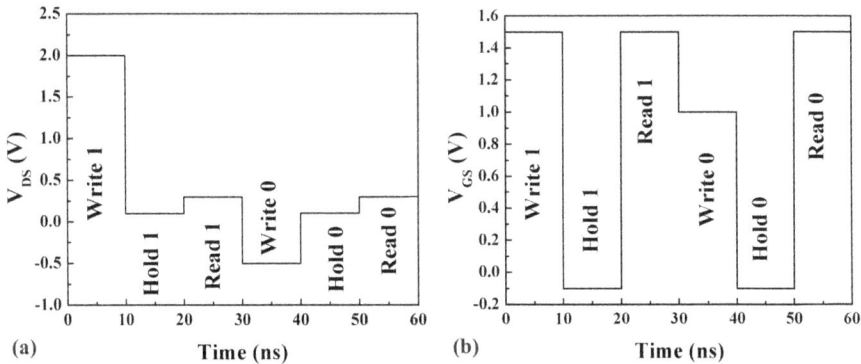

FIGURE 9.2 Transient plot of biasing voltages. (a) Drain voltage V_{DS}, (b) gate voltage V_{GS} given for Write1-Hold1-Read1-Write0-Hold0-Read0 operation sequence for a time period of 10 ns

Write1Write1-Hold1-Read1-Write0-Hold0-Read0 with respect to the time in ns. The time period is 10 ns for each operation.

9.4 RESULT AND DISCUSSION

The electron-hole pairs are generated through impact ionization. The holes are accumulated in the extended channel region. Figure 9.3 shows the I_{DS}-V_{DS} characteristics of the proposed work with and without impact ionization with different isolation dielectric oxides. It can be observed from the graph the effect of impact ionization on the channel potential, and the kink effect observed in the graph shows the storage of extra holes in the floating body. This shows how the impact ionization can elevate the hole concentration, thus enhancing the DRAM performance.

The high k isolation dielectric HfO$_2$-based device offers the lowest off-state current, highest on-state current, and steepest subthreshold slope. The potential across the extended region is more effectively controlled by gate bias due to the high permittivity of the isolation dielectric, which enforces the capacitive coupling through the isolation dielectric oxide. This improves the current drivability and is highly immune to short-channel effects.

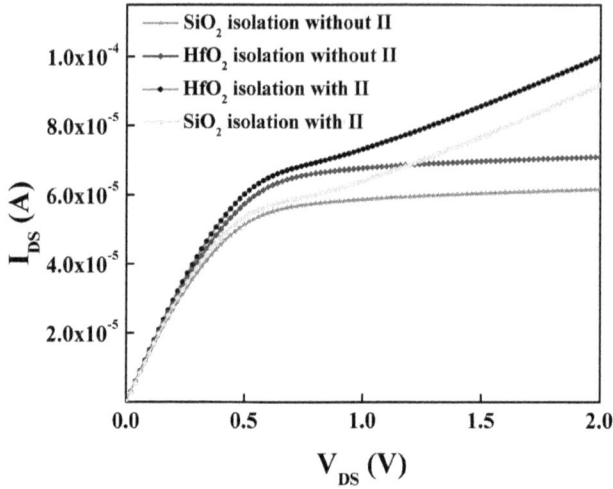

FIGURE 9.3 I_{DS}-V_{DS} characteristics of the proposed work with and without impact ionization with different isolation dielectric oxides

The carrier generation, recombination, and diffusion control the retention characteristics of the device. Hold bias can practically control it. The hole density is highest after the program operation (write1) i.e., 1.651e20 cm^{-3}. Holes are stored in the source side of the channel. The majority of holes are saturated in an extended channel. This is also because of the high electric field near the channel-drain junction of approximately 2.953e6V/cm and the impact ionization rate of 6.254e31 cm^{-3}/s on the drain side. Figure 9.4 shows the contour plots of the impact ionization and electric field of the proposed device. Figure 9.4(a), and 4 (b) shows the impact ionization after the write1 operation, which is highest at the drain-to-channel junction. As the drain potential is high, the rate of impact ionization is very high. The contour plot of the electric field profile can be seen in Figure 9.4 (c), which is at the channel region. The electric field is highest at the drain-to-channel junction. However, some holes are also accumulated in the source side of the device. The positive drain potential attracts electrons, and holes are left at the source side, and some are in the channel region. The rate of recombination is also very high at the source side i.e., 2.253e31 cm^{-3}/s. These values are summarized in Table 9.4 for SiO$_2$ isolation dielectric oxide.

The drive current of the capacitorless 1T-DRAM is modulated by the hole concentration of the device. This can be confirmed from Figure 9.5, which shows the variation of hole concentration after the DRAM operations (W1-H1-R1-W0-H0-R0) with respect to a time period in ns. After the program operation, the hold operation is performed. The holes are accumulated in the channel region. The majority of the holes are stored in the extended channel region due to impact ionization. The density of holes after the hold bias is applied increases. Due to this, the drain current is also enhanced. It is very high i.e., approximately 9.16e-5 A. This can be seen in Figure 9.6, which shows the read current after the DRAM operations

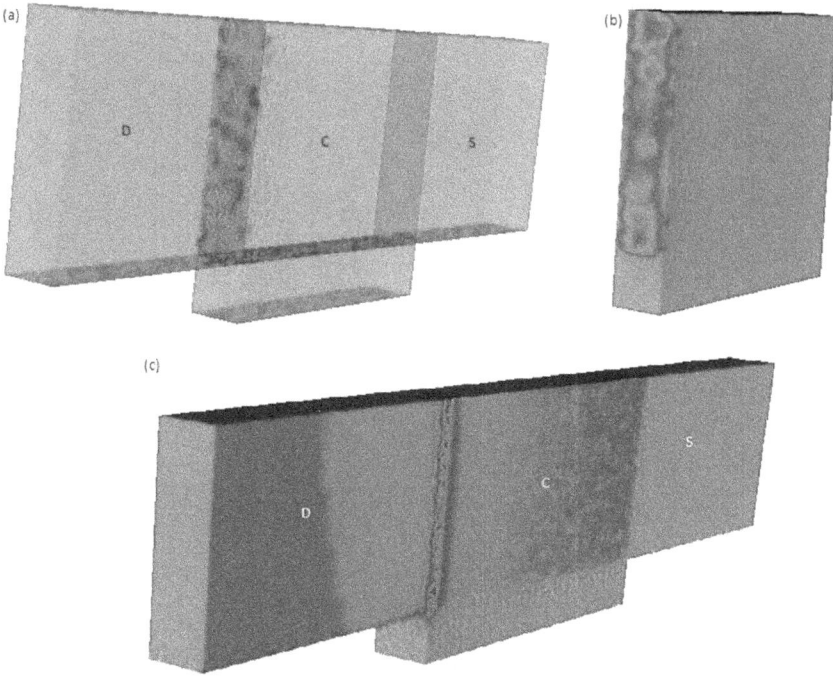

FIGURE 9.4 Contour plots of (a) impact ionization which can be seen at the channel-drain junction, (b) zoom view of impact ionization at junction profile, and (c) electric field of the proposed work after the program operation

(W1-H1-R1-W0-H0-R0) with respect to a time period in ns. Then hold bias is applied after 10 ns. A large number of holes are already accumulated after the program operation by impact ionization. Then read bias is applied, in which some holes start escaping. When the erase operation is performed, the hole density reduces to 4520/cm^3. This is because of the negative drain voltage applied. Then hold bias is applied and the hole concentration starts increasing. This is because of the thermal generation and tunneling of the carriers. Figure 9.7 shows the increment in the hole

TABLE 9.4
Some important parameters after the program operation of capacitorless 1T-DRAM

Parameters	Values
Hole density	1.651e20 cm^{-3}
Electric field	2.953e6 V/cm
Impact ionization	6.254e31 cm^{-3}/s
Recombination	2.253e31 cm^{-3}/s

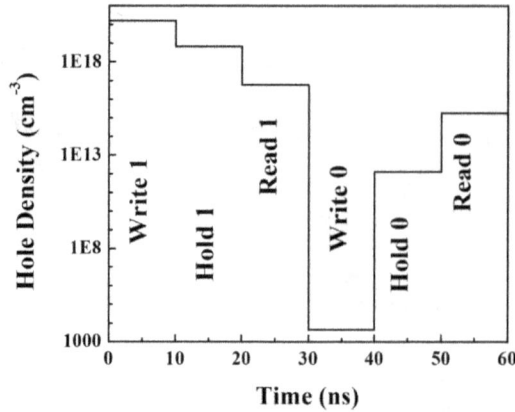

FIGURE 9.5   Hole density variation for different operations (Write1-Hold1-Read1-Write0-Hold0-Read0) for the proposed device

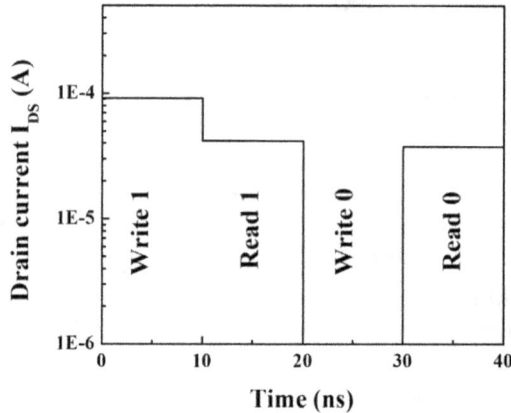

FIGURE 9.6   Drain current of the device for different operations (Write1-Hold1-Read1-Write0-Hold0-Read0) for the proposed device

concentration with an increase in the hold time. After a certain time, the hole concentration starts saturating.

The hold voltage we are taking is VGS= -0.1 V, VDS= 0 V. The hold voltage should be selected correctly. If we are considering a large negative gate voltage for hold bias voltage, this will increase the band-to-band tunneling, hence the leakage after state, thus, increasing the read0 current. This also increases the charge accumulation after the write1 operation.

9.5   CONCLUSION

In 1T-1C DRAM, to increase performance, the capacitor needs to store more charges, and for this, the size of the capacitor cannot be reduced. In addition, the

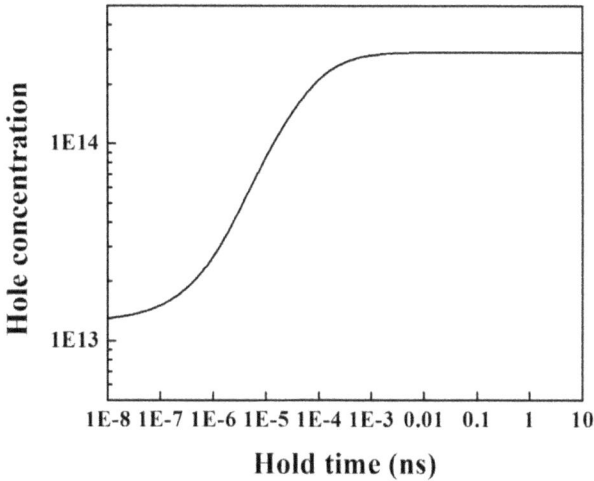

FIGURE 9.7 Hole concentration after the erase operation with respect to hold time for the proposed device

fabrication and scaling of capacitors face many challenges. So the best way for the scaling of DRAM is to suppress the capacitor and find an alternative method for the storage of charges. Hence the concept of capacitorless 1T-DRAM was introduced. The capacitorless 1T-DRAM appears to be the most promising option considering the speed, capacity, and high performance of its embedded memory. One other advantage of this design is that it has less structural complexity and even requires minimum die area. Hence the cost of fabrication is reduced. To achieve a high sensing margin and retention time, a novel solution has been proposed in this chapter. To extend the channel region as in the PDSOI- FinFET, the whole fin was extended downwards. In this proposed solution, only the channel region is extended, which would be a buffer layer for the holes, and the holes are stored in this extended region. Using this concept, the retention characteristics of the device are improved, although there is still a trade-off between the scaling of the memory and the retention period. As the device is scaling down, the retention time is decreasing because the recombination rate is increasing in the floating body. Structural and material optimization could be helpful by providing new models and engineering techniques to overcome this issue.

In the near future when new multi-gate structures will be used, capacitorless 1T-DRAM will be the most promising embedded memory node with low power consumption.

REFERENCES

1. S. Rajendran and R. M. Lourde, "Ab-initio study on FinFETs and their application in IoT aided robotics," *Proc. - 2016 6th Int. Symp. Embed. Comput. Syst. Des. ISED 2016*, pp. 38–42, 2017, doi: 10.1109/ISED.2016.7977051.

2. C. W. Sohn, C. Y. Kang, R. H. Baek, D. Y. Choi, H. C. Sagong, E. Y. Jeong, C. K. Baek, J. C. Lee, J. C. Lee, and Y. H. Jeong, "Device design guidelines for nanoscale FinFETs in RF/analog applications," *IEEE Electron Device Lett.*, 33(9), pp. 1234–1236, 2012, doi: 10.1109/LED.2012.2204853.

3. S. Cristoloveanu, K. H. Lee, M. S. Parihar, H. E. Dirani, J. Lacord, S. Martinie, C. L. Royer, J. C. Barbe, X. Mescot, P. Fonteneau, P. Galy, F. Gamiz, C. Navarro, B. Cheng, M. Duan, F. A. Lema, A. Asenov, Y. Taur, Y. Xu, T. T. kim, J. Wan, and M. Bawedin, "A review of the Z2-FET 1T-DRAM memory: Operation mechanisms and key parameters," *Solid State Electron.*, 143(December), pp. 10–19, 2018, doi: 10.1016/j.sse.2017.11.012.

4. M. H. Kryder and C. S. Kim, "After hard drives-what comes next?," *IEEE Trans. Magn.*, 45(10), pp. 3406–3413, 2009, doi: 10.1109/TMAG.2009.2024163.

5. S. Hong, "Memory technology trend and future challenges," *2010 Int. Electron Devices Meet.*, pp. 292–295, 2010, doi: 10.1109/IEDM.2010.5703348.

6. A. C. Cell, S. Okhonin, M. Nagoga, J. M. Sallese, and P. Fazan, "A capacitor-less 1T-DRAM cell," *IEEE Electron Device Lett.*, 23(2), pp. 85–87, 2002.

7. V. Makarov, V. Sverdlov, and S. Selberherr, "Emerging memory technologies: Trends, challenges, and modeling methods," *Microelectron. Reliab.*, 52(4), pp. 628–634, 2012, doi: 10.1016/j.microrel.2011.10.020.

8. A. Spessot and H. Oh, "Dynamic random access memory," *VLSI Handb.*, 67(4), pp. 1033–1049, 2020, doi: 10.1201/9781420049671-56.

9. N. Rodriguez, S. Cristoloveanu, and F. Gamiz, "Novel capacitorless 1T-DRAM cell for 22-nm node compatible with bulk and SOI substrates," *IEEE Trans. Electron Devices*, 58(8), pp. 2371–2377, 2011, doi: 10.1109/TED.2011.2147788.

10. F. Gamiz, "Capacitor-less memory: Advances and challenges," *Joint Int. EUROSOI workshop and Int. conf. on ultimate and integration on Silicon (EUROSOI-ULIS)*, pp. 68–71, 2016.

11. Wann H. J., Hu C., "A capacitorless DRAM cell on SOI substrate". In Proceedings of IEEE International Electron Devices Meeting 1993 Dec 5 pp. 635–638, 1993, doi: 10.1109/IEDM.1993.347280.

12. M. G. Ertosun and K. C. Saraswat, "Investigation of capacitorless double-gate single-transistor dram: with and without quantum well," *IEEE Trans. Electron Devices*, 57(3), pp. 608–613, 2010, doi: 10.1109/TED.2009.2038651.

13. J. S. Shin, H. Bae, J. Jang, D. Yun, J. Lee, E. Hong, D. H. Kim, and D. M. Kim, "A novel double HBT-based capacitorless 1T DRAM cell with Si/SiGe heterojunctions," *IEEE Electron Device Lett.*, 32(7), pp. 850–852, 2011, doi: 10.1109/LED.2011.2142390.

14. J. T. Lin, W. H. Lee, P. H. Lin, S. W. Haga, Y. R. Chen, and A. Kranti, "A new electron bridge channel 1T-DRAM employing underlap region charge storage," *IEEE J. Electron Device Soc.*, 5(1), pp. 59–63, 2017, doi: 10.1109/JEDS.2016.2633274.

15. H. Jeong, K. W. Song, I. H. Park, T. H. Kim, Y. S. Lee, S. G. Kim, J. Seo, K. Cho, K. Lee, H. Shin, J. D. Lee, and B. G. Park, "A new capacitorless IT DRAM cell: Surrounding gate MOSFET with vertical channel (SGVC cell)," *IEEE Trans. Nanotechnol.*, 6(3), pp. 352–356, 2007, doi: 10.1109/TNANO.2007.893575.

16. J. S. Shin, H. Choi, H. Bae, J. Jang, D. Yun, E. Hong, D. H. Kim, and D. M. Kim, "Vertical-gate Si/SiGe double-HBT-based capacitorless 1T DRAM cell for extended retention time at low latch voltage," *IEEE Electron Device Lett.*, 33(2), pp. 134–136, 2012, doi: 10.1109/LED.2011.2174025.

17. A. Pal, A. Nainani, S. Gupta, and K. C. Saraswat, "Performance improvement of one-transistor DRAM by band engineering," *IEEE Electron Dev. Lett.*, 33(1), pp. 29–31, 2012, doi: 10.1109/LED.2011.2171912.

18. J. Wan, C. Le Royer, A. Zaslavsky, and S. Cristoloveanu, "Progress in Z2-FET 1T-DRAM: Retention time, writing modes, selective array operation, and dual bit storage," *Solid State Electron.*, 84, pp. 147–154, 2013, doi: 10.1016/j.sse.2013.02.010.
19. A. Pal, A. Nainani, Z. Ye, X. Bao, E. Sanchez, and K. C. Saraswat, "Electrical characterization of GaP-silicon interface for memory and transistor applications," *IEEE Trans. Electron Devices*, 60(7), pp. 2238–2245, 2013, doi: 10.1109/TED.2013.2264495.
20. K. R. A. Sasaki, T. Nicoletti, L. M. Almeida, S. D. dos Santos, A. Nissimoff, M. Aoulaiche, E. Simoen, C. Claeys, and J. A. Martino, "Improved retention times in UTBOX nMOSFETs for 1T-DRAM applications," *Solid State Electron.*, 97, pp. 30–37, 2014, doi: 10.1016/j.sse.2014.04.031.
21. J. T. Lin and P. H. Lin, "Multifunction behavior of a vertical MOSFET with trench body structure and new erase mechanism for use in 1T-DRAM," *IEEE Trans. Electron Devices*, 61(9), pp. 3172–3178, 2014, doi: 10.1109/TED.2014.2336533.
22. C. Navarro, M. Duan, M. S. Parihar, F. A. Lema, S. Coseman, J. Lacord, K. Lee, C. S. Matarin, B. Cheng, H. E. Dirani, J. C. Barbe, P. Fonteneau, S. Kim, S. Cristoloveanu, M. Bawedin, C. Millar, P. Galy, C. L. Royer, S. Karg, H. Riel, P. Wells, Y. T. Paul, A. Asenov, and F. J. G. Perez, "Z2-FET as capacitor-less eDRAM cell for high-density integration," *IEEE Trans. Electron Devices*, 64(12), pp. 4904–4909, 2017, doi: 10.1109/TED.2017.2759308.
23. A. Lahgere and M. J. Kumar, "1-T capacitorless DRAM using bandgap-engineered silicon-germanium bipolar I-MOS," *IEEE Trans. Electron Devices*, 64(4), pp. 1583–1590, 2017, doi: 10.1109/TED.2017.2669096.
24. J. Park, M. S. Cho, S. H. Lee, H. D. An, S. R. Min, G. U. Kim, Y. J. Yoon, J. H. Seo, S. H. Lee, J. Jang, J. H. Bae, and I. M. Kang, "Design of capacitorless DRAM based on polycrystalline silicon nanotube structure," *IEEE Access*, 9, pp. 163675–163685, 2021, doi: 10.1109/ACCESS.2021.3133572.
25. S. H. Lee, W. D. Jang, Y. J. Yoon, J. H. Seo, H. J. Mun, M. S. Cho, J. Jang, J. H. Bae, and I. M. Kang, "Polycrystalline-silicon-mosfet-based capacitorless dram with grain boundaries and its performances," *IEEE Access*, 9, pp. 50281–50290, 2021, doi: 10.1109/ACCESS.2021.3068987.
26. E. X. Zhang, D. M. Fleetwood, J. A. Hachtel, C. Liang, R. A. Reed, M. L. Alles, R. D. Schrimpf, D. Linten, J. Mitard, M. F. Chisholm, and S. T. Pantelides, "Total ionizing dose effects on strained Ge pMOS FinFETs on bulk Si," *IEEE Trans. Nucl. Sci.*, 64(1), pp. 226–232, 2017, doi: 10.1109/TNS.2016.2635023.
27. M. N. Reddy and D. K. Panda, "A comprehensive review on FinFET in terms of its device structure and performance matrices," *Silicon*, 2022, doi: 10.1007/s12633-022-01929-8.
28. S. K. Vandana, J. K. Das Mohapatra, K. P. Pradhan, A. Kundu, B. K. Kaushik, B. K. Kaushik, "Memoryless nonlinearity in IT JL FinFET with spacer technology: Investigation towards reliability," *Microelectron. Reliab.*, 119(February), p. 114072, 2021, doi: 10.1016/j.microrel.2021.114072.
29. C. Navarro, S. Navarro, C. Marquez, J. L. Padilla, P. Galy, F. Gamiz, "3-D TCAD Study of the implications of channel width and interface states on FD-SOI Z2-FETs," *IEEE Trans. Electron Devices*, 66(6), pp. 2513–2519, 2019, doi: 10.1109/TED.2019.2912457.
30. K. Lee, S. Kim, J. H. Lee, B. G. Park, and D. Kwon, "Ferroelectric-metal field-effect transistor with recessed channel for 1T-DRAM application," *IEEE J. Electron Devices Soc.*, 10(February), pp. 13–18, 2022, doi: 10.1109/JEDS.2021.3127955.
31. S. Kim, S. J. Choi, D. Il Moon, and Y. K. Choi, "Carrier lifetime engineering for floating-body cell memory," *IEEE Trans. Electron Devices*, 59(2), pp. 367–373, 2012, doi: 10.1109/TED.2011.2176944.

32. S. W. Ryu, J. W. Han, Y. K. Kim, and Y. K. Choi, "Investigation of isolation-dielectric effects of PDSOI FinFET on capacitorless 1T-DRAM," *IEEE Trans. Electron Devices*, 56(12), pp. 3232–3235, 2009, doi: 10.1109/TED.2009.2033412.

33. E. Yoshida and T. Tanaka, "A capacitorless 1T-DRAM technology using gate-induced drain-leakage (GIDL) current for low-power and high-speed embedded memory," *IEEE Trans. Electron Devices*, 53(4), pp. 692–697, 2006, doi: 10.1109/TED.2006.870283.

34. J. W. Han, D. Il Moon, D. H. Kim, and Y. K. Choi, "Parasitic BJT read method for high-performance capacitorless 1T-DRAM mode in unified RAM," *IEEE Electron Dev. Lett.*, 30(10), pp. 1108–1110, 2009, doi: 10.1109/LED.2009.2029353.

35. G. Giusi, M. A. Alam, F. Crupi, and S. Pierro, "Bipolar mode operation and scalability of double-gate capacitorless 1T-DRAM cells," *IEEE Trans. Electron Devices*, 57(8), pp. 1743–1750, 2010, doi: 10.1109/TED.2010.2050104.

36. X. Sun, V. Moroz, N. Damrongplasit, C. Shin, and T. J. K. Liu, "Variation study of the planar ground-plane bulk MOSFET, SOI FinFET, and trigate bulk MOSFET designs," *IEEE Trans. Electron Devices*, 58(10), pp. 3294–3299, 2011, doi: 10.1109/TED.2011.2161479.

37. E. Suzuki, K. Ishii, S. Kanemaru, T. Maeda, T. Tsutsumi, T. Sekigawa, K. Nagai, and H. Hiroshima, "Highly suppressed short-channel effects in ultrathin SOI n-MOSFET's," *IEEE Trans. Electron Devices*, 47(2), pp. 354–359, 2000, doi: 10.1109/16.822280.

38. S. K. Saha, "FinFET devices for VLSI circuits and systems," 2020, doi: 10.1201/9780429504839.

39. A. Razavieh, P. Zeitzoff, and E. J. Nowak, "Challenges and limitations of CMOS scaling for FinFET and beyond architectures," *IEEE Trans. Nanotechnol.*, 18, pp. 999–1004, 2019, doi: 10.1109/TNANO.2019.2942456.

40. M. Jurczak, N. Collaert, A. Veloso, T. Hoffmann, and S. Biesemans, "Review of FINFET technology," *Proc. IEEE Int. SOI Conf.*, 2009, doi: 10.1109/SOI.2009.5318794.

41. J. P. Colinge and A. Chandrakasan, "FinFETs and other multi-gate transistors," 2008, doi: 10.1007/978-0-387-71752-4.

42. D. Bhattacharya and N. K. Jha, "FinFETs: From devices to architectures," *Digit. Analog. Digit. IC Des*, 2014, pp. 21–55, 2015, doi: 10.1017/CBO9781316156148.003.

43. K. Garg and D. Kapoor, "Single gate devices to multigate devices: A review," *Int. J. Innov. Eng. Manag.*, 4(1), pp. 75–80, 2015.

44. S. Rajendran and R. M. Lourde, "Fin FETs and their application as load switches in micromechatronics," *Proc. 2015 IEEE Int. Symp. Nanoelectron. Inf. Syst. iNIS 2015*, pp. 152–157, 2016, doi: 10.1109/iNIS.2015.51.

45. X. Huang, W. C. Lee, C. Kuo, D. Hisamoto, L. Chang, J. Kedzierski, E. Anderson, H. Takeuchi, Y. K. Choi, K. Asano, V. Subramanian, T. J. King, J. Bokor, and C. Hu, "Sub-50 Nm P-Channel FinFET," *IEEE Trans. Electron Devices*, 48(5), pp. 880–886, 2001.

46. D. Hisamoto, W. C. Lee, J. Kedzierski, H. Takeuchi, K. Asano, C. Kuo, E. Anderson, T. J. King, J. Bokor, and C. Hu, "FinFET — A Self-Aligned Double-Gate MOSFETscalable to 20 nm," *IEEE Trans. Electron Devices*, 47(12), pp. 2320–2325, 2000.

47. R. K. Nirala, S. Semwal, and A. Kranti, "A critique of length and bias dependent constraints for 1T-DRAM operation through RFET," *Semicond. Sci. Technol.*, 37(10), 2022, doi: 10.1088/1361-6641/ac8c67.

48. V. Rajakumari and K. P. Pradhan, "Demonstration of an UltraLow energy PD-SOI FinFET based LIF neuron for SNN," *IEEE Trans. Nanotechnol.*, 21, pp. 434–441, 2022, doi: 10.1109/TNANO.2022.3195698.

49. B. Kumar and R. Chaujar, "Numerical simulation of analog metrics and parasitic capacitances of GaAs GS-GAA FinFET for ULSI switching applications," *Eur. Phys. J. Plus*, 137(1), 2022, doi: 10.1140/epjp/s13360-021-02269-z.

50. M. Daga and G. P. Mishra, "Subthreshold performance improvement of underlapped FinFET using workfunction modulated dual-metal gate technique," *Silicon*, 13(5), pp. 1541–1548, 2021, doi: 10.1007/s12633-020-00550-x.

51. M. Rathi and G. P. Mishra, "Improved switching current ratio with Workfuncion modulated junctionless FinFET," *Silicon*, 89(18), p. 01234567, 2022, doi: 10.1007/s12633-022-01969-0.

52. S. Kesherwani, M. Daga, and G. P. Mishra, "Design of Sub-40nm FinFET based label free biosensor," *Silicon*, 2022, doi: 10.1007/s12633-022-01936-9.

53. J. W. Han, S. W. Ryu, C. J. Kim, S. Kim, M. Im, S. J. Choi, J. S. Kim, K. H. Kim, G. S. Lee, J. S. Oh, M. H. Song, Y. C. Park, J. W. Kim, and Y. K. Choi, "Partially depleted SONOS FinFET for unified RAM (URAM) - Unified function for high-speed 1T DRAM and nonvolatile memory," *IEEE Electron Dev. Lett.*, 29(7), pp. 781–783, 2008, doi: 10.1109/LED.2008.2000616.

54. J. W. Han, S. W. Ryu, S. Kim, C. J. Kim, J. H. Ahn, S. J. Choi, J. S. Kim, K. H. Kim, G. S. Lee, J. S. Oh, M. H. Song, Y. C. Park, J. W. Kim, and Y. K. Choi, "A bulk FinFET unified-RAM (URAM) cell for multifunctioning NVM and capacitorless 1T-DRAM," *IEEE Electron Dev. Lett.*, 29(6), pp. 632–634, 2008, doi: 10.1109/LED.2008.922142.

55. S. W. Ryu, J. W. Han, C. J. Kim, S. J. Choi, S. Kim, J. S. Kim, K. H. Kim, J. S. Oh, M. H. Song, G. S. Lee, Y. C. Park, J. W. Kim, and Y. K. Choi, "Refinement of unified random access memory," *IEEE Trans. Electron Devices*, 56(4), pp. 601–608, 2009, doi: 10.1109/TED.2008.2012292.

56. S. J. Choi, J. W. Han, C. J. Kim, S. Kim, and Y. K. Choi, "Improvement of the sensing window on a capacitorless 1T-DRAM of a FinFET-based unified RAM," *IEEE Trans. Electron Devices*, 56(12), pp. 3228–3231, 2009, doi: 10.1109/TED.2009.2033011.

57. N. Collaert, M. Aoulaiche, A. De Keersgieter, B. De Wachter, L. Altimime, and M. Jurczak, "Substrate bias dependency of sense margin and retention in bulk FinFET 1T-DRAM cells," *Solid State Electron.*, 65–66(1), pp. 205–210, 2011, doi: 10.1016/j.sse.2011.06.018.

58. M. S. Cho, H. J. Mun, S. H. Lee, J. Jang, J. H. Bae, and I. M. Kang, "Simulation of capacitorless dynamic random access memory based on junctionless FinFETs using grain boundary of polycrystalline silicon," *Appl. Phys. A Mater. Sci. Process.*, 126(12), pp. 1–10, 2020, doi: 10.1007/s00339-020-04125-w.

59. T. Gong, Y. Wang, H. Yu, Y. Xu, P. Jiang, P. Yuan, Y. Wang, Y. Chen, Y. Ding, Y. Yang, Y. Wang, and Q. Luo, "Investigation of endurance behavior on HfZrO-based charge-trapping FinFET devices by random telegraph noise and subthreshold swing techniques," *IEEE Trans. Electron Devices*, 68(7), pp. 3716–3719, 2021, doi: 10.1109/TED.2021.3082814.

60. T. Tanaka, E. Yoshida, and T. Miyashita, "Scalability study on a capacitorless 1T-DRAM: From single-gate PD-SOI to double-gate FinDRAM," *IEDM Tech. Dig. Int. IEEE Electron Devices Meet.*, pp. 919–922, 2004, doi: 10.1109/iedm.2004.1419332.

61. E. Yoshida, T. Miyashita, and T. Tanaka, "A study of highly scalable DG-FinDRAM," *IEEE Electron Dev. Lett.*, 26(9), pp. 655–657, 2005, doi: 10.1109/LED.2005.853666.

62. User mannual, "Cogenda genius TCAD device simulator," Version 2.1.

10 Literature review of the SRAM circuit design challenges

Ismahan Mahdi, Yasmine Guerbai,
Yassine Meraihi, and Bouchra Nadji

CONTENTS

DOI: 10.1201/9781003359234-10

10.1 INTRODUCTION

The very large-scale integration (VLSI) method has been employed by research-ers for a long time. It refers to the process of assembling millions of transistors into a single chip to form an integrated circuit. The development of novel tech-nologies due to breakthroughs in VLSI reduces design limits while also further enhancing circuit frequency [65]. The tendency of miniaturization has moved on to electronic devices. All modern smart devices come in small, transportable, and compact sizes. The memory and processor are the two circuits that are most frequently seen in these devices. Memory is becoming more and more necessary for the majority of designs.

In today's development, memory takes up more than 85–90% of the chip space. SRAM and DRAM, two memory technologies, provide substantial performance for solid-state drives (SSDs). Therefore, there is a requirement for robust as well as effi-cient memory for multiple integrated devices. SRAM [1, 2, 54] plays a crucial role in VLSI applications due to its low power and high performance. Leakage problems, process instabilities, and SCEs (short channel effects) are all caused by reliability challenges in complementary metal oxide semiconductor (CMOS) design [3, 65]. SRAM is much quicker, more reliable, and uses less power [4], although it has been restricted by CMOS scaling, causing process changes [5, 6].

The major issue in CMOS devices is that supply voltage scaling causes thresh-old voltage scaling. Moore's law has changed CMOS scaling into a nano-scale system [7]. As a consequence, CMOS scaling has reached its limit, with FinFETs [8], tunnel FET (TFET) [9], and carbon nano tubes (CNTs) [10] as potential solu-tions. Within these alternatives, FinFET technology [11, 12] is selected as the best solution for CMOS. FinFET has various benefits over bulk CMOS, includ-ing higher speed, higher drive current per transistor footprint, reduced leakage, no random dopant fluctuation, less power consumption, improved mobility, and transistor scaling.

To reduce the leakage current and the power consumption, several low-power methods are used such as variable threshold CMOS (VTCMOS), multi-threshold CMOS (MTCMOS), self-controlled voltage level (SVL), stacking, and power gating. SRAM is first designed using classic CMOS. Moreover, this causes issues such as increased leakage current and excessive power loss, which degrades SRAM per-formance. Memory must have a low leakage current, a fast access time, and a low power dissipation. As a result, FinFET-based SRAM cells are recommended over CMOS-based SRAM cells [13]. When compared to CMOS-based design architec-tures, FinFET design reduces SCEs [14]. Additionally, it is important to reduce the leakage characteristics of SRAM cells in order to increase cell durability [15].

This chapter is organized as follows: the next section discusses FinFET tech-nology and SRAM architecture. The purpose is described in Section 10.3 and the outline of various FinFET SRAM cells. Section 10.4 discusses the assessment met-rics. Section 10.5 then covers the analytical results of FinFET SRAM using various technologies. Section 10.6 is a comparison of many similar works. We finish with a conclusion in Section 10.7 which is followed by references.

10.2 BASIC CONCEPTS AND RELATED TERMINOLOGIES

10.2.1 FIN FIELD-EFFECT TRANSISTOR TECHNOLOGY

FinFET technology is one of the most appropriate types of FET [16]. This allows for quicker execution and simulation of transistor applications in both analog and digital domains. FinFET appears to be a viable solution for future nanoelectronics because of its low susceptibility, high efficiency, cheap production costs, and low power requirements [17, 18]. FinFETs can be used to replace bulk CMOS transistors [19]. Because of its low leakage current and low standby power, this technique is suitable for the construction of memory circuits [20].

The FinFET structure is considered. It is also known as a substrate-based multi-gate device (MGD). A double gate system is formed by placing the gate on two, three, or four sides of the channel. On the surface of silicon, the source or drain area forms a "fin". FinFET is also known as a multi-gate transistor. The FinFET model incorporates the following zones: a gate-oxide region, a poly-silicon area with high doping, a silicon fin with low doping, and a contact zone with high doping between the source and drain. Figure 10.1 shows a diagrammatic illustration of a FinFET [21].

FinFET architecture provides a variety of design alternatives. It performs in a wide range of modes, including IG, TG, hybrid, and low power. The combination of IG and low-power mode is commonly referred to as the hybrid mode [22]. In terms of fabrication, FinFET devices are identical to CMOS devices. However, FinFETs provide high-performance benefits at relatively low power. FinFET devices are utilized to reduce SCEs and gate-dielectric leakage currents. FinFET is a promising solution for bridging the technological gap between bulk CMOS and new devices like graphene FETs. Hence, FinFET technology is offered as an innovative approach for developing an ultra-low leakage SRAM cell.

It is important to mention some of the advantages of FinFETs which are:

- Channel doping insensitive
- Superior SCE control and better matching
- Higher revenue and reduced cost
- More compact and more efficient in driving current
- Density scale above flat devices (up to 20nm of the substrate)
- Significant effective channel size
- Smaller threshold and source-drain leakage

10.2.2 SRAM MEMORY ARCHITECTURE

SRAM is a type of memory that stores data in a static state until the memory is powered up. SRAM does not require a periodic refresh. It is a volatile memory, which means data is lost if it is not powered. To save each bit, SRAM employs bi-stable latching circuitry. SRAM is employed in CPU memory caches, desktops, PCs, and disks. Figure 10.2 represents the architecture of SRAM memory [23].

The SRAM array is composed of a sense amplifier, a set of SRAM bit cells, a pre-charging circuit, a write driver, a word line driver, and an address decoder. A

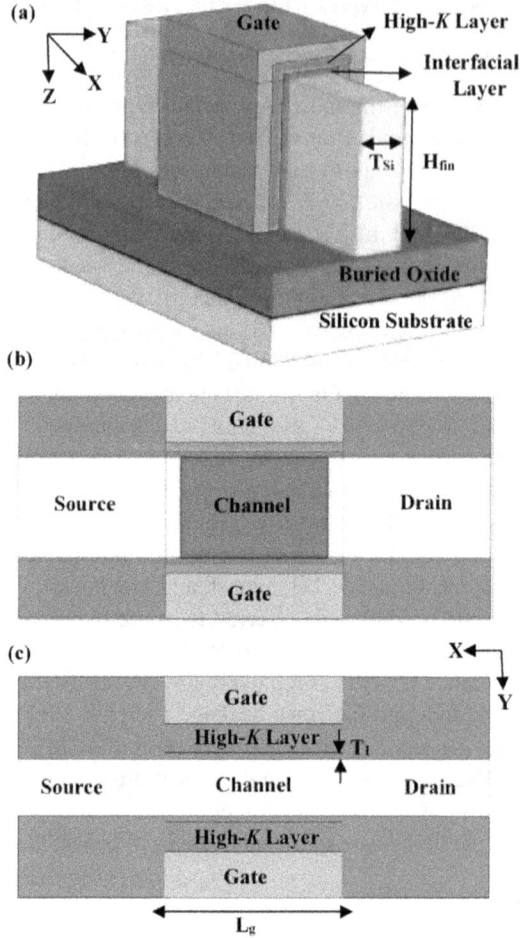

FIGURE 10.1 Diagrammatic illustration of a FinFET in 3D (a) and 2D view (b, c) [21]

large number of words are kept in a single row and are retrieved at the same time. The address word is subdivided into row and column addresses. The SRAM cell is also known as a 1-bit SRAM cell or bit-cell because it has a latch circuit with dual main operating states. The data in the storage cell can be described as a logic "1" or a logic "0". Each SRAM cell has three operating states: hold state, read state, and write state.

SRAM memory cells are typically constructed using basic cross-coupled inverters connected back-to-back, as well as two access transistors [24]. When a word line is enabled for read or write operations, the access transistors that link the cell to the complementary bit line columns are set ON. Some features of this circuit architecture include low static power dissipation, medium power usage, low leakage current, and reduced time required to access data [25, 26].

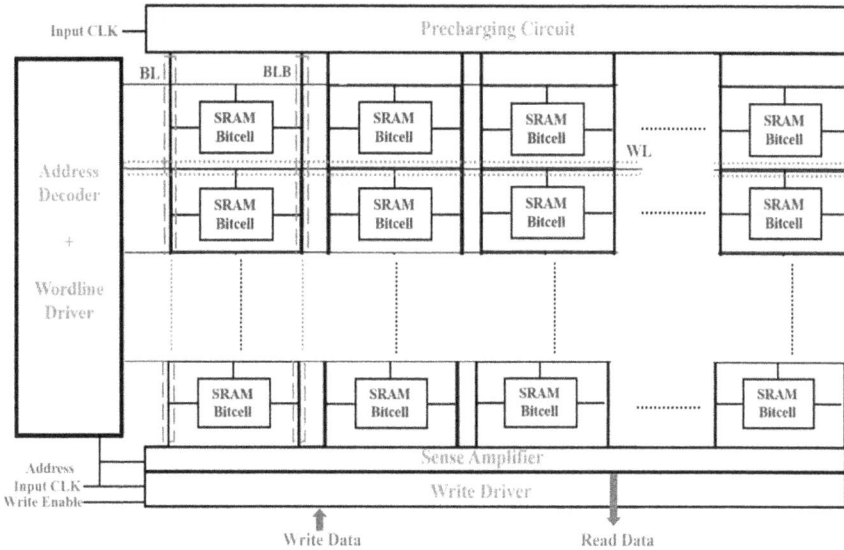

FIGURE 10.2 Conventional SRAM memory architecture [23]

10.3 FINFET-BASED SRAM CELLS

The primary objective of this study is to explore different SRAM memory designs based on FinFET technology. Work feature innovation, flexibility, corner impact, and volume inversion are among FinFET's distinguishing characteristics. The FinFET-based layout provides enhanced performance, lower costs, and increased circuit functionalities.

In order to preserve an adequate production outcome, an advanced circuit approach is required to fill the gap between power, area, durability, speed, and reliability. FinFET-based architectures are presented as a viable solution to bulk devices. Many ways have already been developed, mostly to decrease static power dissipation. Therefore, these technologies can only reduce the changes in leakage current. So, new FinFET-based SRAM cells are used to enhance cell stability and reduce leakage current.

10.3.1 FinFET-6t sram cell structure

Two cross-coupled (2-CC) inverters and dual access FinFET transistors are included in the 6T (six transistors) SRAM [27]. The 2-CC inverters are composed of four transistors known as M1, M2, M3, and M4. Each bit in SRAM is saved on these four FinFET transistors. M5 and M6 are the two access FinFETs, and the source terminals are joined to BL (bitline) and $\overline{BL}$ (or BLB). Because it takes up less space, the 6T SRAM cell is a popular basic used cell. When WL (word line) = 1, the dual access FinFETs are activated, and the bit lines are coupled to a latch that executes a read or write operation. When WL = 0, the access transistors are turned off, and BL and $\overline{BL}$ are disconnected from the latch. A schematic illustration of a FinFET-based SRAM.6T cell is shown in Figure 10.3 [27].

Read, write, and hold are the three basic functions [28, 29] in the SRAM memory. WL is connected to the ground in hold mode. As a result, transistors M1 and M2 turn off, separating the latching circuit from the bit lines. The remaining transistors M3, M4, M5, and M6 constitute a latch structure that maintains stored data until the bit lines are disconnected. In read mode, pre-charge the bit lines to VDD, connect WL to VDD, and turn on the transistors M1 and M2. If Q = 1 and $\overline{Q}$ = 0, the transistors M3 and M6 are turned off, and the transistors M4 and M5 are turned on. As a result, the voltage level of BL remains constant at VDD while the voltage level of $\overline{BL}$ decreases. In write mode, WL is connected to VDD and turns on the transistors M1 and M2.

10.3.2 FINFET-7T SRAM CELL STRUCTURE

Figure 10.4 shows a model of a 7T-FinFET SRAM cell. The 7T (seven transistors) cell design with two 2-CC inverters and four transistors M3, M4, M5, and M6, in

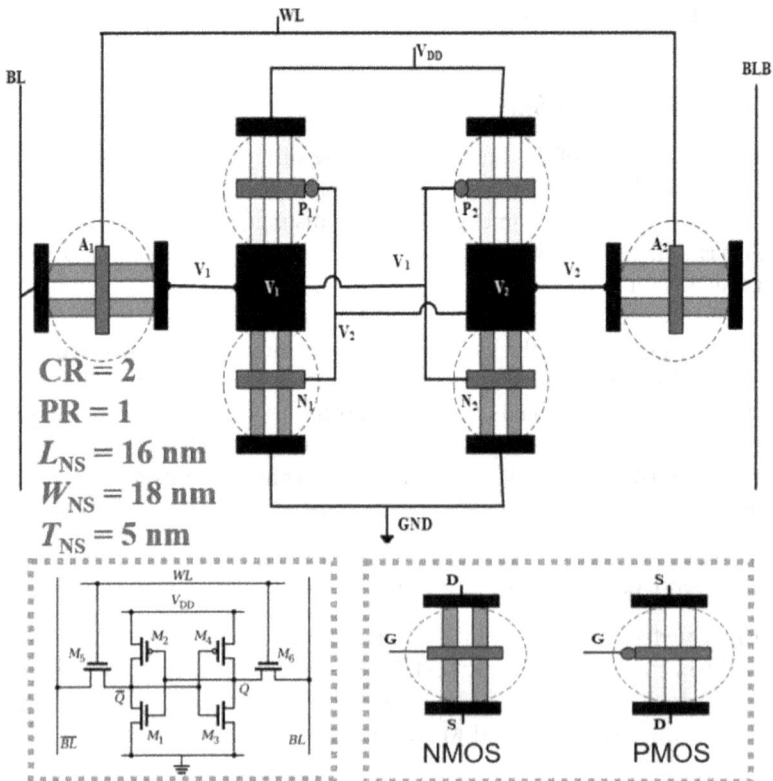

FIGURE 10.3 FinFET 6T SRAM cell structure [27]

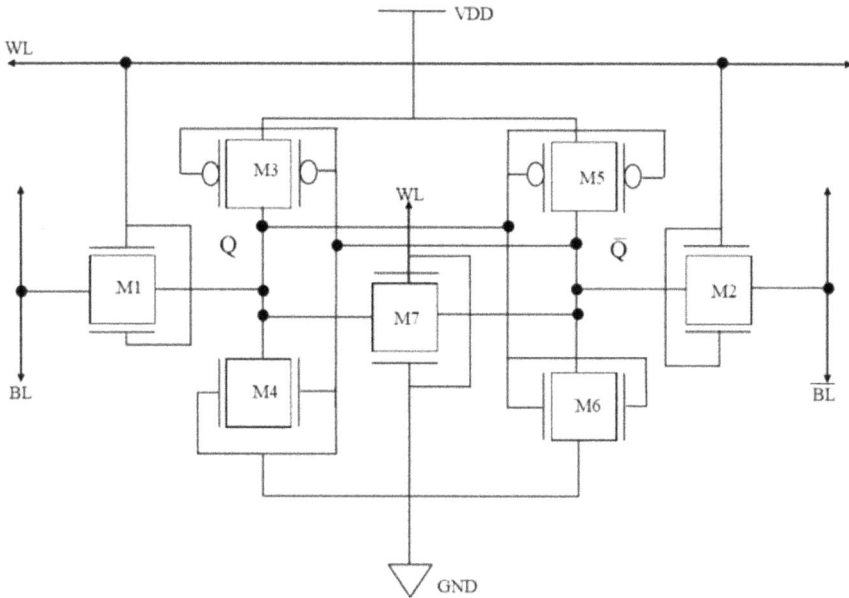

FIGURE 10.4   FinFET 7T SRAM cell structure

addition to an extra transistor M7 are connected to the WL [30, 31]. Dual access transistors M1 and M2 are also connected to the BL and $\overline{BL}$, respectively. The leakage issue in the 6T cell is fixed by introducing the 7T FinFET structure. The M1 and M2 transistors are connected to WL to carry out the read and write operations. The dual bit lines are used as input or output nodes for reading and writing in order to recognize data from SRAM cells utilizing a sense amplifier.

In hold mode, the WL is switched off, and the transistors M3 and M4 become inactive. A sub-threshold leakage current passes through the transistors in the off state because of logic 0 in the SRAM cell. Furthermore, the extra transistor M7 provides both feedback connection and disconnection, and the SRAM cell is entirely dependent on the $\overline{BL}$ to complete write operations [32, 33].

10.3.3   FINFET-8T SRAM CELL STRUCTURE

To solve the restrictions of the 6T cell, the 8T (eight transistors) FinFET SRAM cell is proposed [34]. The major aspect is that the read and write functions are not separated. The cell with the lowest static noise margin (SNM) in reading mode may have better writing capabilities. As a result, if the read and write functions are properly isolated, circuit designers have complete flexibility in improving read and write procedures. Figure 10.5 presents a schematic illustration of a FinFET-8T SRAM cell.

The 8T structure is designed to split read and write operations to provide increased stability while permitting low-voltage operations [34, 35]. The 8T configuration

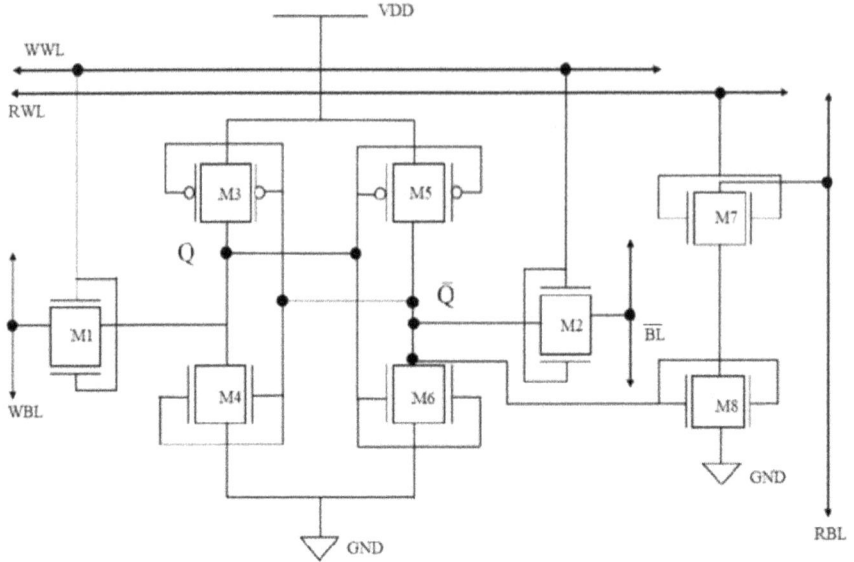

FIGURE 10.5 FinFET 8T SRAM cell structure

indicates that the integration of two FETs to a 6T cell structure provides a read proce-
dure that does not disrupt the cell's internal nodes. As a consequence, this operation
needs a distinct read word line (RWL) and write word line (WWL). RWL is activated,
and RBL (read bitline) is pre-charged to operate in reading mode. If 1 is kept at Q, the
M6 transistor goes on and produces a low resistance channel for the cell current flow
via RBL to ground (GND) as recognized by the sense amplifier [36, 37].

10.3.4 FINFET-9T SRAM CELL STRUCTURE

The 9T (nine transistors) FinFET SRAM cell is mostly composed of dual sub-sections [38–41]. The highest part of 9T SRAM is identical to the 6T cell architecture,
which constitutes M1, M2, M3, M4, and Q. This principal sub-section is used to hold
information. The other part of 9T SRAM has two bitline access transistors M5 and
M6 and one read access transistor M9. The data contained in the cell determines
how transistors M8 and M7 operate. M9 is dependent on a distinct read signal (RD).
Write bitline (WBL) and $\overline{\text{WBL}}$ control the write access transistors, which conduct
write access. Furthermore, read access transistors perform read operation that is
regulated by read word line (RWL). There is a schematic illustration of a 9T FinFET
SRAM cell architecture in Figure 10.6.

   To rectify the leakage issue observed on the 8T cell RBL, a 9T cell architecture is
developed. This allows data to alter during read procedures. The 8T cell is restricted
to low-density applications that can be handled with the 9T structure (by introducing
an M9 transistor between the M7 and M8). As a result, the stack effect phenomenon

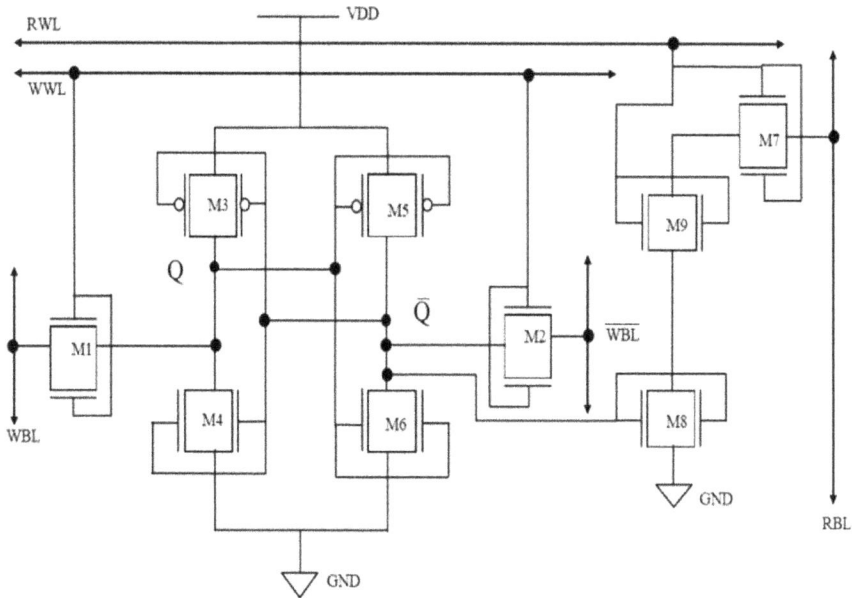

FIGURE 10.6 FinFET 9T SRAM cell structure

dramatically reduces leakage in BL. Stacking occurs in 9T SRAM when OFF state transistors are coupled in series. As a consequence, the top transistor source voltage in the stack will be slightly greater than the lower transistor source voltage. The greater voltage of the top transistor raises the threshold voltage. This increase in threshold voltage will minimize leakage.

10.3.5 FINFET-10T SRAM CELL STRUCTURE

The 10T (ten transistors) SRAM cell consists of four pull-ups, four pull-downs, and two access transistors. This 10T cell is implemented to reduce power dissipation and leakage current. The dual threshold voltage method is applied via transistors in the read line, which improves the current ON/OFF ratio. RWL is connected to the sources of M10 and M9 transistors. WWL, BL, and $\overline{BL}$ are also associated with the access FinFETs. The write range is increased by the use of transistors M7 and M8. The static current is reduced by assuming that access transistors are twice the size of pull up transistors. Except for access transistors, all other transistors in the 10T architecture are limited to the shortest feasible gate length [42–46].

Compared to the 9T cell design, FinFET-based 10T cell design reduces leakage current by employing transistors M7, M8, M9, and M10. The two access transistors are employed to interconnect the nodes for the read process. Because there is no transfer of read current by storage nodes, read stability is well managed. To execute a write operation, node Q saves 1 and node $\overline{Q}$ saves 0. By providing a high supply voltage, node Q is discharged through the access and the pull-up transistor, causing it to be forced down to "0".

10.3.6 FINFET-11T SRAM CELL STRUCTURE

SRAM 11T (11 transistors) is intended to reduce energy usage. Static power consumption is a major challenge in SRAM engineering. This 11T design aims to reduce static power consumption while enhancing performance in the sub-threshold zone. The transistors M2, M4, M5, and M6 have the same properties as the 6T SRAM cell. Furthermore, the size of the transistors M2 and M3 is downscaled to be similar to the size of the PMOS transistor. The 11T cell has WL, BL, RWL, and read/write interfaces [47, 48]. Since it has series-connected drivers, which are supplied by BL, $\overline{BL}$, and read buffers, this cell has a low loss of power.

10.3.7 FINFET-12T SRAM CELL STRUCTURE

The SRAM 12T (12 transistors) cell contains the transistors M3, M7, M4, M5, M8, and M6, as well as dual read or write ports, M9, M1, M11, M10, M2, and M12. WWL1, WWL2, BL, and $\overline{BL}$ are all column-based, whereas RWL and VGND (virtual ground) are row-based. The decrease in power and current justifies the 12T design [49, 50]. It is created for low-voltage use. The 12T bit-cell switches off the supply voltage of the left or right half-cell during the writing operation to reduce the pull-up network. This architecture accelerates the writing process without the need for additional timing control or peripheral write help circuits.

10.3.8 FINFET-13T SRAM CELL STRUCTURE

The fluctuation impacts of an intrinsic parameter reduce the SRAM cell's stability properties. Among them are random dopant variation, line-edge roughness, and gate-oxide-thickness difference. The 13T SRAM cell is intended to provide greater SM and improved performance. The majority of these cells separate functions like read (R) and write (W) to achieve higher NM. The 13T (13 transistors) design [51] consists of a CC Schmitt Trigger (ST) inverter, dual transistors in the read line, and one MAL (W-Access transistor), as well as MAR1 (R-Access transistor). The suggested description of the ST13T SRAM cell comprises a design change with the transmission gate (TG) usage in the access line. TG passes over the voltage range, i.e. (between "0" and "1"), boosting device performance properly. FinFETs enhance power consumption by overcoming the leakage issues of planar devices and delivering superior efficiency [52, 66].

10.4 PERFORMANCE ASSESSMENT METRICS

10.4.1 STATIC NOISE MARGIN (SNM)

The standard technique for measuring SRAM bit-cell stability is SNM. It is determined by the cell ratio (CR), supply voltage, and pull-up ratio (PR). SRAM cell

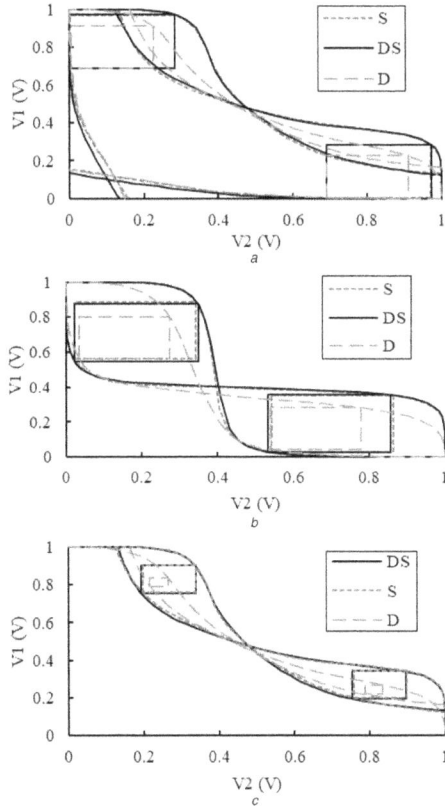

FIGURE 10.7 SNMs of drain/source-JLSiNT FET (a), drain-JLSiNT FET (b), and source-JLSiNT FET (c) based 6T SRAM cell [53]

stability depends on good SNM. CR is the ratio of the sizes of the driver transistor to the load transistor during the read operation. The ratio of the sizes of the load transistor to the access transistor is referred to as PR during a write operation.Figure 10.7 shows SNMs or butterfly curves of the 6T SRAM structure [53].

The driver transistor affects 70% of the SNM value. SNM determines both read and write margins and is proportional to the threshold voltages of NMOS and PMOS devices. If CR starts to rise, the DT size grows as well. Additionally, as the current increases, so does the cell speed. SNM is therefore obtained in order to modify the CR value. We acquired various SNMs in different SRAM cell technologies for several CR values. This is given by the formula below:

$$SNM = \frac{1}{\sqrt{2}} * \min\left[\max_{-\sqrt{2}<u<0} \left(|V1-V2|\right), \max_{0<u<\sqrt{2}} \left(|V1-V2|\right) \right] \qquad (1)$$

10.4.2 TEMPERATURE

The most essential metric to measure high temperature affects the device's performance when it is switched on or off. Power dissipation commonly causes a rise in device temperature. If a temperature rise is detected, it may cause a circuit fault, affecting power, performance, and reliability. Temperature can have a significant impact on other design characteristics such as access time. Furthermore, when the temperature rises, the leakage current may also rise exponentially in a FinFET device. As a consequence, temperature is regarded as the most important performance metric in VLSI circuit design.

10.4.3 POWER AND DELAY

Power and delay are key parameters in SRAM circuit design. The main advantage of SRAM based on FinFET technology is its low access time and low energy consumption. In SRAM, column height, as well as line delays, have a significant impact on propagation delay. In conclusion, the segmentation technique reduces the delay. It is known that oversizing the FinFET device minimizes the delay. Leakage currents must be reduced using increased transistor threshold voltage to reduce power delay.

10.4.4 POWER DELAY PRODUCT (PDP)

Power delay product (PDP) is determined using the transient analysis performance of SRAM cells. It is a parameter for measuring a circuit's energy usage. PDP is defined as the product of gate delay and average power. Furthermore, PDP supports processors that run at a lower frequency. For read and write operations, an optimal SRAM cell requires a smaller PDP. The transistor size is chosen to get the lowest possible PDP by optimizing the transistor size, which then reduces the delay without boosting the power usage.

10.4.5 READ NOISE MARGIN (RNM)

Read noise margin (RNM) is used to assess the reliability of SRAM cells, and the RNM is proportional to CR. To get better RNM, the pull-down FinFET should be larger than the access transistor. The pull-up ratio is determined by the size of the transistor. So, RNM increases as the pull-up ratio value goes up. Furthermore, the read margin is exactly proportional to the CR. RNM technique analysis is comparable to SNM one. The readability of an SRAM cell is described by RNM, which is based on voltage transfer curves (VTCs). RNM is calculated using the transistor's current model. Pull-down transistor upsizing improves RNM, resulting in an increase in access FinFET gate length. To avoid unintentionally writing 1 into an SRAM cell, a careful FET device is required.

10.4.6 WRITE NOISE MARGIN (WNM)

To write data into an SRAM cell we use write noise margin (WNM). It is the highest bitline voltage (BLV) that can twist the cell state of a FinFET -SRAM when the $\overline{BL}$ voltage is set high. WNM is proportional to PR, and it improves as the PR value rises. WNM voltage is the maximum noise voltage (NV) present at BL during a complete write operation. Only when noise voltage surpasses the WNM voltage does a write fail happens, and superior stability is represented by higher WNM. The use of an access FinFET allows for a faster discharge of 1 and hence a faster write 0. Thus, WNM improves with strong access at the read margin.

10.5 ANALYTICAL RESULTS OF FINFET SRAM IN DIFFERENT TECHNOLOGIES

This section discusses the analysis of different SRAM cells in several nanometer technologies. The performance study of 6T SRAM in 22 nm technology is shown in Figure 10.8, and the tool utilized is a predictive technology model library. Figure 10.9 shows a comparison between 6T SRAM CMOS technology and 6T SRAM FinFET technology [59]. Figure 10.10 compares the performance of 6T SRAM in planar and FinFET technologies [60]. Figure 10.11 compares 7T FinFET SRAM technology to different SRAM cells [31]; Tanner was used as the simulation tool. Table 10.1 shows a comparison of 7T SRAM in the H-Spice tool [20].

Figure 10.12 illustrates a comparison between FinFET 7T SRAM and 8T SRAM [61] using the Cadence Virtuoso tool. Figure 10.13 shows a comparison of 9T FinFET SRAM technology [62] in Cadence software. Figure 10.14 compares 10T SRAM and

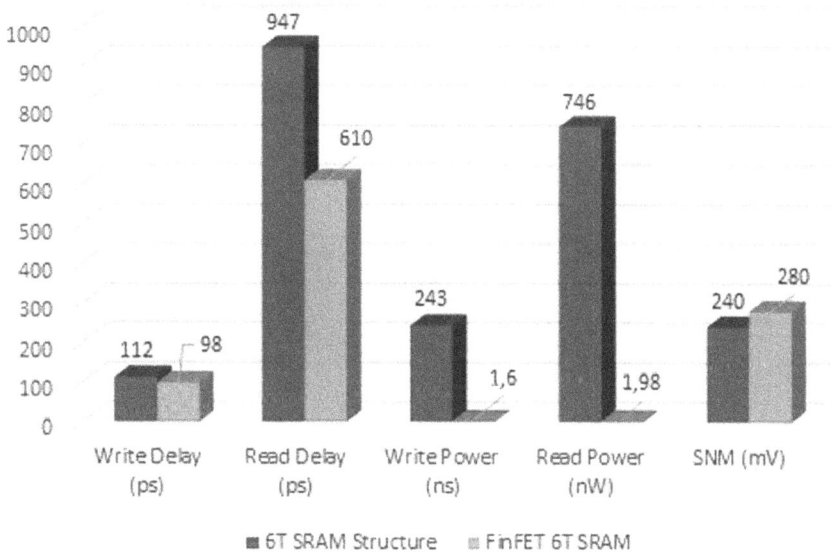

FIGURE 10.8 Performance of 6T SRAM in 22 nm technology (T = 25 °C) [29]

(a)

■ 45 nm   ■ 32 nm   ▣ 16 nm

(b)

■ 16 nm   ■ 7 nm   ▣

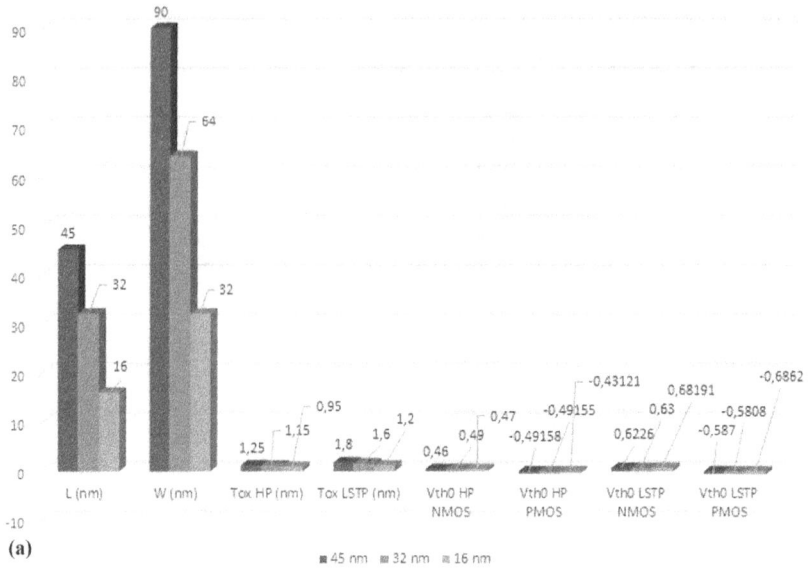

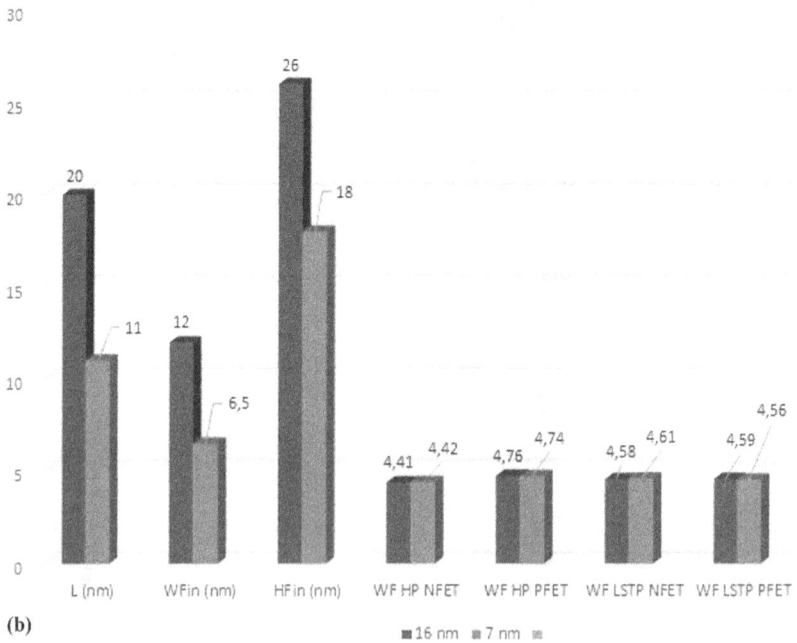

FIGURE 10.9   Comparison of 6T SRAM in CMOS (a) and FinFET (b) technology [59]

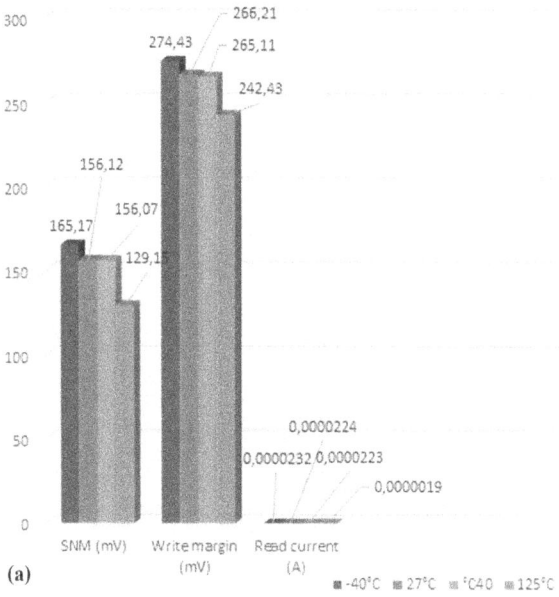

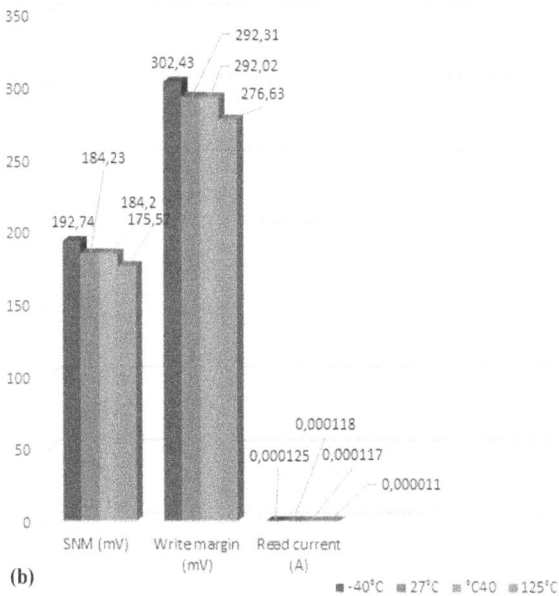

FIGURE 10.10 Comparison of 6T SRAM in planar (MOSFET) (a) and FinFET (b) technology [60]

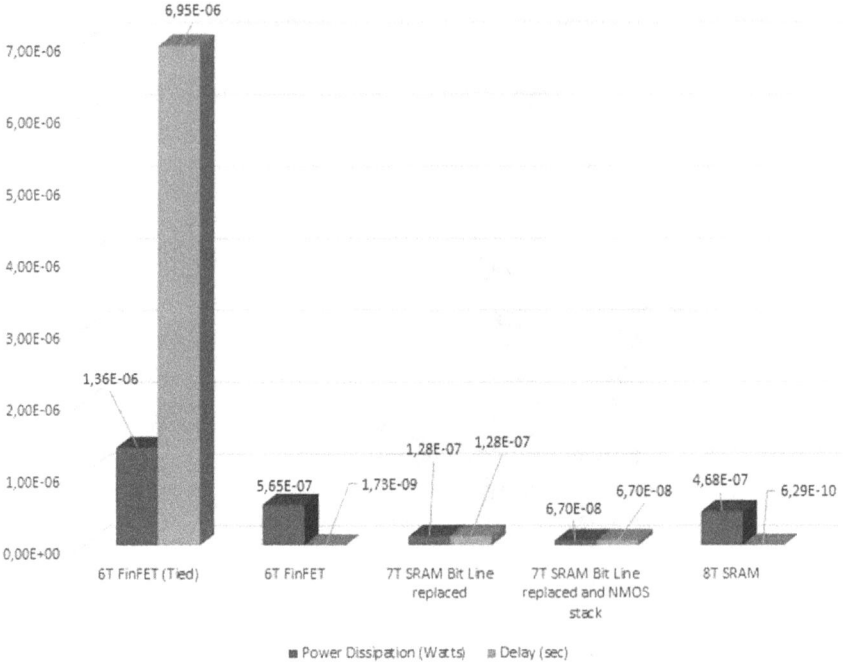

FIGURE 10.11 Comparison of 7T SRAM with other SRAM cells technologies (45 nm) [31]

6T FinFET SRAM at 7 nm technology [63] in the Cadence Virtuoso tool. Figure 10.15 compares the 11T FinFET SRAM to different cells on 10 nm technology [64]. Table 10.2 compares 12T FinFET SRAM in 32 nm technology [50-56] with the H-Spice simulator. Figure 10.16 compares 13T FinFET SRAM in 22 nm technology [51] using the Cadence Virtuoso tool (V.6.1) [].

10.6 ANALYTICAL RESULTS OF FINFET SRAM IN DIFFERENT TECHNOLOGIES

The used technology, device name, employed technique, and key characteristics of various FinFET SRAM cells are compared in this section. The FinFET-based SRAM comparison is shown in Table 10.3.

TABLE 10.1
Comparative analysis of 7T SRAM [20]

Technology	Average power	Delay (ns)	Power delay
CMOS (22 nm)	0.169	0.129	0.0219
CMOS (16 nm)	0.139	0.223	0.0311
FinFET (22 nm)	0.023	0.090	0.0021
FinFET (16 nm)	0.015	0.002	0.0028

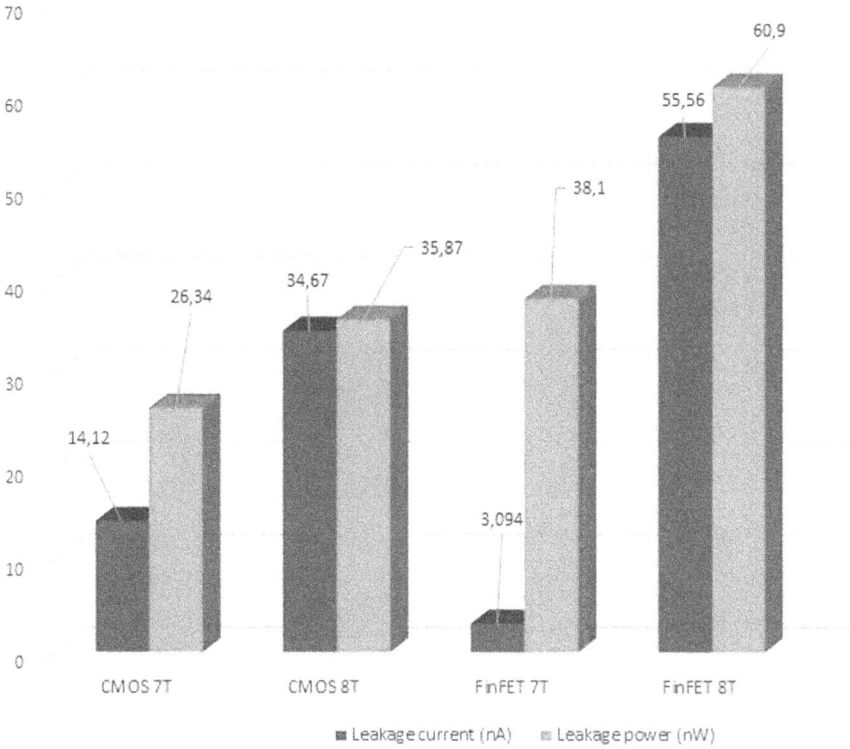

FIGURE 10.12 Comparison of FinFET 7T and 8T SRAM [61]

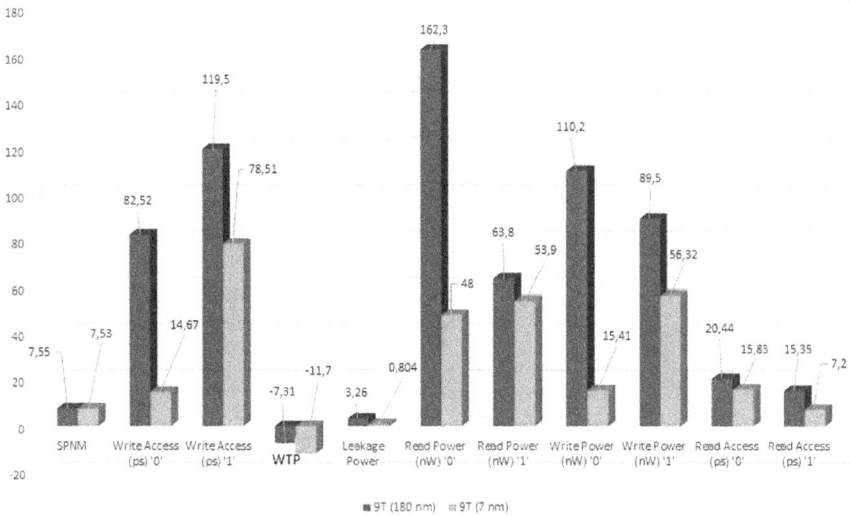

FIGURE 10.13 Comparison of 9T FinFET at 180 nm and 7 nm technologies [62]

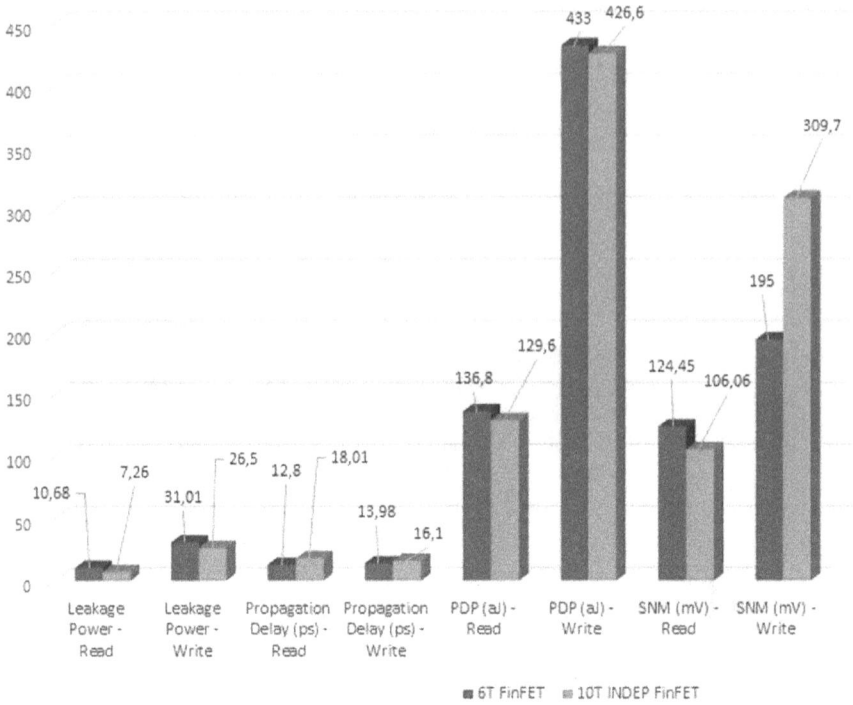

FIGURE 10.14 Comparison of 10T and 6T FinFET at 7 nm technology [63]

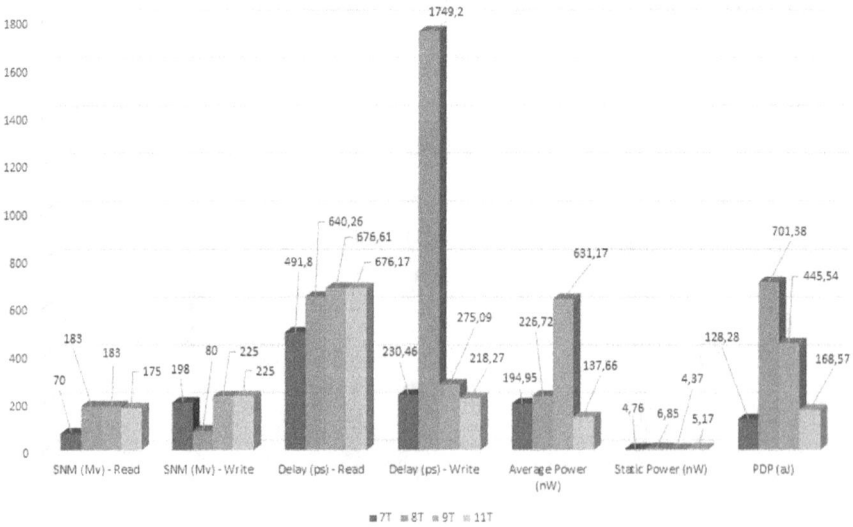

FIGURE 10.15 Comparison of 11T FinFET SRAM with other cells at 10 nm technology [64]

TABLE 10.2
12T FinFET SRAM cell comparison at 32 nm technology [50]

	Average power		
Type	Hold (nW)	Read (nW)	Write (nW)
12 T CMOS	180.79	161.30	201.61
12 T FinFET	14.06	45.64	60.06

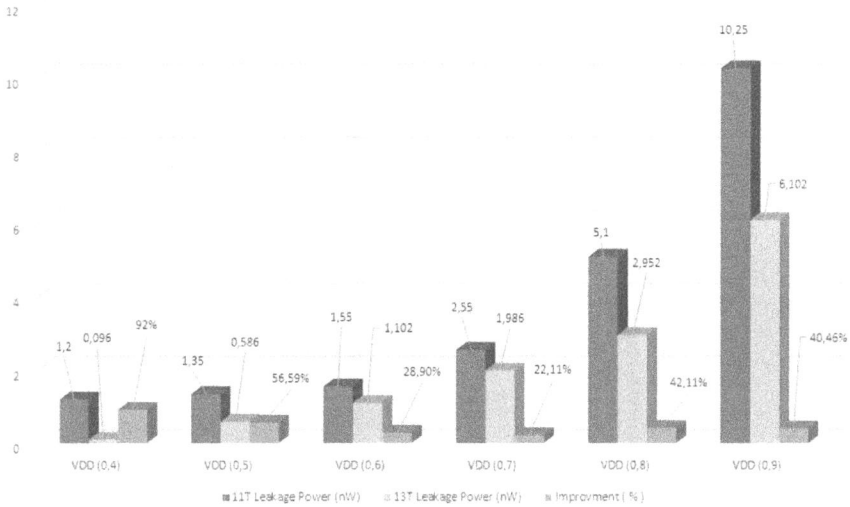

FIGURE 10.16   13T FinFET SRAM cells comparison at 22 nm technology [51]

10.7   CONCLUSION

Different SRAM designs based on FinFET technology were examined in this litera-ture review. This chapter shows that it is optimal to design SRAM using FinFET, as it provides lower static power consumption and latency than CMOS SRAM cells, and the delay is also minimized in both read and write operations. FinFETs offer several advantages over bulk MOSFETs, including the fact that FinFETs were designed with a process fabrication flow identical to the typical SOI CMOS process, while DG MOSFETs have a complex manufacturing method.

Compared to other DG MOSFET architectures, FinFET offers a high package density. FinFET-based SRAM models are offered to remove SCEs. Relative to typi-cal MOSFET-based SRAM cells, these models demonstrate a considerable reduction in leakage current and power dissipation. As a result, this chapter helps to improve knowledge of the behavior of FinFET-based SRAM in which low power, high speed, low leakage, and high performance are required.

TABLE 10.3
Some FinFET SRAM Cells technologies comparison

Reference	Technology	Device name	Employed technique	Characteristics
[31]	45 nm	7TFinFETSRAM	Power voltage scaling	RBL charging is not required, and the voltage supply is reduced
[32]	14 nm	7T FinFET SRAM	Near threshold voltage (NTV) process	In the NTV zone, an appropriate design option for high performance and low power usage is available
[33]	20 nm	7T FinFET SRAM	NTV process	High write and read margins, as well as a short write time
[37]	16 nm	8T FinFET SRAM	Supply voltage-floating	Enhancement of reading and writing abilities
[39]	22 nm	9T FinFET (Single-ended SRAM)	NTV and yield estimation methods	The minimum operating voltage is 0.3 V
[40]	22 nm	9T FinFET SRAM	Multi-threshold	Three times lower power consumption
[41]	22 nm	FinFET (9T SRAM)	Power gated 9T SRAM	Requires less power for both read and write operations, and a small area for each bit
[44]	32 nm	10T FinFET SRAM	Built-in feedback system and back gate biasing	Positive feedback and back gate biasing improve stability
[45]	16 nm	MOSFET and 10T FinFET SRAM	NTV technique	FinFET-based SRAM will be utilized without SCE below 32nm
[46]	32 nm	CMOS and 10T FinFET SRAM	Supply voltage reduction	At 400 mV, increases RSNM by 20% and reduces WSNM by 5%
[57]	20 nm	8T FinFET SRAM	Supply voltage	The FinFET technology has a reduced sub-threshold swing (SS)
[58]	32 nm	8T FinFET SRAM	Back gate voltage	Minimizing the discharge activity in the write operation

Furthermore, this FinFET SRAM architecture is well adapted for a variety of electronic applications such as mobile technologies, embedded systems, CPUs, processors, DSPs, SD-RAMs, and so on. They are often employed in various low-power CMOS circuit applications. In the future, a multi-fin FinFET device might be employed to improve the device's driving efficiency. This enables the electronic revolution to be faster and more reliable. Moreover, the manufacturing procedures necessary to achieve such tight criteria are a topic that can be investigated in the future. To improve performance, the transistor count may be minimized while developing the SRAM architecture.

REFERENCES

1. Agrawal, R., & Tomar, V. K. "Analysis of low power reduction techniques on cache (SRAM) memory". In *2018 9th International Conference on Computing, Communication and Networking Technologies (ICCCNT)*. IEEE, 1–7 (2018).
2. Pasandi, G., Mehrabi, K., Ebrahimi, B., Fakhraei, S. M., Afzali-Kusha, A., & Pedram, M. "Low-power data encoding/decoding for energy-efficient static random access memory design". *IET Circuits, Devices and Systems*, 13(8), 1152–1159 (2019).
3. Lin, Z., Wu, X., Li, Z., Guan, L., Peng, C., Liu, C., & Chen, J. "A pipeline replica bit-line technique for suppressing timing variation of SRAM sense amplifiers in a 28-nm CMOS process". *IEEE Journal of Solid-State Circuits*, 52(3), 669–677 (2016).
4. Chiu, P. F., Chang, M. F., Wu, C. W., Chuang, C. H., Sheu, S. S., Chen, Y. S., & Tsai, M. J. "Low store energy, low VDDmin, 8T2R nonvolatile latch and SRAM with vertical-stacked resistive memory (memristor) devices for low power mobile applications". *IEEE Journal of Solid-State Circuits*, 47(6), 1483–1496 (2012).
5. Mocuta, A., Weckx, P., Demuynck, S., Radisic, D., Oniki, Y., & Ryckaert, J. "Enabling CMOS scaling towards 3nm and beyond". In *2018 IEEE Symposium on VLSI Technology*. IEEE, 147–148 (2018).
6. Yusop, N. S., Nordin, A. N., Khairi, M. A., & Hasbullah, N. F. "The impact of scaling on single event upset in 6T and 12T SRAMs from 130 to 22 nm CMOS technology". *Radiation Effects and Defects in Solids*, 173(11–12), 1090–1104 (2018).
7. Zhang, K. "Circuit design in nano-scale CMOS technologies". In *2018 IEEE Asian Solid-State Circuits Conference (A-SSCC)*. IEEE, 1–4 (2018).
8. Ensan, S. S., Moaiyeri, M. H., Moghaddam, M., & Hessabi, S. "A low-power single-ended SRAM in FinFET technology". *AEU – International Journal of Electronics and Communications*, 99, 361–368 (2019).
9. Liautard, R., Trojman, L., Arevalo, A., & Procel, L. M. "TFET and FinFET hybrid technologies for SRAM cell: Performance improvement over a large VDD-range". In *2019 IEEE Fourth Ecuador Technical Chapters Meeting (ETCM)*. IEEE, 1–5 (2019).
10. Jia, X., & Wei, F. "Advances in production and applications of carbon nanotubes". In *Single-Walled Carbon Nanotubes*. Yan Li and Shigeo Maruyama eds. Springer, 299–333 (2019).
11. Birla, S. "FinFET SRAM cell with improved stability and power for low power applications". *Journal of Integrated Circuits and Systems*, 14(2), 1–8 (2019).
12. Birla, S. "Variability aware FinFET SRAM cell with improved stability and power for low power applications". *Circuit World* 45 (4) (2019).
13. Panchal, J., & Vishal, R. *Design and Implementation of 6T SRAM Using Finfet with Low Power Application*, 5 (7) 2395–0056 (2017).

14. Narendar, V., & Mishra, R. A. "Analytical modeling and simulation of multigate FinFET devices and the impact of high-k dielectrics on short channel effects (SCEs)". *Superlattices and Microstructures*, 85, 357–369 (2015).

15. Tripathi, T., Chauhan, D. S., & Singh, S. K. "A novel approach to design SRAM cells for low leakage and improved stability". *Journal of Low Power Electronics and Applications*, 8(4), 41 (2018).

16. Jurczak, M., Collaert, N., Veloso, A., Hoffmann, T., & Biesemans, S. "Review of FinFET technology". In *2009 IEEE International SOI Conference*. IEEE, 1–4 (2009).

17. Dadoria, A. K., Khare, K., Gupta, T. K., & Singh, R. P. "Leakage reduction by using FinFET technique for nanoscale technology circuits". *Journal of Nanoelectronics and Optoelectronics*, 12(3), 278–285 (2017).

18. Guo, D., Karve, G., Tsutsui, G., Lim, K. Y., Robison, R., Hook, T., Vega, R., et al. "FinFET technology featuring high mobility SiGe channel for 10nm and beyond". In *2016 IEEE Symposium on VLSI Technology*. IEEE, 1–2 (2016).

19. Yakimets, D., Bardon, M. G., Jang, D., Schuddinck, P., Sherazi, Y., Weckx, P., Miyaguchi, K., et al. "Power aware FinFET and lateral nanosheet FET targeting for 3nm CMOS technology". In *2017 IEEE International Electron Devices Meeting (IEDM)*. IEEE, 20–24 (2017).

20. Garg, M., & Singh, B. "A comparative performance analysis of CMOS and finfet based voltage mode sense amplifier". In *2016 8th International Conference on Computational Intelligence and Communication Networks (CICN)*. IEEE, 544–547 (2017).

21. Tayal, S., & Ashutosh, N. "Comparative analysis of high-K gate stack based conventional & junctionless FinFET". 2017 14th IEEE India Council International Conference (INDICON), Roorkee, India, 2017, pp. 1–4, doi: 10.1109/INDICON.2017.8 487675.

22. Rajprabu, R., Arun Raj, A., & Rajnarayanan, R. "Performance analysis of CMOS and FinFET logic". *IOSR Journal of VLSI and Signal Processing (IOSR-JVSP)*, 2(1), 1–6 (2013).

23. Mohammed, M. U., Athiya, N., Liaquat, A., & Masud, H. C. "FinFET based SRAMs in Sub-10nm domain". *Microelectronics Journal*, 114, 105–116, (2021) ISSN 0026-2692, https://doi.org/10.1016/j.mejo.2021.105116.

24. Tiwari, N., Neema, V., Rangra, K. J., & Sharma, Y. C. "Performance parameters of low power SRAM cells: A review". *i-Manager's Journal on Circuits & Systems*, 6(1), 25 (2017).

25. Gavaskar, K., & Ragupathy, U. S. "An efficient design and analysis of low power SRAM memory cell for ULTRA applications". *Asian Journal of Research in Social Sciences and Humanities*, 7(1), 962–975 (2017).

26. Bhaskar, A. "Design and analysis of low power SRAM cells". In *2017 Innovations in Power and Advanced Computing Technologies (i-PACT)*. IEEE, 1–5 (2017).

27. Tayal, S., Smaani, B., Rahi, S. B., Upadhyay, A. K., Bhattacharya, S., Ajayan, J., Jena, B., et al. "Incorporating bottom-up approach into device/circuit co-design for SRAM-based cache memory applications". *IEEE Transactions on Electron Devices*, 69(11), 6127–6132 (2022). https://doi.org/10.1109/TED.2022.3210070.

28. Gupta, M. K., Weckx, P., Cosemans, S., Schuddinck, P., Baert, R., Yakimets, D., Jang, D., et al. "Device circuit and technology co-optimisation for FinFET based 6T SRAM cells beyond N7". In *2017 47th European Solid-State Device Research Conference (ESSDERC)*. IEEE, 256–259 (2017).

29. Banu, R., & Shubham, P. "Design and performance analysis of 6T SRAM cell in 22nm CMOS and FinFET technology nodes". In *2017 International Conference on Recent Advances in Electronics and Communication Technology (ICRAECT)*. IEEE, 38–42 (2017).

30. Asli, R. N., & Taghipour, S. "A near-threshold soft error resilient 7T SRAM cell with low read time for 20 nm FinFET technology". *Journal of Electronic Testing*, 33(4), 449–462 (2017).

31. Sneha, G., Krishna, B. H., & Kumar, C. A. "Design of 7T FinFET based SRAM cell design for nanometer regime". In *2017 International Conference on Inventive Systems and Control (ICISC)*. IEEE, 1–4 (2017).

32. Yang, Y., Jeong, H., Song, S. C., Wang, J., Yeap, G., & Jung, S. O. "Single bit-line 7T SRAM cell for near- threshold voltage operation with enhanced performance and energy in 14 nm FinFET technology". *IEEE Transactions on Circuits and Systems. Part I: Regular Papers*, 63(7), 1023–1032 (2016).

33. Ansari, M., Afzali-Kusha, H., Ebrahimi, B., Navabi, Z., Afzali-Kusha, A., & Pedram, M. "A near-threshold 7T SRAM cell with high write and read margins and low write time for sub-20 nm FinFET technologies". *Integration*, 50, 91–106 (2015).

34. Neelima, C. H., Ravinder, T., & Sudha, D. "Design of 8TSram using FinFET technology". *Test Engineering and Management*, 82, 3168–3171 (2020).

35. Guler, A., & Jha, N. K. "Three-dimensional monolithic FinFET-based 8T SRAM cell design for enhanced read time and low leakage". *IEEE Transactions on Very Large Scale Integration (VLSI) Systems*, 27(4), 899–912 (2019).

36. Alias, N. E., Hamzah, A., Tan, M. L. P., Sheikh, U. U., & Riyadi, M. A. "Low-power and high performance of an optimized FinFET based 8T SRAM cell design". In *2019 6th International Conference on Electrical Engineering, Computer Science and Informatics (EECSI)*. IEEE, 66–70 (2019).

37. Monica, M., & Chandramohan, P. "A novel 8T SRAM cell using 16 nm FinFET technology". *SASTech-Technical Journal of RUAS*, 16(1), 5–8 (2017).

38. Sharma, N. "Ultra low power dissipation in 9T SRAM design by using FinFET technology". In *2016 International Conference on ICT in Business Industry & Government (ICTBIG)*. IEEE, 1–5 (2016).

39. Yang, Y., Park, J., Song, S. C., Wang, J., Yeap, G., & Jung, S. O. "Single-ended 9T SRAM cell for near-threshold voltage operation with enhanced read performance in 22-nm FinFET technology". *IEEE Transactions on Very Large Scale Integration (VLSI) Systems*, 23(11), 2748–2752 (2014).

40. Moradi, F., & Tohidi, M. "Low-voltage 9T FinFET SRAM cell for low-power applications". In *2015 28th IEEE International System-on-Chip Conference (SOCC)*. IEEE, 149–153 (2015).

41. Oh, T. W., Jeong, H., Kang, K., Park, J., Yank, Y., & Jung, S. O. "Power-gated 9T SRAM cell for low-energy operation". *IEEE Transactions on Very Large Scale Integration (VLSI) Systems*, 25(3), 1183–1187 (2016).

42. Ichihashi, M., Woo, Y., Karim, M. A. U., Joshi, V., & Burnett, D. "10T differential-signal SRAM design in a 14- nm FinFET technology for high-speed application". In *2018 31st IEEE International System-on-Chip Conference (SOCC)*. IEEE, 322–325 (2018).

43. Singh, A., Sharma, Y., Sharma, A., & Pandey, A. "A novel 20nm FinFET based 10T SRAM cell design for improved performance". In *International Symposium on VLSI Design and Test*. Springer, 523–531 (2019).

44. Yadav, N., Dutt, S., Pattnaik, M., & Sharma, G. K. "Double-gate FinFET process variation aware 10T SRAM cell topology design and analysis". In *2013 European Conference on Circuit Theory and Design (ECCTD)*. IEEE, 1–4 (2013).

45. Kaur, N., Gupta, N., Pahuja, H., Singh, B., & Panday, S. "Low power FinFET based 10T SRAM cell". In *2016 Second International Innovative Applications of Computational Intelligence on Power, Energy and Controls with Their Impact on Humanity (CIPECH)*. IEEE, 227–233 (2016).

46. Pal, S., Bhattacharya, A., & Islam, A. "Comparative study of CMOS-and FinFET-based 10T SRAM cell in Subthreshold regime". In *2014 IEEE International Conference on Advanced Communications, Control and Computing Technologies*. IEEE, 507–511 (2014).

47. Birla, S. "FinFET SRAM cell with improved stability and power for low power applications". *Journal of Integrated Circuits and Systems*, 14(2), 1–8 (2019). https://doi.org/10.29292/jics.v14i2.57.

48. Maabi, S., Sayyah, E. S., Moaiyeri, M. H., & Hessabi, S. "A low-power hierarchical FinFET-based SRAM". *The CSI Journal Computer Science and Engineering*, 13, 54–60 (2016).

49. Yadav, N., Shah, A. P., & Vishvakarma, S. K. "Stable, reliable, and bit-interleaving 12T SRAM for space applications: A device circuit co-design". *IEEE Transactions on Semiconductor Manufacturing*, 30(3), 276–284 (2017).

50. Kishore, K. K., & Radha, B. L. "Design and implementation of 12-T SRAM Cell in 32nm FinFET technology". *International Journal of Engineering Research in Electronics and Communication Engineering (IJERECE)*, 4(8) (2017), 216–220.

51. Saxena, S., & Mehra, R. "Low-power and high-speed 13T SRAM cell using FinFETs". *IET Circuits, Devices and Systems*, 11(3), 250–255 (2016).

52. Guo, X., & Stan, M. R. "Design and aging challenges in FinFET circuits and Internet of things (IoT) applications". In *Circadian Rhythms for Future Resilient Electronic Systems*. Springer, 143–189 (2020).

53. Tayal, S., & Nandi, A. "Enhancing the delay performance of junction-less silicon nanotube based 6T SRAM". *Micro and Nano Letters*, 13(7), 965–968 (2018). https://doi.org/10.1049/mnl.2017.0867.

54. Ebrahimi, B., Zeinolabedinzadeh, S., & Afzali-Kusha, A. "Low standby power and robust FinFET based SRAM design". In *2008 IEEE Computer Society Annual Symposium on VLSI*. IEEE, 185–190 (2008).

55. Patil, S., & Bhaaskaran, V. K. "Optimization of power and energy in FinFET based SRAM cell using adiabatic logic". In *2017 International Conference on Nextgen Electronic Technologies: Silicon to Software (ICNETS2)*. IEEE, 394–402 (2017).

56. Kishor, M. N., & Narkhede, S. S. "Design of a ternary FinFET SRAM cell". In *2016 Symposium on Colossal Data Analysis and Networking (CDAN)*. IEEE, 1–5 (2016).

57. Farkhani, H., Peiravi, A., Kargaard, J. M., & Moradi, F. "Comparative study of FinFETs versus 22nm bulk CMOS technologies: SRAM design perspective". In *2014 27th IEEE International System-on-Chip Conference (SOCC)*. IEEE, 449–454 (2014).

58. Kim, Y. B., Kim, Y. B., & Lombardi, F. "New SRAM cell design for low power and high reliability using 32nm independent gate FinFET technology". In *2008 IEEE International Workshop on Design and Test of Nano Devices, Circuits and Systems*. IEEE, 25–28 (2008).

59. Almeida, R. B., Marques, C. M., Butzen, P. F., Silva, F. R. G., Reis, R. A., & Meinhardt, C. "Analysis of 6 T SRAM cell in sub-45 nm CMOS and FinFET technologies". *Microelectronics Reliability*, 88, 196–202 (2018).

60. Kumar, A. A., & Chalil, A. "Performance analysis of 6T SRAM cell on planar and FinFET technology". In *2019 International Conference on Communication and Signal Processing (ICCSP)*. IEEE, 0375–0379 (2019).

61. Kushwah, R. S., & Akashe, S. "Analysis of leakage reduction technique on FinFET based 7T and 8T SRAM cells". *Radio-Electronics and Communications Systems*, 57(9), 383–393 (2014).

62. Vijapur, P. R., & Uma, B. V. "Comparative analysis of novel 9T static random access memory at different technologies of FinFET". *International Journal of Pure and Applied Mathematics*, 118(24) (2018), 1–7.

63. Mushtaq, U., & Sharma, V. K. "Design and analysis of INDEP FinFET SRAM cell at 7-nm technology". *International Journal of Numerical Modelling: Electronic Networks, Devices and Fields* 33, (5), (2020), 33:e2730.

64. Ensan, S. S., Moaiyeri, M. H., & Hessabi, S. "A robust and low-power near-threshold SRAM in 10-nm FinFET technology". *Analog Integrated Circuits and Signal Processing*, 94(3), 497–506 (2018).

65. Smaani, B., Rahi, S. B., & Labiod, S. "Analytical compact model of nanowire junctionless gate-all-around MOSFET implemented in Verilog-A for circuit simulation". *Silicon*, 14(16), 10967–10976 (2022). https://doi.org/10.1007/s12633-022-01847-9.

66. Labiod, S., Smaani, B., Tayal, S., Rahi, S. B., Sedrati, H., & Latreche, S. "Mixed-mode optical/electric simulation of silicon lateral PIN photodiode using FDTD method". *Silicon* (2022). https://doi.org/10.1007/s12633-022-02081-z.

11 Challenges and future scope of gate-all-around (GAA) transistors
Physical insights of device-circuit interactions

Shobhit Srivastava and Abhishek Acharya

CONTENTS

DOI: 10.1201/9781003359234-11

11.1 INTRODUCTION

No doubt, FinFET technology is the slogger of today's semiconductor world. But as demand for further scaling with a desire for ultra-low-power and high-speed applications results in undesired short-channel effects, a new transistor is required. Here gate-all-around (GAA) devices come into existence. The GAA structure helps to mitigate unwanted short-channel effects by enhancing channel controllability. In GAAFETs, the channel surrounds all of its sides through a high-κ and interfacial oxide layer. Thanks to science and technological innovation, the GAAFET family brings together different transistors and their competitive benefits. This chapter tries to answer why and how 3D devices emerge. In addition to the limitation of FinFET (a 3D device, gate surrounded by three sides), it further talks about the scope and challenges of different competitive GAAFET members (nanowire FET, nanosheet FET, junctionless nanosheet FET, complementary FET, and forksheet FET) of the GAAFET family. It is worth mentioning that a smaller benefit of the device performance exerts a massive performance enhancement on circuit-level applications. However, the advantages of device enhancement concurrently exaggerate the limitation of devices at circuit-level applications. So, an elaborated idea of GAAFETs holding the benefits and challenges at the circuit is also discussed here.

11.2 THE TRANSITION FROM PLANER FETS TO 3D FETS

The shape and material of MOSFET change a lot from time to time, but it has had the same basic structures since its invention: the gate region, the channel region, the source, and the drain region. In the device, the source, drain, and channel are silicon regions are doped with atoms of other elements to produce either a region with an abundance of negative mobile charge (n-type) or positive mobile charge (p-type). CMOS technology requires all types of transistors that make up today's computer chips. How transistor structure, shape, and size have changed are shown in Figure 11.1 in a sequential manner.

The earlier workhorse "planer FET" drove the industry for more than 30 years from its birth. But the demand for miniaturization or to follow Moore's law pushes hard the technology to shrink down the device size. The adverse effects of scaling down the planer MOS have given birth to newer silicon-on-insulator (SOI) technology.

11.2.1 BENEFITS OF SOI OVER BULK MOS

Silicon-on-insulator (SOI) technology uses the idea of fabricating layered silicon-insulator-silicon substrates to reduce parasitic capacitance and improve performance [1]. In SOI-based devices, the thin semiconductor (mostly silicon) layer is above an insulator, generally silicon dioxide (SOI) or sapphire (SOS). The insulator choice mainly depends on the intended application. Sapphire performs well in radiation-sensitive and RF applications, while SiO_2 reduces the short-channel effects in other microelectronics devices.

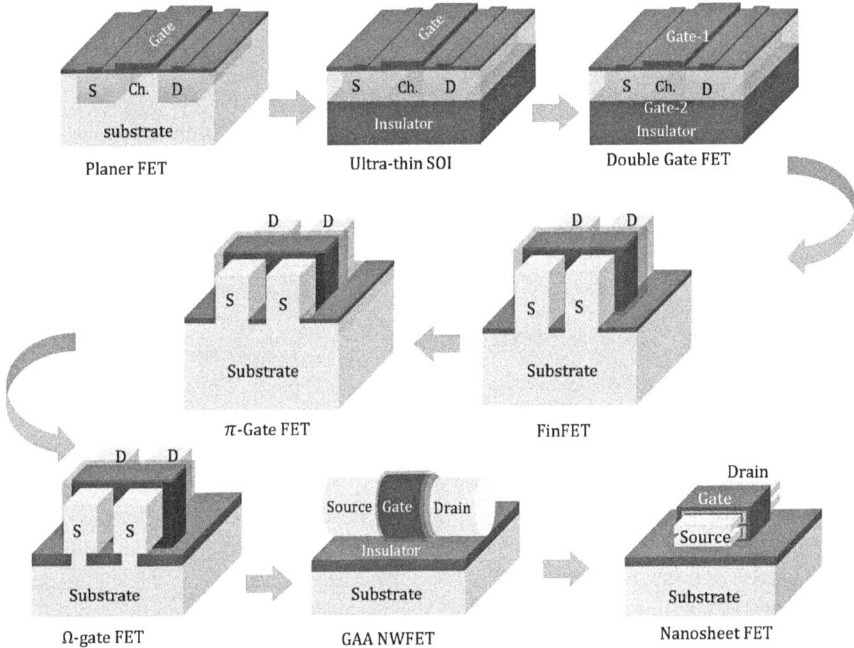

FIGURE 11.1   Structural evolution of transistors from planer to 3D technology

The benefits of SOI relative to conventional silicon (bulk CMOS) are as follows:

- Parasitic capacitance reduces due to isolation from the bulk silicon, improving power consumption performance.
- Resistant to the latch-up condition due to complete isolation of the n- and p-well in structures.
- Operable to work at low V_{DD}. Shows higher performance at the same V_{DD}.
- Reduction in temperature dependency.
- Comparatively better wafer utilization gives a high yield due to high packing density.
- Reduction in antenna issues.
- Consideration of body or well traps is less necessary, as it does not show any significant impact on device performance due to the separation of the device from bulk.
- Higher power efficiency because of low leakage current owing to good isolation.

The only drawback of SOI technology over conventional technology is its increased manufacturing cost. As of 2012, only AMD and IBM used the SOI approach for high-end processors, while other manufacturers like Intel, Global Foundries, and TSMC used an older approach of silicon wafers to fabricate devices on chips.

11.2.2 BENEFITS OF DUAL GATE OVER SOI

Farrah and Steinberg were the first to coin the idea of a thin-film double-gated transistor (TFT) in 1967. Toshihiro Sekigawa patented the idea of a double-gate MOSFET in 1980, where he demonstrated that the limitation of short-channel effects could be considerably reduced by sandwiching an SOI device between two connected gate electrodes. The use of DG-MOSFET in logic gate design gives significant improvements over conventional single-gate CMOS design [2]. DG-MOSFET shows a good response for high drive current comparatively to FinFETs.

11.2.3 THE EMERGENCE OF 3D TECHNOLOGY

As the dimension scaled toward the nanometer level, control of the channel region from two side gates was insufficient to eliminate the short-channel effects. Scientists tried hard to reduce this unwanted effect by changing channel, oxide, and metal contact material.

Meanwhile, in 1996, Indonesian engineer E. Leobandung, while working at Minnesota University, came up with the idea of cutting a wider MOS channel into many narrower channels to do further device scaling and improve drive current by enhancing channel width [3]. This led to a structure that is what the latest Fin-FET seems to be. The growth in control of electrostatics in the channel by gate placing different sides of the channel is shown in Figure 11.2.

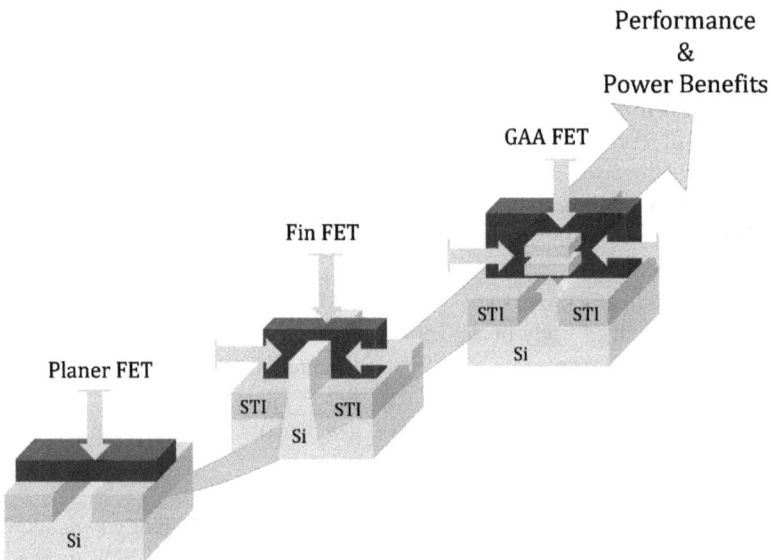

FIGURE 11.2 Improvement in electrostatic control in the channel through technological evolution [4]

Later a group led by TSMC's Chenming Hu and Hisamoto made the tabulated quantum leap between 1998 to 2004; Tsu-Jae King Liu, a dean and a Carlson professor at the UC Berkeley College of Engineering gave this information during a conference on VLSI Technology Symposium [5].

They coined the name "FinFET" (fin-based field-effect transistor) in the year 2000 [5] to narrate a non-planar, multi-gate transistor. The key points with FinFET over conventional devices (planer) are mentioned here.

(a) **Advantages of FinFETs**
  • Offers good channel controllability even at a low voltage.
  • Reduces leakage current associated with OFF condition by reducing DIBL.
  • Size shrinking makes it able to operate at a lower operating voltage.
  • Smaller dimensions make it a power-budget device on the chip.
  • Intrinsic ($\sim 1 \times 10^{15}$) doping of the channel causes a few dopant-induced variations.
  • Comparatively, low retention voltage makes it suitable for memory design.
  • Short-channel effects are reduced.
(b) **Disadvantages of FinFETs**
  • Low driving current.
  • Increased parasitic capacitance.
  • Distributed parasitics make its estimation more complicated.
  • Because of the 3D structure, it demands a high aspect ratio.
  • The feasibility of body biasing is no more.

11.3 GATE-ALL-AROUND TRANSISTOR FAMILY

Table 11.1 shows that up to 10 nm FinFET had been working satisfactorily until 2002. Scientists started looking to improve FinFET characteristics by changing gate metal, oxide, and the channel region to make scaling continue.

GAAFET is very much similar to FinFET technology with only the exception of gates surrounding the channel region. As shown in Figure 11.2, GAAFET surrounds the channel all over the side through a high-k metal gate; it provides a high electrostatic control in the channel region. This controllability has become a key achievement for low-power digital applications. The benefit of GAAFETs has been effectively demonstrated using both theoretical and experimental methods. Furthermore, the limitation of smaller ON current owing to small dimensions can be improved using III-V materials with higher mobility. Additionally, InGaAs nanowires, which have greater electron mobility than silicon, are successfully created. GAAFETs are the successor to FinFETs because they can work at sizes below 7 nm; even IBM demonstrates 5 nm process technology.

Note: As of 2020, Intel and Samsung planned for mass production of the multi-bridge channel (MBC) FET at the 3 nm node. In contrast, TSMC continues to use

TABLE 11.1

Performance improvement of FinFET through device design parameter

Breakthrough	Inventor	Technological node	Year
n-type FinFET	Group led by Digh Hisamoto and Chenming Hu.	17 nm	1998
p-type FinFET	Digh Hisamoto, Xuejue Huang, Chenming Hu, Wen Chin Lee, Leland Chang, Charles Kuo, Hideki Takeuchi, and Erik Anderson.	Sub-50 nm	1999
FinFET	Chenming Hu, Nick Lindert, P. Xuan, Yang-Kyu Choi, S. Tang, Erik Anderson, D. Ha, Jeffrey Bokor, and Tsu-Jae King Liu.	15 nm	2001
10-nm FinFET	Shibly Ahmed, Cyrus Tabery, Scott Bell, Jeffrey Bokor, Tsu-Jae King Liu, David Kyser, Chenming Hu, Leland Chang, and Bin Yu.	10 nm	2002
High-κ FinFET	D. Ha, Yang-Kyu Choi, Hideki Takeuchi, W. Bai, Tsu-Jae King Liu, M. Ameen, and A. Agarwal.	10nm	2004

FinFETs at the 3 nm technology nodes despite having gate-all-around transistors at the research level.

11.3.1 THE NANOWIRE FET

Voltage scaling has been utilized with planar transistors for generations to reduce power consumption. However, use was eventually constrained by short-channel effects. The answer was Fin-FETs, which allowed for additional voltage scaling. Regrettably, restrictions arise once again. The channel must have a gate completely surrounding it in order to have optimum electrostatics control, termed a "gate-all-around" (GAA) FET. Typically, GAAs are nanowires. For this technological advancement, a research work of 1988 regarding vertical surrounding gate transistor (SGT) [Figure 11.3], done by the Toshiba research team, including H. Takato, F. Masuoka, and K. Sunouchi, became the foundation for GAAFET [6]. SGT works similarly to a planer transistor in ON current with a very low OFF current (~10^{-14}) even in PMOS, ensuring excellent electrostatic in the μm regime.

Later, in 2003, Yi Cu. et al. fabricated a high-performance silicon nanowire FET (NWFET) of a 10~20 nm diameter with a gate length of 800 to 2000 nm [7]. The scaling trend again causes lower driving capability. Targeting this issue, a twin 10 nm diameter silicon NWFET was designed in 2005, which gave a way to improve driving capability by adding nanowires horizontally and vertically, as shown in Figure 11.4 (a). The twin NWFET [inset of Figure 11.4 (a)] gives ON current in a

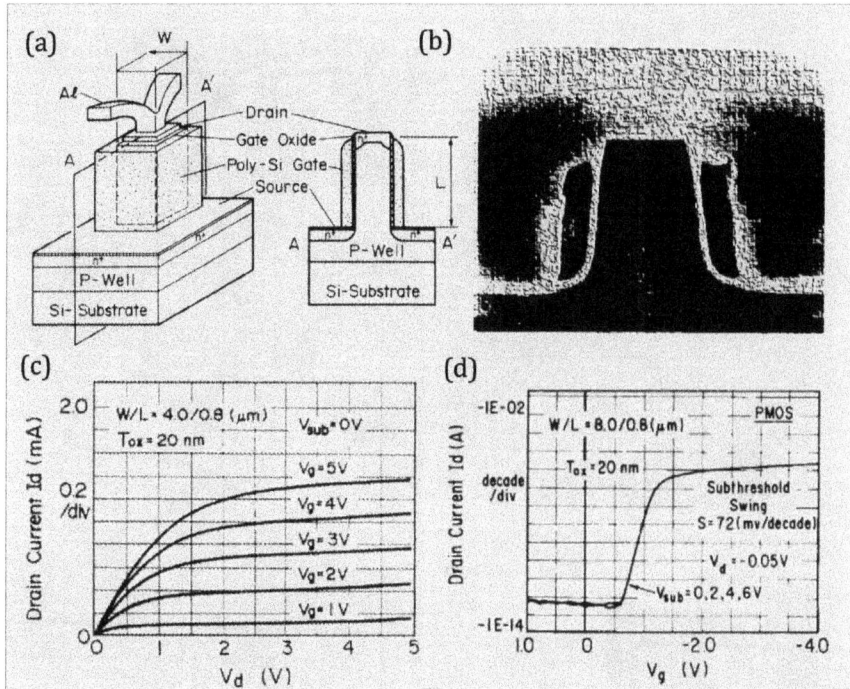

FIGURE 11.3   (a) 3D schematic of vertical surrounding gate transistor (SGT) and 2D cross-sectional cut of SGT across the plane a-a' (inset), (b) SEM cross-sectional view, (c) I-V characteristic of SGT, and (d) subthreshold characteristic of PMOS SGT [6]

range of ~2mA/µm, no roll-off to the threshold voltage, a subthreshold swing of ~70 mV/dec, and ~20 mV/V of drain-induced barrier lowering, as shown in Figure 11.4 (b), (c) [8]. But horizontal and vertical stacking of NWs gives parasitic increment, which ultimately causes a burden on device performance. This led to a structure with no horizontal stacking, only vertical stacking, termed stacked nanosheet FET.

11.3.2 THE NANOSHEET TRANSISTOR (NSFET)

However, the integration complexities of nanowires outweigh the benefits. This leads to the creation of a unique version of GAA, with all the advantages but minimal complications. Researchers proudly introduced NSFET around the year 2017 [Figure 11.5(a)]. The 2-dimensional cross-sectional SEM view of the NSFET cut at mid of channel is shown in Figure 11.5(b). The key advancement with NSFETs was the device drive enhancement by sheet width in contrast to FinFETs, where only one fin is allowed [9].

Further increment in ON current can be feasible by stacking the multiple channels vertically. Recent research advancement shows that a new way to increase the ON current in NSFET is by interbridging the staked sheet [10, 11], shown in Figure

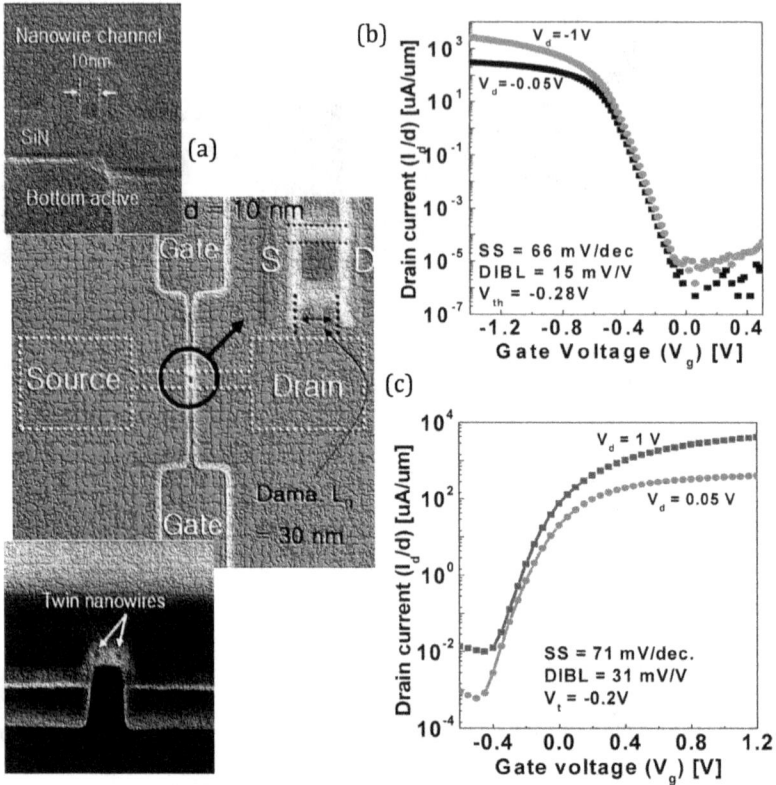

FIGURE 11.4   (a) Top SEM view of the nanowire at dia=10nm and gate length Lg=30nm, with cross-sectional SEM view of nanowire covered by SiN (inset) and SEM of twin nanowire (inset), transfer characteristics of (b) n+ poly-Si gated n-type twin silicon NWFET, and (c) TiN metal gated p-type twin silicon NWFET [8]

11.5(c), (d). This architecture can further be helpful in the mitigation of device self-heating [12, 13] (Figure 11.6).

Some points worth highlighting with regard to NSFETs are mentioned here:

*   One advantage of NSFET is that additional area is not required to improve speed. FinFETs need fins to be laterally added, while NSFETs can be vertically stacked.
*   NSFETs are compatible with FinFET design. The designer can replace FinFETs with NSFETs without changing the footprint.
*   Performance (power, speed, area, accuracy, etc.) can also be improved without an area increase that suits all applications, including AI, automation driving, 5G, and high-performance computing.
*   NSFETs are the most advanced technology that provides solutions from low-power to high-performance applications.

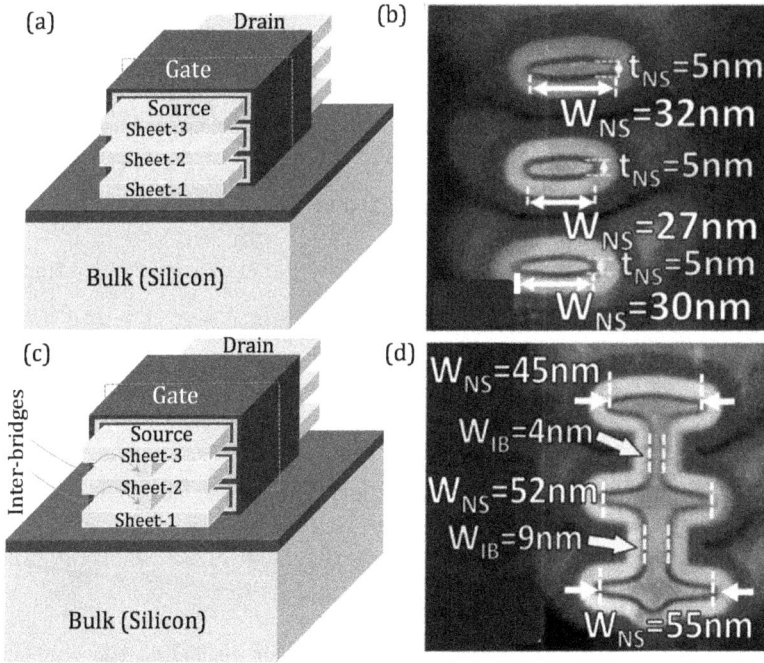

FIGURE 11.5 (a) 3D structure of 3-stack NSFET, (b) 2D cut-plane cross-sectional SEM view of NSFET at mid of channel [9], (c) 3D structure of 3-stack TreeFET, and (d) 2D cut-plane SEM view of TreeFET at mid of channel [10]

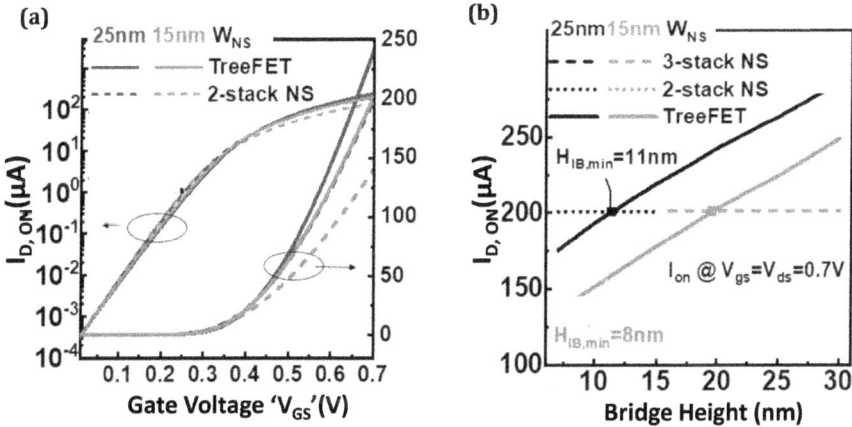

FIGURE 11.6 (a) Transfer characteristic of 2-stack NSFET & TreeFET, (b) comparative analysis of ON current variation of 2-stack NSFET and 2-stack TreeFET with 3-stack NSFET [11]

- The key advantage of NSFETs is their short-channel control, which plays a vital role in threshold voltage (V_{TH}) variation. Nanosheets offer fewer V_{TH} variations, which is essential to achieving good performance.
- The remarkable thing is that it can be fabricated with minimal deviation from FinFET (manufacturing methodology).
- Excellent electrostatics control. By varying the width and height of the sheet, we can optimize capacitance and resistance with minimal compromise to accuracy.
- Stacked NSFET offers versatile design options as per the consumer's requirements.
- It maintains tight control over leakage current by I_{OFF} of the order of 10^{12}.
- It shows a better subthreshold swing over FinFET.

NEGATIVE POINTS TO THE NSFET

It would be wrong to say that with perfection, there are no negatives. NSFET has some drawbacks. Industry researchers and scientists are working hard to drive NSFET toward the ideal. Some of the drawbacks points of the nanosheet transistor are mentioned below.

- NSFET has a self-heating effect due to compact sheets that may cause the cross-talk or falsely trigger itself.
- Decreasing the sheet width causes a decay in the speed of operation, as I_{ON} is directly proportional to the width.
- Increasing the width leads to an increment in I_{ON} current, but a simultaneous increase in capacitance becomes a bottleneck to improved performance.
- Stacking multiple devices needs some optimized adjacent distance to make isolation from cross-talk.
- Reduction in technological nodes through NSFET is limited for analog/RF applications.

It is worth highlighting that further scaling is restricted due to the limitation created by junctions. This restriction of device scaling can be reduced to some extent by junctionless devices.

11.3.3 JUNCTIONLESS NSFET

For the time being, all the existing FETs are formed by selectively introducing the dopant atoms into the bulk semiconductor, which forms the junctions. Scaling results in a device dimension drop-down under the 10-nm node and demands extremely high doping gradients. To hold the laws of diffusion and the statistical nature of the distribution of dopant atoms, these junctions cause increasing challenges in the fabrication industry.

In 2010, Jean-Pierre Colinge and his colleagues designed and proposed a new type of field-effect transistor that has no junctions at all [Figure 11.7], and they found that its electrical characteristics were comparable to the trending junction FETs with

(a)   (b)

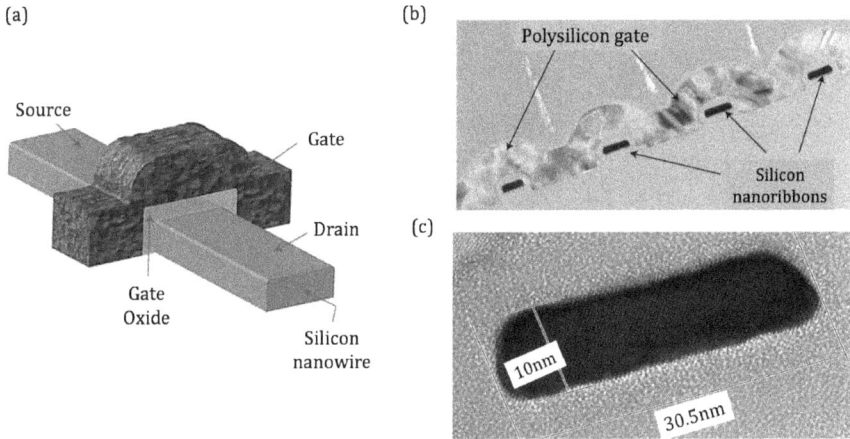

FIGURE 11.7   (a) Schematic of n-type junctionless nanowire transistor, (b) TEM image of five parallel silicon-gated nanoribbons with a common polysilicon gate, and (c) zoomed view of a single nanowire device [14]

more benefits [Figure 11.8] [14]. In the junctionless gated FETs, the silicon nanowire (uniformly doped n-type) acts as a channel, and the gate material is of p-type polysilicon, as shown in Figure 11.7(a). In the case of p-channel FETs, the opposite dopant polarities are used [15, 16].

Si-based junctionless nanowire FET shows an ideal subthreshold slope of 60mV/ dec, {theoretically, the lowest value of SS $=(k_B T/q) \ln (10)$ at T= 300 K} to classical FETs with extremely low leakage currents (~1×10^{15}A) and lesser mobility degradation due to gate voltage and temperature comparatively [Figure 11.8(a)].

When gate voltage ($V_{.G.}$) in JLFET is at $V_{.G.} < V_{TH}$ condition, the channel region is depleted of the electrons, resulting in an OFF current; as the gate voltage increases to the threshold voltage ($V_{.G.} = V_{TH}$), a string-like channel of n-type silicon joins the source and drain and drain current starts rising. At the above threshold situation ($V_{.G.} \geq V_{TH}$), the induced channel starts expanding in areas as soon as a situation of flat band energy has been reached ($V_{.G.} = V_{FB} \gg V_{TH}$). The channel region simply becomes a resistor, as shown in Figure 11.9.

Still, the limitation of the subthreshold region and the self-heating effect persists in the nanoscale devices.

11.3.4 TUNNEL JUNCTION NSFET

The problem mentioned above seeks a new device that gives good channel controllability with the steep subthreshold slope, which can be targeted by tunneling phenomena [17, 18]. Regarding this area-scaled nanosheet tunnel-FET (AS-NSTFET) came into existence [19], as shown in Figure 11.10, with its improved transfer characteristics [Figure 11.10(b), (c)]. AS-NSTFET improves ON current by utilizing area/ line tunneling rather than point tunneling [20] with an excellent subthreshold swing of 20mV/dec and a very small OFF current.

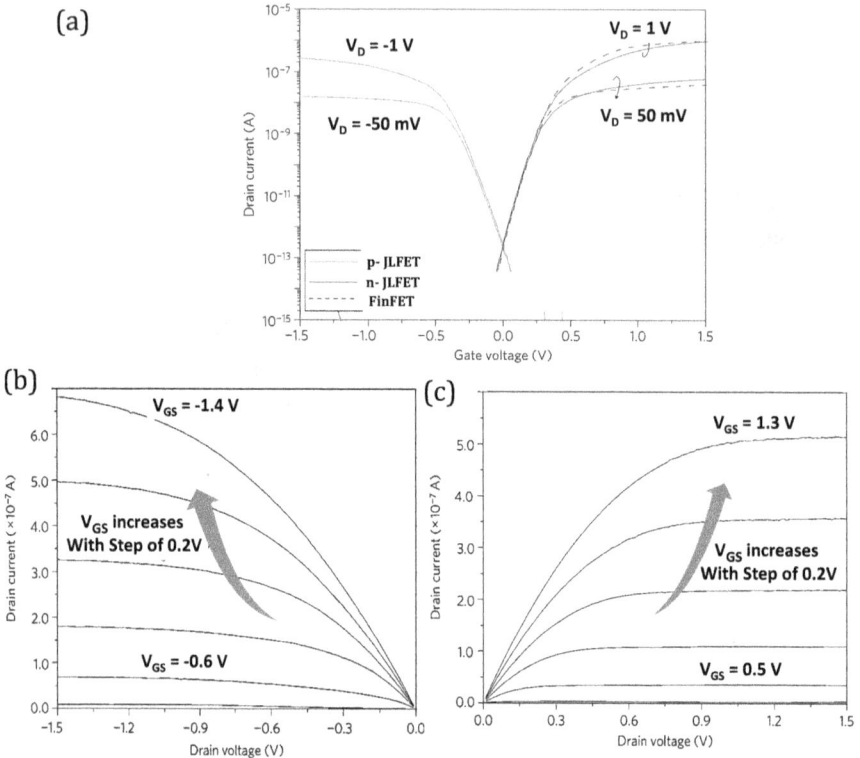

FIGURE 11.8 (a) Comparative analysis of transfer characteristics for junctionless FET (JLFET) with conventional Trigate FET. I_{Off} is below the observing limit of the measuring instrument (1×10^{15} A) with the I_{ON}/I_{OFF} ratio larger than 1×10^6, (b) the output characteristic of p-channel, and (c) n-channel junctionless FET [14]

In order to achieve a higher ON current, the gate is extended over to the source region, which gives the delayed $V_{T., ON}$ without much increment in the ON current due to point tunneling. In order to invoke area/line tunneling, an epi-layer-based TFET [Figure 11.10(d)] was designed, which gives improved ON current, as shown in Figure 11.10(c).

The logic circuits today rely on pairing two types of transistors – NMOS and PMOS. A separate interconnect is required to make such a pair. This restricts the improvement in packing density. Sheet stacking-based FET shows extraordinary compactness for such a combination using forksheet FET, a complementary field-effect transistor (CFET).

11.3.5 FORKSHEET FET

Forksheet FET (FS-FET) generally consists of multiple vertically stacked sheets controlled by a fork-gated structure [Figure 11.11(a)] [21]. By adding a dielectric

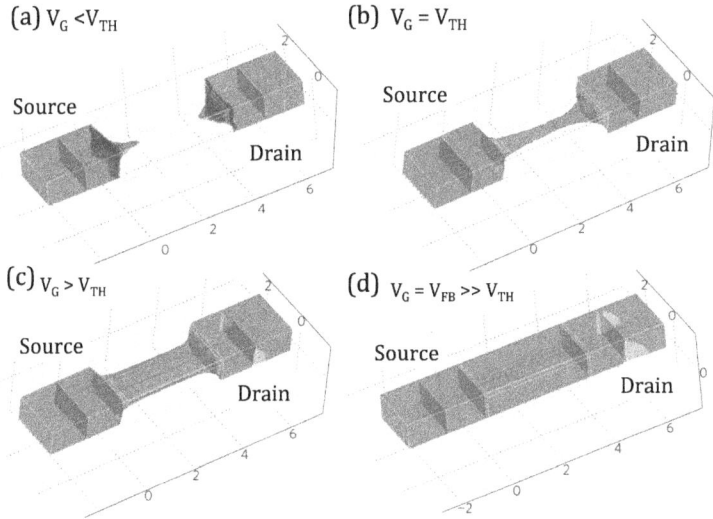

FIGURE 11.9   The charge inversion created by gate voltage at different gate voltages [14]

wall between the pMOS and nMOS, the gate edge self-aligns with the device by avoiding overlay margin and enables patterning simplicity. As a result, the p-gate trench becomes physically isolated from the n-gate trench, which simplifies the work function of metal fill during the RMG process. The further process flow used for making the forksheet is like the one used for manufacturing nanosheet FETs. This makes it an attractive natural extension of the gate-all-around nanosheet FET flow. A highly magnified TEM image of a two-sheet stacked forksheet is shown in Figure 11.11(b).

The I_D-V_{GS} curves for both nMOS and pMOS forksheet FETs give excellent sub-threshold swing [Figure 11.11(c)]. A comparison of the gate pitch path of the fork-sheet FET shows that it requires less area on the chip than NSFET and FinFET [Figure 11.12]. FS-FET holds the advantage of robustness over the other two due to the dielectric wall that exists between nMOS and pMOS [22].

Though forksheet FET gives more compactness than NSFET, it lacks electro-static controllability due to reduced channel control in the z-direction (see Figure 11.12) compared to the typical NSFET. But still, most of the channel controllability is provided by gate metals present along the y-direction rather than the z-direction, hence still providing reasonable electrostatic control. Further scaling results in a new structural modification of NSFET, where pMOS and nMOS sit one over the other.

11.3.6   COMPLEMENTARY FET (CFET)

Complementary FETs are a more compact version of a gate-all-around (GAA) transistor [23]. Traditional GAAFET stack several sheets/wires vertically or horizontally

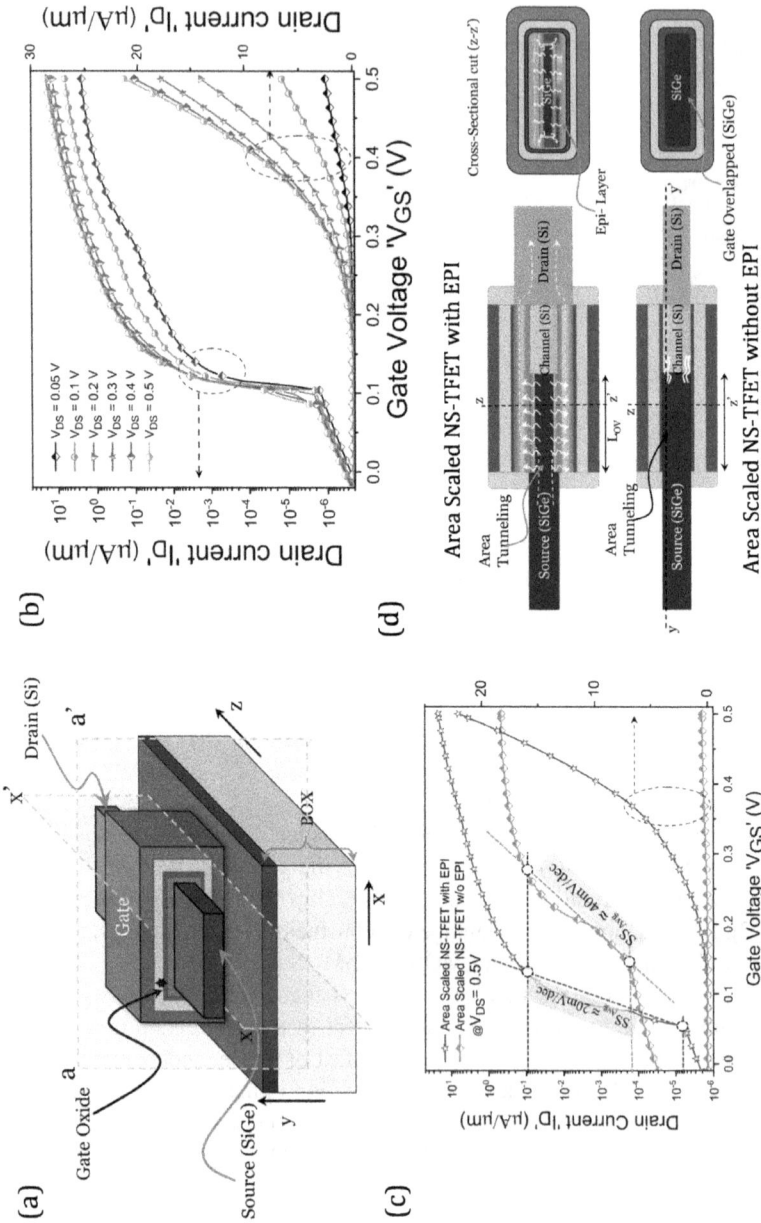

FIGURE 11.10 3D-schematic of area-scaled nanosheet-TFET, (b) transfer characteristic of AS NS-TFET, (c) Id-Vg of the devices NS-TFET with epitaxial layer and without epitaxial layer, and (d) 2-D slice of AS NS-TFET showing area tunneling in NSFET with an epitaxial layer and gate-induced area tunneling in NSFET without an epitaxial layer [19]

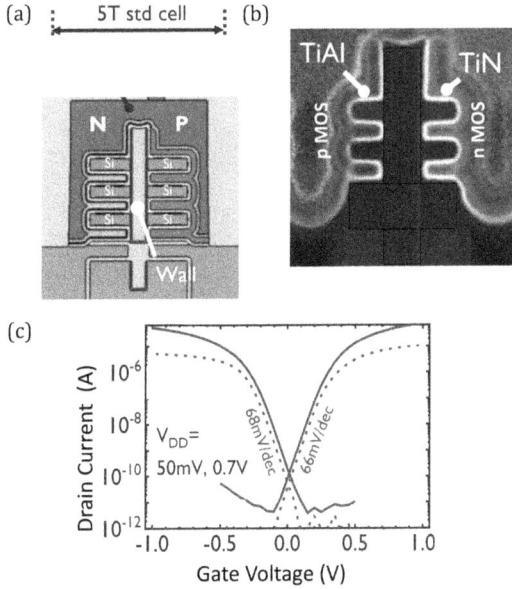

FIGURE 11.11   (a) Three layered forksheet transistors, (b) SEM image of FS-FET, and (c) an approximate symmetric transfer characteristic for p-type FS-FET and n-type FS-FET [21]

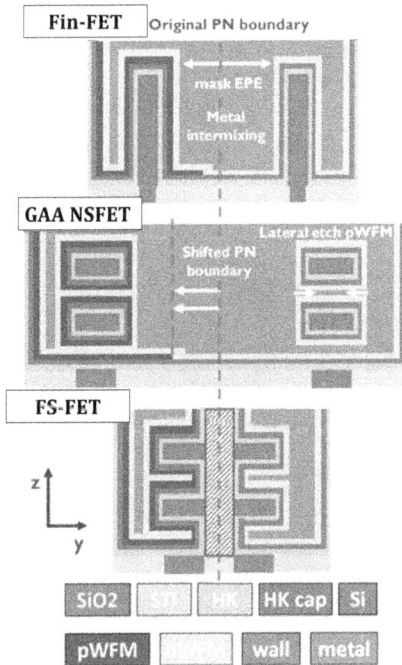

FIGURE 11.12   Comparative analysis of reduced gate pitch of forksheet with NSFET and FinFET [22]

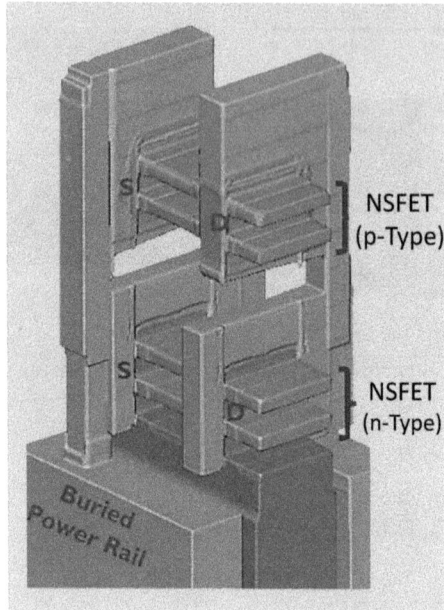

FIGURE 11.13   3D-schematic of complementary FET [24]

in order to enhance drivability. Any logical circuit demands a CMOS combination of nMOS and pMOS due to its great benefits. Scaling beyond 2 nm led to a new idea to stack nMOS and pMOS wires/sheet stacks on each other, as shown in Figure 11.13. This "folding" of the nMOS and pMOS eliminates the nMOS to pMOS separation bottleneck by reducing the cell active area footprint [24].

The key achievement of CFET is the area without compromising electrostatic controllability [Figure 11.14]. CFET offers the same electrostatic control as a traditional GAA device.

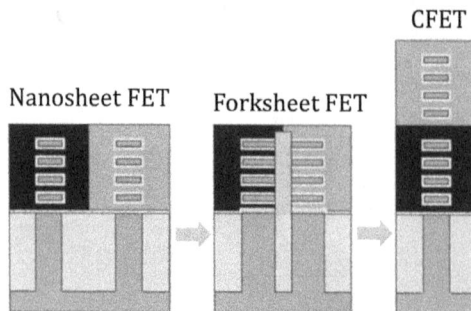

FIGURE 11.14   A regular degradation in chip area requirement from NSFET to CFE.

11.4 CHALLENGES AND FUTURE SCOPE WITH THE GAAFET FAMILY

Each member of the GAA family has advantages with a few hidden limitations. The challenging points of the GAA members discussed in Section 10.3 hold critical scope for the innovation of novel transistors with improved reliability and compactness. Further scaling of the fundamental technological node of individual novel GAA members urges an in-depth observation of device reliability and short-channel effect. For example, increasing the height of fins or placing more fins to get a high ON current in FinFET again becomes a limitation of fragileness/scaling, respectively. This led to research on the current hike concerning the height and width of fins, which led to the invention of nanosheet FET [25]. On keeping scaling in mind, further I_{ON} improvement became feasible by stacking the sheet vertically, which again gives rise to the degradation of drivability because of the self-heating effect (SHE) [26, 27]. The improvement in device characteristics created by SHE is made by incorporating hetero-dielectric structure at the gate in NSFETs [28].

Stacking the sheet vertically in NSFET is also limited to a number because, beyond that limit, the current does not increase proportionally owing to increase in path resistance from source to drain for the bottom sheet. NSFET suffers from another limitation of bias temperature instability due to scaling. PBTI/NBTI is the consequence of traps (defects) charging and discharging under stress conditions where oxygen vacancies act as the major traps in the gate dielectric. Larger width/height shows the aggravated PBTI effects but less threshold voltage (V_{TH}) variation [29]. Furthermore, using a vertical combo spacer can help optimize the electrothermal behavior of GAA transistors below 7 nm. A 118% and 18% enhancement in I_{ON} and I_{OFF} can be achieved using HfO_2 in place of SiO_2, respectively [30]. Though CFET shows extraordinary compactness of the device on the wafer. It has a limitation of characteristic asymmetric behavior due to different hole and electron mobility, which cannot be compensated by the width increment of pMOS, as pMOS is on the top of nMOS in CFET.

11.5 DEVICE-CIRCUIT INTERACTION

Phenomenologically, quantum effects are going to be more pronounced at the latest advanced node, causing unusual and unexpected changes in the behavior of nanoscale devices. These restrictions might not be the deal-breaker for circuit size reduction. Electron momentum starts to impact the structure of the traces when IC interconnections are so small that there is essentially nothing left of them to carry current. The limitation of IC size scaling might not be entirely established by quantum mechanics but by reliability loss caused by metal migration too, which not only limits the scaling but also alters the behavior of nano-devices based on parasitic increment, electric field alteration, and process limitation [31].

The limit is already being reached and began to have an impact a decade ago when the smallest-geometry processes were specified with a mean time between failures (MTBF) of less than a decade, which is now moving toward less than five years. Will anyone buy an electronic item/laptop knowing that the MTBF of the chip is less than five years old? The answer is a big "no", considering that simultaneous device and circuit interaction is mandatory for cart IC-size scaling. These effects are quantified in further sub-sections based on digital/analog applications.

11.5.1 DIGITAL DESIGN PERSPECTIVES

Memory storage is a fundamental performance and energy bottleneck in approximately all computing systems. Storage as a digital application has billions of transistors per chip. The increment of storage capacity without further increment of chip size demands the accommodation of more transistors within the same size, provided they can be reliable with a good life span [32]. This ongoing work in combating scaling challenges of NAND flash memory is briefly discussed here. 3D NAND flashes have been investigated as a significant challenger because of their potential to replace traditional 2D-floating gate cells. Despite the inherent problems of process complexity and poor data retention, there has recently been considerable progress toward mass manufacturing. Several challenges, such as materials, cell architecture, and process, still need to be explored for a better future. The NSFET-based 6T SRAM cell shown in Figure 11.15(a) is used to study how the suggested device topology improves SRAM performance.

Delay and stability are two essential performance characteristics of the optimized NSFET-based 6T SRAM that are analyzed with five crucial parameters: RAT, HSNM, WAT, RSNM, and WSNM. Butterfly curves for different modes (hold and read/write) are shown in Figure 11.15 (b), (c), and (d), respectively. The scaling down of the channel length further degrades the SNM because of DIBL [33], while scaling into thickness enhances the stability of SRAM at the same time. Again here, DIBL plays a role in pull-up/pull-down FETs at a lower thickness of channel for stability improvement.

The latest advance non-volatile (NV) memories, like PRAM, ReRAM, and STT-MRAM, have gone through explosive research in the few years.

At the same time, DRAM technology is also experiencing difficult technology scaling challenges to maintenance and enhancement of its capacity, energy efficiency, and reliability, and it is significantly more costly than conventional techniques. Some promising research and design directions to overcome challenges handled by memory scaling are discussed regarding these issues.

- Enabling new DRAM building blocks, functionality, interfaces, and enhanced integration of the DRAM with the other communicating system (DRAM-System co-design).
- Designing a memory system using emerging non-volatile memory with taking merits of multiple different technologies (hybrid-memory architecture).
- Enabling predictable performance with quality of the system (QoS) to target sharing the memory system for high-priority applications (QoS-aware memory systems).

Beyond 20 nm, DRAM is expected to scale down in iterations under the 1xnm regime, such as 1xnm (for 16nm to 19nm), 1ynm (for 14nm to 16nm), and 1znm

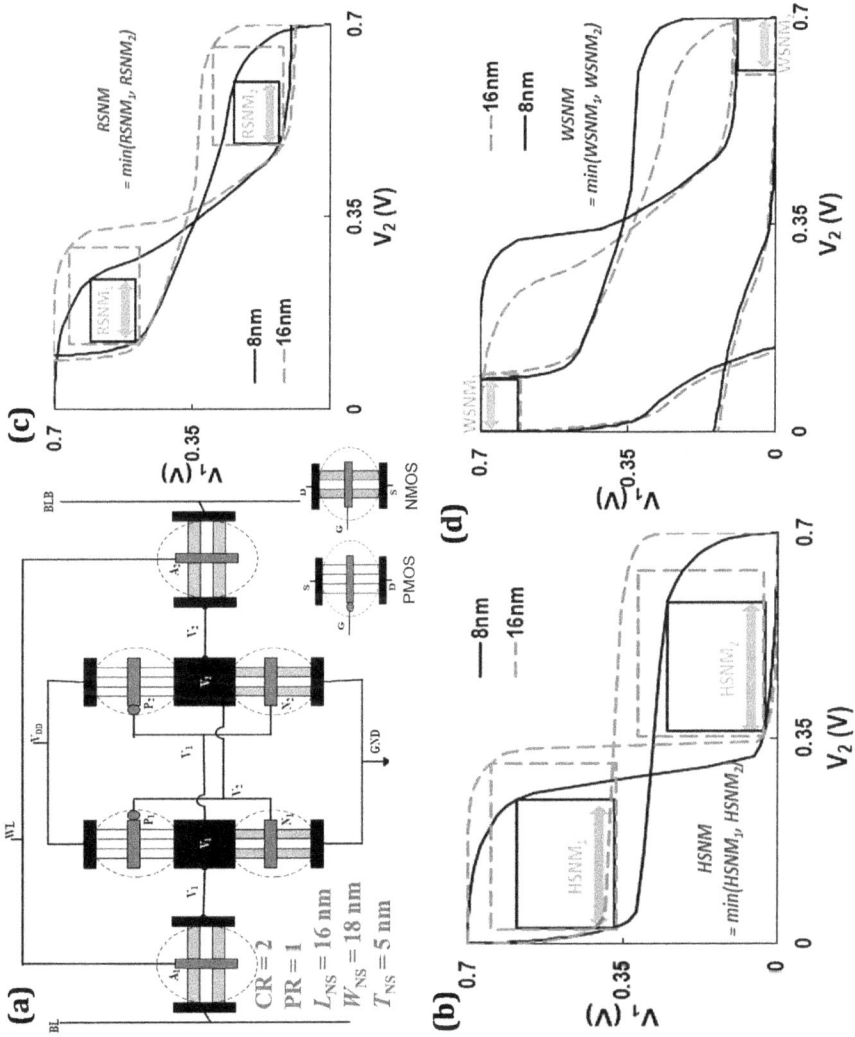

FIGURE 11.15 (a) NSFET-based 6T SRAM with butterfly curves during hold mode (b), read mode (c), and write mode (d) due to variation in channel length [33]

(for 12nm to 14nm). DRAM technology in the 1xnm range confronts significant hurdles, such as obtaining adequate store capacitance and sensing margin. FET cell capacitors with new materials should be explored to ease the challenges with error detection and correction methodology. Along with this, DRAM has been facing issues of performance degradation due to scaling, and it requires advanced 3D transistors (GAA devices), which offer reliable and speedy performance with low power. Without using conventional device geometric scaling, a 3D integration with TSV offers a novel option for high density, high speed, low power, and broader bandwidth. However, GAA devices have their own challenges of reliability and high manufacturing costs, which need to be overcome before they can be commercially used [34].

Due to its fast read/write speeds and superior cycle durability, STT-MRAM is thought to be the only non-volatile memory that can match the performance of DRAM. PRAM and ReRAM are promising candidates to replace conventional NOR/NAND flash and pioneer the field of memories.

11.5.2 ANALOG DESIGN PERSPECTIVES

As discussed above, there are too many scopes and challenges with GAA family members related to memory-based applications, but what about analog circuits? When it comes to obtaining optimal performance and functionality, analog design is difficult to achieve. In a system, most of the tests and chip failures come from analog design. Recently, a report by Cadence Design Systems showed that approximately 95% of field failures occur from analog blocks in the design because the analog circuits not only demand benefits over speed, power, and area but also have a huge demand like bandwidth and gain improvement, less signal distortion, sensitivity to power supply variations, and other noise sensitivities like phase noise and noise figure.

High voltage devices (HV devices) face a current limiting issue due to quasi-saturation before the occurrence of channel pinch-off. With an increment in the gate bias, huge drain potential drops occur in the drift region (below the drain terminal). These push the transistors to operate in the ohmic region [Figure 11.16(b)], ultimately reducing the overall transconductance. For example, these phenomena (limitation of I_{ON} & QS effect) are observed in a test circuit made of STI-DeMOS, as shown in Figure 11.16(c). The high voltage high-speed (GHz) level shifter made up of DeMOS for 5V operation gives a 15% improvement in the speed compared to other counterparts as in [35].

11.6 CIRCUIT-RELATED RELIABILITY ISSUES

This section elaborates on why this is happening and what would be the right approach to handle these challenges strategically. Analog circuits deal with higher voltage swings, temperature, and electric currents, which means a designer must manage thermal stress optimally without extra demand of area on the chip. That can be possible by GAA devices using vertical stacking of sheets to get high drivability.

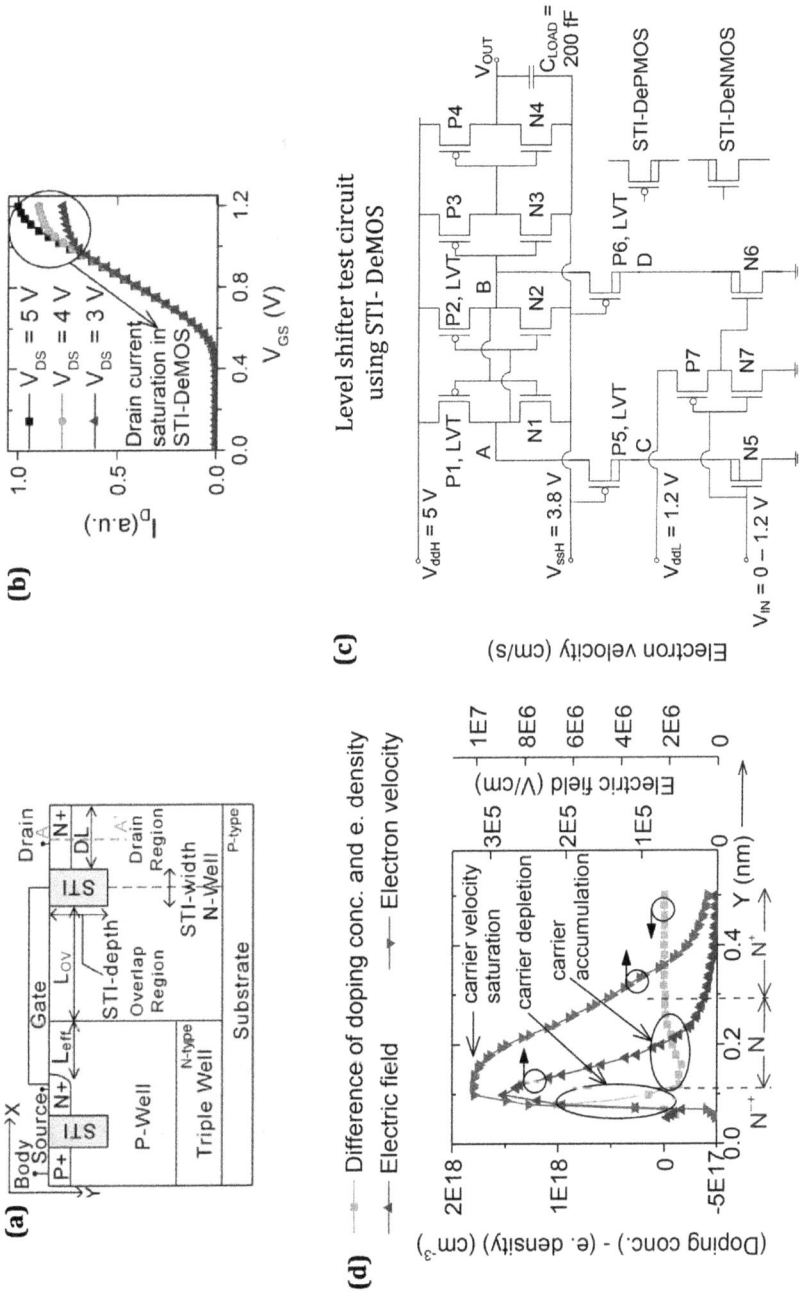

FIGURE 11.16 (a) 2D STI-DeMOS, here gate overlap and drain region combined are termed as the drift region, (b) I-VGS characteristics showing current saturation at higher gate voltage, (c) STI-DeMOS based level shifter circuit, and (d) different operating region of STI-DeMOS biased at V_{DS} = 5 V and V_{GS} = 1.2 V, across the AA' cross-section in Figure 11.16(a) [35]

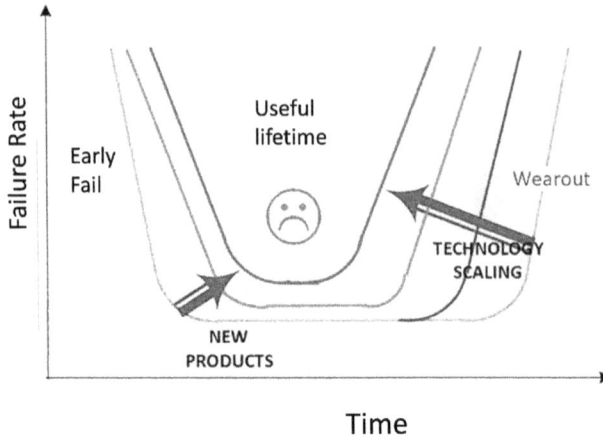

FIGURE 11.17   Bathing-tub curve showing failure rate with respect to the circuit design lifetime [36]

In the case of reliable operation, GAA members are struggling with self-heating effects which reduce the signal noise margin. And things got more sophisticated as the number of application types increased from a circuit. At that time, success totally depends on choosing the right set of parameters to optimize. An optimized analog design takes a high degree of process iteration and time from designers to pass the project on time. Processing challenges in modern CMOS technologies have led to several reliability concerns, resulting in deteriorated and over-lifetime product performance, as shown in Figure 11.17. Reliability effects today include those related to transistor aging, such as bias temperature instability (BTI) and hot carrier injection (HCI), as well as interconnect degradation [36].

Though this is not a comprehensive list, some common reliability-related challenge issue and their diminution are discussed below.

11.6.1   TIME-DEPENDENT DIELECTRIC BREAKDOWN (TDDB)

The breakdown caused by high electric fields when passing through the gated oxide is known as TDDB. This high-intensity electric field further generates traps and ultimately results in stress-induced gate leakage current due to either an open or a short. This is basically a time-dependent voltage, and stresses are applied to the device/circuit with higher voltages exposed for a longer time, causing greater defects per million impact.

TDDB can be handled by electrical checks to confirm that all internal and external nodes in the circuit schematics/layouts avail the defined constraints. Furthermore, modulation in the gate and channel currents at higher gate voltage (V_G) must be correctly modeled in the file containing technology parameters while verifying its consequences on the performance of the circuit. From being circuit design perspective, this breakdown can be handled by selecting lower voltage

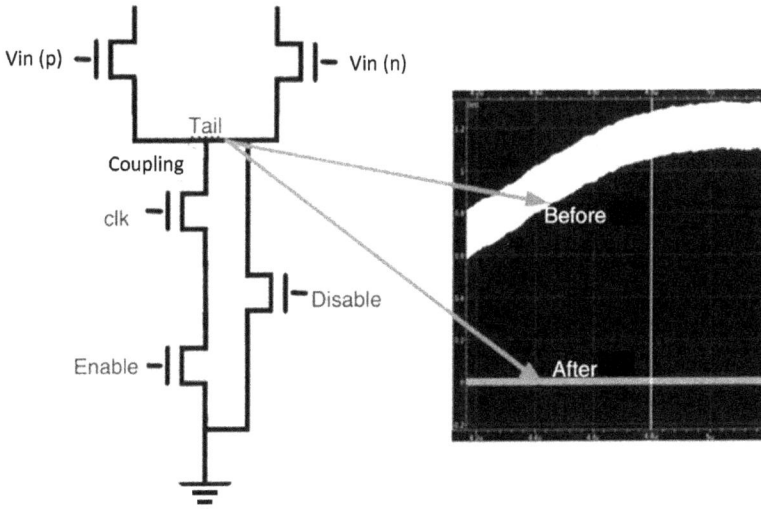

FIGURE 11.18 Undriven nodes handling through clock gates

operating cells, which can be feasible through GAA devices. Stacked topologies, which help to reduce HCI effects, can also be effective in the reduction of TDDB. Techniques like slowing down slew rates and power gating prevent the exposure time of the circuit to high voltages, hence restricting degradations. In the circuit, TDDB can further be prevented by avoiding undriven nodes that can be altered by capacitive coupling. Weakly stacked or undriven nodes are prone to undershoot and can generate another concern with float nodes. It is always recommended to do electrical rule checks so they can be easily identified and if required, a network should be installed to discharge the accumulated a charge leaker network should be added, as shown in Figure 11.18.

11.6.2 HOT CARRIER INJECTION (HCI)

Advanced transistors (like GAAFET), due to short-channel, current flows because of a large lateral E-field, generate electron-hole pairs (EHP) through impact ionization. Some EHP-generated charge carriers dive into the gate, resulting in transistor degradation. As aging is modeled, HCI directly impacts threshold voltage shift, mobility degradation, and transconductance reduction. Generally, high load and slow slew rates cause severe HCI degradation. Hence, this accounts for the higher time margin. The high power supply for long time periods is the forecaster of HCI failure in analog circuits. In order to get sustained post-aging performance, the guard bands should be applied at the time of design. Further, HCI degradation can also be controlled in analog circuits by the V_{DS} supply voltages, which can be achieved by connecting the diode-making circuit/device to limit drain-source voltage, as depicted in Figure 11.19.

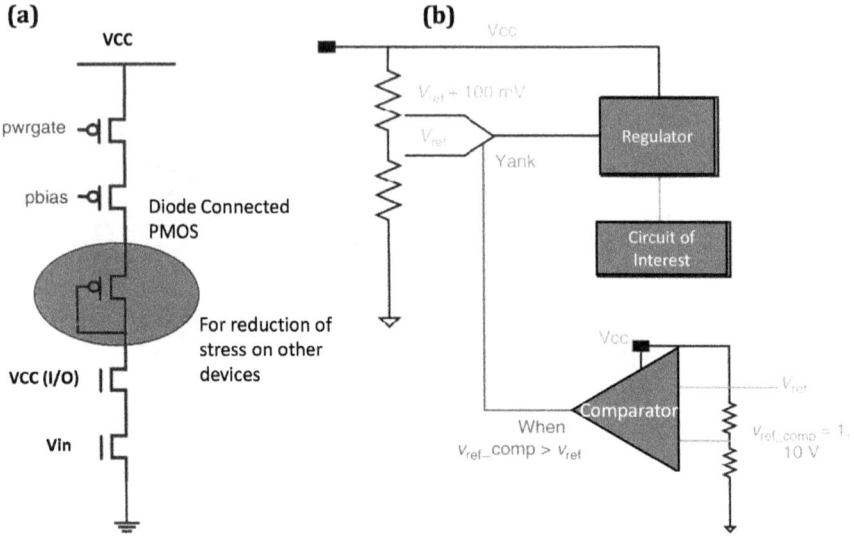

FIGURE 11.19   HCI mitigation from high VDS by (a) stacking a diode-forming device and (b) by controlled supply [36]

11.6.3   BIAS TEMPERATURE INSTABILITY (BTI)

A phenomenon occurs at stressed bias voltage applied to the gate. It results in degraded device performance, like an undesired threshold voltage increment of the device when the circuit is imposed over long periods [37]. This BTI degradation can be easily explained with the help of the atomistic trap-based BTI (ATB) model, as shown in Figure 11.20(a). This gets worse in short-channel devices. PMOS suffers from negative bias temperature instability (NBTI), while positive bias temperature instability (PBTI) occurs in NMOS devices with positive stress. It is observed that

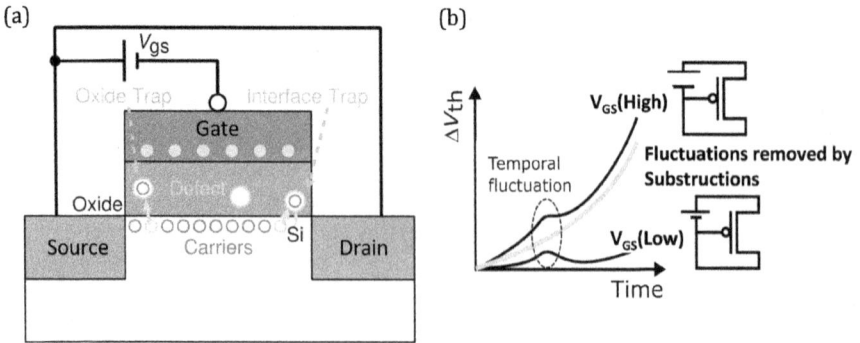

FIGURE 11.20   (a) Explanation of NBTI degradation with ATB model, and (b) degradation in threshold voltage due to BTI as time spent [37]

PMOS devices are more susceptible to BTI; recovery is irreversible even with the removal of stress. A contributor to temperature instability is that the carriers are trapped in the gate oxide due to defects/breaking of silicon-to-hydrogen bonds, causing charge accumulation. Both BTI and HCI are proportionally dependent on the V_{GS} or V_{DS}, consequently degrading the threshold voltage (Figure 11.20(b)).

The gate stress and high voltages worsen the condition by increasing the failure rate. BTI is aggravated at higher temperatures and also depends on the ON time periods of a transistor.

11.6.4 DESIGN TECHNIQUES FOR RELIABILITY

In summary, we provide a few design techniques that need to be discussed to wrap up the scope of challenges.

- Use GAA devices to form lower supply designs whenever possible.
- Make use of standard design guidelines, like good slopes having lower fanout in logic paths.
- The power gating approach to limit currents has high performance and activity factors.
- Paths must be optimized, like optimum dc paths and clock networks.
- Optimize and manage the dc and high supply currents in the system design.
- Create a path to discharge undriven nodes.
- The low-frequency clock may solve long-term effects such as HCI and BTI.
- Utilize duty-cycle correction, chopping, and offset cancellation, as well as other approaches that regulate overall variance to manage currents across process skews, temperatures, and voltages.
- Meticulously perform the floor planning, layout for reliability verification, and degradation-aware standard cells.

These techniques can provide designs that are simpler to close from the perspective of reliability and reduce the process of design cycles in cleaning up reliability flows. Ultimately, the failure rate can be lowered, and the design's durability can be enhanced.

11.7 CONCLUSION

It can be concluded that the scaling of devices is the focus of industry demand. As the device structure changes from a planer to 3D structure, the demand for further scaling with a desire for ultra-low power and high-speed applications leads to undesired short-channel effects. In addition to the limitation of FinFET (3D FET), its further advancement comes with the scope and challenges of different members of the GAAFET family, viz. nanowire FET, nanosheet FET, junctionless nanosheet FET, complementary FET (CFET), and forksheet FET. This chapter describes the accomplishments and issues related to GAA-based circuit-level design. Design techniques need to be taken care of in relation to reliability issues when dealing with circuits containing 3D devices to benefit from the challenges. Device and circuit design challenges need to be quantified simultaneously to target demands.

REFERENCES

1. G. K. Celler and S. Cristoloveanu, "Frontiers of Silicon-On-Insulator," *Journal of Applied Physics*, 93(9), p. 4955, April 2003, doi: https://doi.org/10.1063/1.1558223.
2. S. K. Gupta, G. G. Pathak, D. Das and C. Sharma, "Double Gate MOSFET and Its Application for Efficient Digital Circuits," *2011 3rd International Conference on Electronics Computer Technology*, pp. 33–36, 2011, doi: 10.1109/ICECTECH.2011.5941650.
3. E. Leobandung and S. Y. Chou, "Reduction of Short Channel Effects in SOI MOSFETs with 35 nm Channel Width and 70 nm Channel Length," *1996 54th Annual Device Research Conference Digest*, pp. 110–111, 1996, doi: 10.1109/DRC.1996.546334.
4. Yong Joo Jeon, "Making Semiconductor History: Contextualizing Samsung's Latest Transistor Technology," *Samsung Newsroom*, May 2019, https://semiconductor.samsung.com/newsroom.
5. D. Hisamoto et al., "FinFET-A Self-Aligned Double-Gate MOSFET Scalable to 20 nm," *IEEE Transactions on Electron Devices*, 47(12), pp. 2320–2325, December 2000, doi: 10.1109/16.887014.
6. H. Takato, K. Sunouchi, N. Okabe, A. Nitayama, K. Hieda, F. Horiguchi and F. Masuoka, "High-Performance CMOS Surrounding Gate Transistor (SGT) for Ultra-High-Density LSIs," *Technical Digest International Electron Devices Meeting*, pp. 222–225, 1988, doi: 10.1109/IEDM.1988.32796.
7. Y. Cui, Z. Zhong, D. Wang, W. U. Wang and C. M. Lieber, "High-Performance Silicon Nanowire Field Effect Transistors," *Nano Letters*, 3(2), pp. 149–152, 2003, doi: 10.1021/nl0258751.
8. S. D. Suk et al., "High Performance 5nm Radius Twin Silicon Nanowire MOSFET (TSNWFET): Fabrication on Bulk si Wafer, Characteristics, and Reliability," *IEEE International Electron Devices Meeting, 2005*. IEDM Technical Digest, pp. 717–720, 2005, doi: 10.1109/IEDM.2005.1609453.
9. D. Jang et al., "Device Exploration of Nanosheet Transistors for Sub-7-nm Technology Node," *IEEE Transactions on Electron Devices*, 64(6), pp. 2707–2713, June 2017, doi: 10.1109/TED.2017.2695455.
10. H. Y. Ye and C. W. Liu, "On-Current Enhancement in TreeFET by Combining Vertically Stacked Nanosheets and Interbridges," *IEEE Electron Device Letters*, 41(9), pp. 1292–1295, September 2020, doi: 10.1109/LED.2020.3010240.
11. C. T. Tu et al., "Experimental Demonstration of TreeFETs Combining Stacked Nanosheets and Low Doping Interbridges by Epitaxy and Wet Etching," *IEEE Electron Device Letters*, 43(5), pp. 682–685, May 2022, doi: 10.1109/LED.2022.3159268.
12. C. J. Tsen, C.-C. Chung and C. W. Liu, "Self-Heating Mitigation of TreeFETs by Interbridges," *IEEE Transactions on Electron Devices*, 69(8), pp. 4123–4128, August 2022, doi: 10.1109/TED.2022.3183967.
13. S. Srivastava, M. Shashidhara and A. Acharya, "Investigation of Self-Heating Effect in Tree-FETs by Interbridging Stacked Nanosheets: A Reliability Perspective," *IEEE Transactions on Device and Materials Reliability*, doi: 10.1109/TDMR.2022.3227942.
14. J. P. Colinge, C. W. Lee, A. Afzalian, N. D. Akhavan, R. Yan, I. Ferain, et al., "Nanowire Transistors without Junctions," *Nature Nanotechnology*, 15, pp. 1–5, February 2010, doi: 10.1038/NNANO.2010.15.

15. J. M. Sallese, F. Jazaeri, L. Barbut, N. Chevillon and C. Lallement, "A Common Core Model for Junctionless Nanowires and Symmetric Double-Gate FETs," *IEEE Transactions on Electron Devices*, 60(12), pp. 4277–4280, December 2013, doi: 10.1109/TED.2013.2287528.

16. L. C. Chen, M. S. Yeh, K. W. Lin, M. H. Wu and Y. C. Wu, "Junctionless Poly-si Nanowire FET With Gated Raised S/D," *IEEE Journal of the Electron Devices Society*, 4(2), pp. 50–54, March 2016, doi: 10.1109/JEDS.2016.2514478.

17. A. Acharya, A. B. Solanki, S. Dasgupta and B. Anand, "Drain Current Saturation in Line Tunneling-Based TFETs: An Analog Design Perspective," *IEEE Transactions on Electron Devices*, 65(1), pp. 322–330, January 2018, doi: 10.1109/TED.2017.2771249.

18. A. B. Acharya, S. Solanki, S. Glass, Q. T. Zhao, B. Anand and B. Anand, "Impact of Gate–Source Overlap on the Device/Circuit Analog Performance of Line TFETs," *IEEE Transactions on Electron Devices*, 66(9), pp. 4081–4086, September 2019, doi: 10.1109/TED.2019.2927001.

19. S. Srivastava, S. Panwar and A. Acharya, "Proposal and Investigation of Area Scaled Nanosheet Tunnel FET: A Physical Insight," *IEEE Transactions on Electron Devices*, 69(8), pp. 4693–4699, August 2022, doi: 10.1109/TED.2022.3184915.

20. M. Schmidt et al., "Line and Point Tunneling in Scaled Si/SiGe Heterostructure TFETs," *IEEE Electron Device Letters*, 35(7), pp. 699–701, July 2014, doi: 10.1109/LED.2014.2320273.

21. P. Weckx et al., "Novel Forksheet Device Architecture as Ultimate Logic Scaling Device towards 2nm," *2019 IEEE International Electron Devices Meeting (IEDM)*, pp. 36.5.1–36.5.4, 2019, doi: 10.1109/IEDM19573.2019.8993635.

22. H. Mertens et al., "Forksheet FETs for Advanced CMOS Scaling: Forksheet-Nanosheet Co-Integration and Dual Work Function Metal Gates at 17nm N-P Space," *2021 Symposium on VLSI Technology*, pp. 1–2, 2021.

23. B. Vincent, J. Boemmels, J. Ryckaert and J. Ervin, "A Benchmark Study of Complementary-Field Effect Transistor (CFET) Process Integration Options Done by Virtual Fabrication," *IEEE Journal of the Electron Devices Society*, 8, pp. 668–673, 2020, doi: 10.1109/JEDS.2020.2990718.

24. A. Veloso et al., "Nanosheet FETs and Their Potential for Enabling Continued Moore's Law Scaling," *2021 5th IEEE Electron Devices Technology & Manufacturing Conference (EDTM)*, pp. 1–3, 2021, doi: 10.1109/EDTM50988.2021.9420942.

25. D. Jang et al., "Device Exploration of Nanosheet Transistors for Sub-7-nm Technology Node," *IEEE Transactions on Electron Devices*, 64(6), pp. 2707–2713, June 2017, doi: 10.1109/TED.2017.2695455.

26. M. J. Kang, I. Myeong and K. Fobelets, "Geometrical Influence on Self-Heating in Nanowire and Nanosheet FETs Using TCAD Simulations," *2020 4th IEEE Electron Devices Technology & Manufacturing Conference (EDTM)*, pp. 1–4, 2020, doi: 10.1109/EDTM47692.2020.9117971.

27. L. Cai, W. Chen, G Du, X. Zhang and X. Liu, "Layout Design Correlated With Self-Heating Effect in Stacked Nanosheet Transistors," *IEEE Transactions on Electron Devices*, 65(6), pp. 2647–2653, June 2018, doi: 10.1109/TED.2018.2825498.

28. Y. S. Song, J. H. Kim, G. Kim, H. M. Kim, S. Kim and B. G. Park, "Improvement in Self-Heating Characteristic by Incorporating Hetero-Gate-Dielectric in Gate-All-Around MOSFETs," *IEEE Journal of the Electron Devices Society*, 9, pp. 36–41, 2021, doi: 10.1109/JEDS.2020.3038391.

29. W. Chen, L. Cai, Y. Li, K. Wang, X. Zhang, X. Liu and G. Du, "Investigation of PBTI Degradation in Nanosheet nFETs With HfO2 Gate Dielectric by 3D-KMC Method," *IEEE Transactions on Nanotechnology*, 18, pp. 385–391, 2019, doi: 10.1109/TNANO.2019.2909951.

30. R. Liu, X. Li, Y. Sun and Y. Shi, "A Vertical Combo Spacer to Optimize Electrothermal Characteristics of 7-nm Nanosheet Gate-All-Around Transistor," *IEEE Transactions on Electron Devices*, 67(6), pp. 2249–2254, June 2020, doi: 10.1109/TED.2020.2988655.
31. S. H. Lee, "Scaling Trends and Challenges of Advanced Memory Technology," *Technical Papers of 2014 International Symposium on VLSI Design, Automation and Test*, pp. 1–1, 2014, doi: 10.1109/VLSI-DAT.2014.6834928.
32. L. Sung et al., "First Experimental Study of Floating-Body Cell Transient Reliability Characteristics of Both N- and P-Channel Vertical Gate-All-Around Devices with Split-Gate Structures," *2022 IEEE International Reliability Physics Symposium (IRPS)*, pp. 7B2-1–7B.2-6, 2022, doi: 10.1109/IRPS48227.2022.9764454.
33. S. Tayal et al., "Incorporating Bottom-Up Approach into Device/Circuit Co-Design for SRAM-Based Cache Memory Applications," *IEEE Transactions on Electron Devices*, 69(11), pp. 6127–6132, November 2022, doi: 10.1109/TED.2022.3210070.
34. M. Yellepeddi, A. Kelkar and J. Waldrip, "Analog Circuit Design Strategies for Reliability Tolerance: Planning for Reliability Effects While Designing Circuits in Modern CMOS Technologies," *IEEE Solid-State Circuits Magazine*, 12(4), pp. 79–85, Fall 2020, doi: 10.1109/MSSC.2020.3021843.
35. P. S. Swain, M. S. Baghini, V. R. Rao, M. Shrivastava and H. Gossner, "Device-Circuit Co-design for High-Performance Level Shifter by Limiting Quasi-Saturation Effects in Advanced DeMOS Transistors," *2016 IEEE International Nanoelectronics Conference (INEC)*, pp. 1–2, 2016, doi: 10.1109/INEC.2016.7589264.
36. G. Gielen et al., "Emerging Yield and Reliability Challenges in Nanometer CMOS Technologies," *Proceedings Date Conference*, pp. 1322–1327, 2008.
37. R. Kishida, T. Asuke, J. Furuta and K. Kobayashi, "Extracting Voltage Dependence of BTI-Induced Degradation Without Temporal Factors by Using BTI-Sensitive and BTI-Insensitive Ring Oscillators," *IEEE Transactions on Semiconductor Manufacturing*, 33(2), pp. 174–179, May 2020, doi: 10.1109/TSM.2020.2983060.

Index

For Product Safety Concerns and Information please contact our EU
representative GPSR@taylorandfrancis.com
Taylor & Francis Verlag GmbH, Kaufingerstraße 24, 80331 München, Germany

www.ingramcontent.com/pod-product-compliance
Lightning Source LLC
Chambersburg PA
CBHW060347220326
41598CB00023B/2830